The Prior Unity of Science and Spirit

Toward a Panpsychic Noetic Quantum Ontology

David Paul Boaz

First Printing, 2023

ISBN-13: 978-1-960583-76-5 print edition
ISBN-13: 978-1-960583-77-2 e-book edition

Waterside Productions
2055 Oxford Ave
Cardiff, CA 92007
www.waterside.com

For
The Precious Copper Mountain Institute Sangha
from whom I have learned so much

Books of David Paul Boaz Dechen Wangdu

The Noetic Revolution:
Toward an Integral Science of Matter, Mind and Spirit
Mindfulness Meditation: The Complete Guide
Buddhist Dzogchen: Being Happiness Itself
The Teaching of the Buddha: Being Happy Now

Contents

Introduction: Toward a Panpsychic Noetic Quantum Ontology vii

I: Primordial Consciousness: Dzogchen Panpsychism/ Kosmopsychism 1

II: Toward a Foundational Noetic Quantum Ontology: Quantum Nonlocality and Buddhist Emptiness 71

Conclusion: From the Prior Unity of Science and Spirit Arises a Panpsychic Noetic Quantum Ontology 335

Appendix A: Let It Be: Brief Course in Basic Mindfulness Meditation 343

Appendix B: The Neuroscience, Logic, and Metaphysics of Mindfulness Meditation 353

Bibliography 365

Introduction: Toward a Panpsychic Noetic Quantum Ontology

Preparing the Ground

What in heaven and earth and the wide world of the Science of Consciousness could "Panpsychic Noetic Quantum Ontology" possibly mean? For any meaningful clarity we must engage recently respectable Western Contemplative Science as it has arisen over at least 5000 years in the mind of the human being in both the East and the West.

In other words, we must ponder the necessity of contemplating our trans-conceptual spiritual belly buttons upon the incoming lifeforce of subtle love-wisdom *jnanaprana* wind of each conscious mindful breath. Better still, we must actually engage such spooky 'psycho-spiritual' practice in the midst of the confusion and distractions of our busy stressed out everyday lifeworld! No cognitive picnic, to be sure.

Well, why bother? Why indeed. We are told through all these many centuries by our human noetic (body mind spirit subject-object unity) primordial wisdom traditions—Hindu, Buddhist, Taoist, Abrahamic monotheistic Hebrew, Christian, Islam—that such view and praxis is perforce the 'innermost secret' cause of awakening the prosaic, narcissistic course conceptual human mind to the indwelling always already present compassionate *Presence* of that greater love-wisdom mind that is non-conceptual primordial Spirit ground of all this arising phenomenal experience of ours.

And that we are told by the wise, is who we actually are—'supreme identity' of our formless, timeless, selfless 'supreme source' ground in which, or in whom this all arises and participates.

But don't believe it! That great love-wisdom mind truth is utterly beyond belief. Without the trans-conceptual noetic direct experience (*yogi pratyaksa*) of that Spirit ground it is little more than prosaic conceptual self-stimulation.

More good news. We shall discover in these pages that this rather off-putting cognitive contemplative mindfulness meditation discipline is the primary cause of our individual and collective human happiness—both relative human flourishing (*eudiamonia, felicitas*), and ultimate harmless Happiness Itself (*paramananda, mahasukha, beatitudo*). As our *relative* human consciousness evolves in its *ultimate* formless timeless all subsuming primordial awareness-consciousness ground we begin to heal the narcissistic negative emotional afflictions that are fear/anger, hatred, random grasping desire, greed, and individual and ethnic pride—the causes of our all too human ethnic hatred, despotism, endless war, and psychic despair. That is why we bother with such an endeavor.

Unless you have already established such a happy psycho-spiritual practice, you—as a self-ego-I—won't want to do it. And should you, as a courageous self-ego-I, *choose* to embark, you will be amazed at how utterly random, scattered, and obsessive the 'wild horse of the mind' can be; and how many lame excuses you can find not to practice the training and discipline of it. Yes, of course the spiritually untrained mind is out of control. But one may freely *choose* to learn to tame it. Results appear almost immediately.

Notice that this near ubiquitous condition of the untrained mind obtains in even the most highly trained intellectual virtuosos—quantum physicists, mathematicians, philosophers, psychiatrists—but not in buddhas, mahasiddhas, and saints.

Thus you begin—with the awakened guidance of the very subtle love-wisdom mind of the *Dzogchen* meditation master—with extremely simple yet profound 'mindfulness of breathing' as Gautama the Buddha called it—'brief moments many times' throughout your day. As you begin to *feel* that non-conceptual clarity and deep inner peace and bliss the scattered 'monkey mind' begins to stabilize and you become a meditation practitioner. And that changes everything! [*Appendix A*]

As we begin our assiduous practice of mindfulness meditation (Boaz 2022, Ch. 8) we shall notice that we must enhance and expand our usually imbalanced human cognition beyond its all too normal habitual objective conceptual cage, to trans-conceptual intuitive, noetic contemplative cognition. In other words, we train the obsessive mostly mindless 'wild horse of the mind' to utilize our complementary *noetic cognitive doublet*—cognitive consciousness processional that is our 'outer' objective conceptual thinking mind (Science), and our 'inner' subjective intuitive non-conceptual noetic mind (Spirit).

The Prior Unity of Science and Spirit is a Panpsychic Noetic Ontology

Our primary task in this book is to demonstrate the ontic prior unity of *relative* quantitative objective Science and the qualitative *ultimate* perfectly subjective Spirit ground in which it arises; that by developing an acausal nondual *Dzogchen* panpsychic/kosmopsychic Noetic Quantum Ontology.

From the prior unity of objective Science (as in the mathematical formalism of quantum physics), and its formless, timeless, nondual (beyond the conceptual subject-object split) perfectly subjective all subsuming Spirit ground naturally arises a panpsychic/cosmopsychic centrist integral Noetic Quantum Ontology that perforce transcends yet includes the mathematical formalism of Irwin Schrödinger's quantum Ψ-wave function. My explanation of the deeper meaning of the prodigious quantum theory constitutes a new 'foundational interpretation' of the Quantum Field Theory/Quantum Electrodynamics (QFT/QED) mathematical formalism and thus dimensionally exceeds the classical local realist/physicalist explanations of the seven main 'foundational quantum ontologies' introduced in Chapter II. We urgently need to know what it is that quantum mechanics means for human destiny beyond its mere mathematical formalism—wondrous bestower of our computers, smart phones, laser communications, and not so wondrous nuclear weapons.

That is my thesis in this brief consciousness exploration of the astounding knowledge paradigm shift that is the inherent subjectivity of quantum theory as it arose from its objective roots in classical physics; and its natural foundation in Max Planck's 1900 'quantum of action' as it arose in the primordial boundless whole in which, or in whom, our phenomenal worlds of space and time arise, participate, and are instantiated.

I shall argue—along with our primary noetic wisdom traditions—that that boundless all subsuming aboriginal 'supreme source' or Base (*gzhi rigpa*), or Basic Space (*dharmadhatu, chöying*) of all arising spacetime phenomena is the formless, timeless primordial emptiness ground (*dharmakaya, Tao*) of all such cosmic phenomena arising therein, including all of us, our Mother Earth, our solar system, our home galaxy, and indeed the entire dimension of all pervading *kosmos*—Reality Being Itself. Broadly construed, that is panpsychism. Pan (everything), psychism (primordial mind). Everything is that enfolded Basic Space (*chöying*) awareness unfolding as all this spacetime stuff.

Moreover, I shall argue that this vast whole is mereologically (part-whole relations) the 'grounding relation' for our foundational integral panpsychic *Dzogchen Kosmopsychic* Noetic Quantum Ontology. Mereologically, the infinite multiplicity of the separate spacetime parts are perforce included and subsumed within the singular formless, timeless, selfless *kosmic* primordial whole that embraces and grounds, and in which, or in whom arises the entire physical cosmos.

Mereology may be understood as the holistic logic that 'proves' that where there exist microcosmic and macrocosmic parts, there necessarily exists a greater whole that includes them. Just so, where there is a great all subsuming whole, there are such constituting parts subsumed within it. Simple logic. That vast unbounded whole that is Reality Being Itself subsumes and pervades all of the spacetime phenomenal participating parts and is instantiated in space and time by them—with not a whit of *ultimate* separation. And we the *conscious* parts abide in an *ultimate* relation of identity with the great whole that is our aboriginal ground. Relative parts and their ultimate whole constitute a prior and present *one truth unity* (*dzog*).

The indwelling primordial Presence of that 'groundless ground' is our 'supreme identity' whose recognition then realization is the primary cause of both relative and ultimate human happiness. That is who we actually are. We awaken to that sublime *kosmic process* through the practice of the psycho-spiritual path. We should feel better already. Again, don't *believe* this. Such trans-conceptual nondual wisdom is far beyond belief. It is rather a matter of contemplative direct experience (*yogi pratyaksa*). As Gautama the Buddha of this present age told, "Don't believe what I teach; come and see (*ehi passika*)."

We shall come to understand that Middle Way Buddhist nondual *Ati Dzogchen* view and practice reveals the fundaments of an already present panpsychic kosmopsychic integral Noetic Quantum Ontology—a foundational ontic 'quantum interpretation' that reaches beyond the twenty or so mere classical realist/physicalist 'interpretations' now on offer. We shall together explore the seven most viable of them. We shall see that none of them transcends the 'scientific' local realist/physicalist ontic ideology enured in 400 years of Metaphysical Local Scientific Realism/Physicalism metaphysical bias. An integral Noetic Quantum Ontology is required to fathom the deeper *ultimate* meaning of the inherent subjectivity of the wondrous quantum theory. It lies beyond its mere *relative* arcane quantum mathematical formalisms.

Yes, we must venture beyond Relative Truth (*samvriti satya*) to engage the Ultimate Truth (*paramartha satya*) that is perfectly subjective ontic one truth unity (*dzog*) of nondual Being Itself—luminous 'clear light innermost secret' always already present Presence of That (*tathata*). As Einstein's pal quantum physicist David Bohm told, "The vast implicate order of the one unbroken whole is the ultimate ground for the existence of everything."

That prodigious understanding completes physics' Relativistic Quantum Field Theory (QFT/QED) as it opens into and adorns the infinite nondual *ultimate* Spirit ground of ever incomplete *relative* provisional fallible Science. That primordial awareness-consciousness whole or ground constitutes the metaphysical foundation, the mereological 'grounding relation' for all subsuming panpsychic *Dzogchen Kosmopsychism*—the *Perfect Sphere of Dzogchen*—in which, or in whom, our centrist integral Noetic Quantum Ontology naturally arises.

Well, how may we come to know and to realize this great truth? That noetic contemplative certainty as to the prior unity of objective Science and its perfectly subjective Spirit ground arises through the meditative practice of 'mindfulness of breathing' (*shamatha*), and penetrating insight of contemplative analysis (*vipashyana*), under the guidance of a qualified *Dzogchen* master, as we have just seen.

Strange Interlude

So now Dear Reader, just for a moment please place your conscious attention-awareness upon the lifeforce wisdom *jnanaprana* wind in your belly, at your heart, in your medial prefrontal cortex just behind your forehead, and at the crown of your head. Breathe naturally for a minute or two and observe without judgment your habitual 'self-referencing' or 'selfing' with its interminable 'Small Mind' thoughts arising and ceasing. Let thoughts naturally pass away on the out breath. Should an obsessive thought not spontaneously pass, invite it to stay for tea. In due course it shall pass like a dark cloud on the pure vast empty sky that is your 'Big Mind'.

Thoughts are inherently insubstantial. They are only diaphanous thoughts. Positive or negative, they possess no more reality than you *choose* to give them. Rest briefly in the peace and clarity of that luminous Basic Space. Human happiness arises from that placement of attentional awareness upon the always already present Presence of the primordial ground that you actually are.

And yes, it requires a bit of practice to move beyond such concepts. But only a bit. And yes, there is no need to *believe* any of this. It is clearly beyond belief. Our thoughts and beliefs, when subjected to the trans-conceptual quiescent breathing of *shamatha,* and the penetrating insight of *vipashyana* are little more than conceptual self-stimulation. [*Appendix A*]

Now 'bracket' or place in abeyance the natural cognitive biases of our human deep cultural background 'global web of belief' as you begin to engage in these pages that indwelling primordial love-wisdom mind Presence that you are now. [*Appendix B*]

I: Primordial Consciousness: Dzogchen Panpsychism/ Kosmopsychism

> Subject and object are only one. The barrier between them does not exist.
>
> —Werner Heisenberg

Panpsychism

There is now abroad in the world human cognosphere a paradigmatic primordial stream of ancient, pre-modern wisdom—both West and East—that continues uninterrupted into Modern and Postmodern Western objectivist scientific metaphysical ontology, and as well, into religious and spiritual ontology. [Ontology is the study of the ultimate nature of Being Itself. Epistemology is how we know it.] That noble Primordial Wisdom Tradition is known in the West as panpsychism.

Panpsychism literally means pan (everything, whole) and psyche (spirit, mind, awareness). Panpsychism may be understood then as all subsuming primordial awareness-consciousness Spirit ground in which, or in whom arises and abides all appearing phenomenal spacetime reality, and the Science that attempts to understand and describe it. Panpshchism is thus the prior and present unity of objective Science and its perfectly subjective nondual all inclusive Spirit ground.

In the West panpsychism has been the default philosophy of mind or consciousness from the ancient Hermetic Greeks, Hebrews, and esoteric Gnostic Hermetic Christians to the present day. It has become in the 21st century important in managing recent versions of 1) the 'foundational interpretations of quantum theory'; and 2) the perennial Mind-Body problem, namely, the 'hard problem of consciousness' of Newton, Locke, Leibniz, Wundt, and recently David Chalmers (1996, 2015).

Chalmers in 2015 used Hegel's dialectical method—thesis, antithesis, synthesis—to analyze the viability of ancient and modern panpsychism. *Thesis*: metaphysical materialism is true; all phenomena are fundamentally physical. *Antithesis*: Dualism is true; not all phenomena are fundamentally physical. *Synthesis*: Panpsychism is necessarily true.

Panpsychism possesses the centrist syncretic truth of a relative Metaphysical Monistic Materialism, a Metaphysical Property Dualism, and even an ultimate antirealist monistic Metaphysical Idealism, as in panpsychic Middle Way Buddhism that is the conceptual foundation of acausal nondual *Ati Dzogchen.* [Boaz 2023] A proto-panpsychic Property Dualism is what Chalmers decides on.

As to the quantum theory, panpsychic interpretations have been offered by David Bohm, A.N. Whitehead, Bertrand Russell, and recently by Jon Schaffer and Shan Goa, as we shall soon see.

Just so, that all-pervading panpsychic ultimate nature of all (pan) appearing reality is all pervading primordial awareness-consciousness (psyche), ultimate *kosmic* 'Big Mind' that embraces and pervades the relative conceptual 'Small Mind' of mere cosmic spacetime reality. That ultimate primordial awareness-consciousness is panpsychic cosmopsychic aboriginal fundamental ground, boundless timeless whole of reality Being Itself.

Panpsychism is then a proto-idealist ontology, historically aligned with philosophical Idealism, that 'Big Mind' or the mental-spiritual dimension of reality—basal timeless consciousness/awareness itself—is the formless fundamental ground, the vast boundless whole in which or in whom the Small Mind of physical and mental form arises and participates. Panpsychism offers a venerable and viable ontological alternative to Metaphysical Scientific Local Realism/Materialism/Physicalism, and to Metaphysical Cartesian Dualism. Appearing reality to sentient beings is then panpsychic or 'pan-mind'.

Panpsychism is mereologically (relation of parts to wholes) interpreted by reductionist realist/physicalists as micropsychism, wherein macrocosmic trees and stars and living beings are derived from more basic microcosmic quarks and leptons.

Conversely, cosmopsychism (Schaffer 2010) is the syncretic view that the great whole of reality is ontologically prior to and greater than the sum of its micropsychic parts. We shall discover that that is the understanding of *Dzogchen Kosmopsychism*, the foundation of our centrist integral Noetic Quantum Ontology. Our Eastern wisdom traditions are replete with cosmopsychism. Hindu Advaita Vedanta, Buddhist *Dzogchen* and Essence *Mahamudra*, *Tao-chia* are profound examples.

In the East panpsychism begins with the earliest Vedas. In the West we see it first in the foundational 6th and 5th century BCE Greek pre-Socratic philosophers—Thales, Parmenides, and Heraclitus. Panpsychism may be seen as the esoteric wisdom orientation of the primary wisdom traditions of our species—Hindu Veda-Vedanta, Middle Way Madhyamaka Buddhism and *Ati Dzogchen*, Taoism, and in the esoteric voice of Abrahamic Monotheism—*Kabbalah*, Gnostic and Hermetic mystical Christianity, Islamic Sufism.

This luminous body of centrist monistic panpsychist basal noetic nondual primordial wisdom (gnosis, *jnana, yeshe*) includes the teaching of the greatest minds and spirits in the global wisdom history of our species.

In the East: Gautama the Buddha, Middle Way Buddhists Nagarjuna, Chandrakirti, Longchen Rabjam (Longchenpa), and the many Mahayana/Vajrayana Buddhist *mahasiddhas*, Taoism founder Lao Tzu, Advaita Vedanta founder Adi Shankara.

Panpsychists in the West include metaphysical monists Moses, Jesus the Christ, St. Augustine, St. Theresa, St. John of the Cross and the Christian Holy Saints, Parmenides, Heraclitus, Plato, Plotinus, Proclus, Zeno and his Stoics, Spinoza, Leibniz (pluralistic panpsychism); the German idealists including Kant, Hegel, and Schopenhauer; American and British idealists George Berkeley, F.H. Bradley, J. Royce, B. Blanshard, William James (Neutral Monism), theist H. Lotze, and the British Cambridge Neoplatonists.

Panpsychic physicists include Sir James Jeans, Bernard d'Espagnat, Niels Bohr, Irwin Schrödinger, Arthur Eddington, and mathematicians A.N. Whitehead, and even Bertrand Russell in his Russellian Monism.

Contemporary philosophers include Alan Wallace, Ken Wilber, David Chalmers, Tim Sprigge, Don Hoffman, David Pearce. And many many more. Good company indeed.

Whence Metaphysics?

Metaphysics, literally 'beyond physics', on F. H. Bradley's Idealism account is "The discovery of bad reasons for what we instinctually know to be true." For Scots radical empiricist polymath David Hume, any text pretending to address metaphysical questions should be "Committed to the flames, for it can contain nothing but sophistry and illusion."

The British empiricists—Lock, Berkeley, Hume—began this anti-metaphysical bias of the Western Tradition that is with us still. The panpsychic Subjective Idealism of Immanuel Kant briefly restored a bit of respect for metaphysical truth, but the early 20th century advent of the "logical empiricism/positivism" cabal that was to become the proto-religious prevailing metaphysical 'web of belief' of the Western mind—Scientific Local Realism/Materialism/Physicalism—soon buried it in a belief system of unscientific purely objectivist observer-independent metaphysical cognitive bias.

Good news. Beginning in the late 20th century metaphysics has again become mainstream in academic philosophy, if not in the hallowed halls of classical Big Science.

Metaphysical questions include the relation of mind to body, the ultimate nature of physical substance and its relation to the mental dimension, causation/causality, ontology/being and theistic and non-theistic nondual God; all inherently metaphysical—beyond physical/physics exoteric materialist "global web of belief" (Quine 1969). That quantum and gravitational cosmology presume to construct propositions as to the *ultimate nature* of any purely physical cosmos/universe, it too is metaphysics at the macrocosmic level—a monistic, physicalist metaphysic that admits of no logical, mathematical, or empirical verification or proof. 'Scientific' metaphysics indeed.

At the microcosmic level of physical reality quantum physics provides ontological (metaphysical) "interpretations"—*opinions* and belief systems—as to the ultimate nature of the physical reality ostensibly described by their mathematical formalisms. For example, is Schrödinger's prodigious infinitely superposed quantum Ψ-wave function reducible to an objectively "real" microphysical phenomenon (an electron); or is it a subjective trans-physical metaphysical process? Are numbers and sets physically real objects, or emblematic ideal Platonic archetypes? Is all this arising physical and mental reality stuff *ultimately* objectively physical; or some spooky subjective panpsychic process; or as the Buddhist Mahayana Two Truths—Relative and Ultimate—view indicates, a *one truth unity* (*dzog*) of both. These are all inherently 'unfalsifiable' metaphysical questions that admit of no 'scientific' empirical evidence or logical/mathematical proof.

All such views express sub-textual metaphysical ontology, not scientific empirical objective fact. Such debates are inherently metaphysical, often in the guise of empirical debates. Again, the unconscious prevailing Western metaphysic is monistic Metaphysical Scientific Local Realism/Physicalism—in short, the ultimate nature of all this arising reality is purely physical. But surely the world is *ultimately* physical, after we scientifically reduce the mental and spiritual dimensions down to purely physical brain structure and function (Metaphysical Functionalism). Is such a radical reduction possible? Probable? Surely Physicalism is just a "brute fact" of nature. But is it? What shall we make of this informal logical fallacy known to logicians as 'begging the question' of ontic Physicalism?

In the spirit of Hume and the British Empiricists twentieth century ideological hostility to metaphysics (literally beyond physics) has continued through the scientific vogues of Logical Positivism/ Logical Empiricism, Quine's Naturalism, Wittgenstein's "ordinary language" philosophy, and the 'scientific reductionist' Physicalism of the Modern and Postmodern Scientific Materialism doctrinaire 'web of belief' (Quine 1969).

Again, this realist/materialist/physicalist metaphysical arc is the prevailing ontic and epistemic ideology that presently pervades all of the sciences, even psychology and neuroscience; and it cognitively trickles down to become the metaphysic of mass mind "Common Sense Realism", Bertrand Russell's "metaphysics of the stone age". Here our appearing realities are all quantifiable and only purely physical. Qualitative love, compassion, intuition, poetry, religion, spirit—all reduced to purely physical brain electrochemistry in the head! Dismal dominate ontic trope indeed.

Moreover, this epistemic Scientific Local Realism and ontic Physicalism of proto-religious triumphal Scientific Materialism ("Scientism" in its most religious fundamentalist raiment) has colonized the Western "common mind" of recent mass culture. And mass materialist culture has embraced modern science's hostility toward metaphysics through its hidden metaphysical valorization and idealization of Scientific Materialism and the so called "scientific method", largely unaware that such a view is itself a purely metaphysical cognitive presumptive presupposition; a 'conformation bias' firmly established in our perceptual and conceptual deep cultural background unconscious 'global web of belief'.

Yes, metaphysical views are *ipso facto* beyond physics and so, in spite of our deep cognitive background Greek materialist web of belief, admit of no "scientific" empirical or logical proof. Ontology (*ontos*/being), the science of the "what" and the "how" of ultimate existence—of primordial Being Itself—that metaphysics is expressed in metaphysical statements of belief. Socratic irony?

So, ontology, what ultimately exists, and how it exists is, *ipso facto* (by definition) metaphysical, beyond the colossal conceptual reach of empirical physics. To habitually, obsessively reduce the panpsychic boundless whole of vast trans-conceptual nondual ultimate reality itself to the subject matter of mere physics is an important brand of "category mistake" that has tragically diminished our human being to Lewis Carroll and Alice's "bag of neurons" relegated to the deterministic darkness of a godless separate entirely physical universe. We have in the West adopted a metaphysical ontology that leaves us out in the cosmic cold. Dreadful metaphysic indeed.

Now that is rapidly changing with the advent of our inherently subjective quantum theory, and the 20th century cognitive cultural infusion of Middle Way Buddhism with its panpsychic/kosmopsychic nondual *Ati Dzogchen* view and practice.

Toward a Noetic Revolution in Science and Spirit

Now the good news! We have upon us in the post-Postmodern human cognosphere the advent of a new post-quantum, post-empirical Kuhnian scientific revolution! Our 400 year old secular European Modern Enlightenment zeitgeist that resulted in reductionist objectivist materialist physics and cosmology has utterly failed to explain, or explain away its many logical and mathematical paradoxes and empirical anomalies—not the least of which is the inherent subjectivity of the Kuhnian 'paradigm shift' represented by 'spooky' (Einstein's term is *spukhaft*) quantum entanglement/nonlocality.

Here is Thomas Kuhn's cognitive process. Progress in "normal science" has resulted through such unanswerable anomalies and formal logical and mathematical paradoxes (especially Russell's Paradox) in a "scientific crisis" and a "paradigm shift" or "gestalt shift" that yields a new "scientific revolution"—a new scientific knowledge paradigm. We have seen that as the old Modernist Materialist empirical physics paradigm has failed us, hitherto prodigal flaky metaphysics has recently returned as respectable cognition into the domain of academic philosophy, and so, haltingly, of philosophy of physics and cosmology. Scientific and cultural revolution indeed. But whence the noetic revolution?

Physicists fear philosophy. But the quantum collapse of old paradigm positivist scientific objective certainty has forced dialogue with philosophers of physics and cosmology, and even Buddhist philosophy, that we may at last discover just what it is that Quantum Electrodynamics (QED) and quantum cosmology actually tells us about the proto-noetic quantum nature of reality, and how to interpret its several competing 'foundational quantum ontologies'. The consummation to be wished is resolution of critical physics anomalies arising through the logical and empirical incommensurability of QED and General Relativity Theory (GRT). Our Noetic Quantum Ontology will help.

A refreshing renascent proto-spiritual *sub specie aeternitatis* (from the view of eternity) metaphysical vehicle has now entered this ontic fray to contend with proto-religious Metaphysical Scientific Local Realism/Materialism/Physicalism for metaphysical hegemony of the Western mind. That syncretic wisdom vehicle is the union of Western monistic panpsychism with Eastern monistic Buddhist panpsychic/kosmopsychic *Dzogchen*, the Great Completion, as we shall soon see.

What Panpsychism is Not

Panpsychism is not panthestic and should not be conflated with pantheism (everything is God), nor with panenthism (God is in everything). Panpsychism is a viable alternative to what is now considered by philosophers of mind and philosophers of physics and cosmology to be the failure of pan-materialism—the reductionist Metaphysical Monistic Physicalism of 20th century science and philosophy—the view that everything is just physical, or is reducible to the functionalist gambit, that is to say, physical brain structure and function. That metaphysical ploy is known to the philosophy of physics trade as 'scientific reductionism'.

Panpsychism is also a viable alternative to Metaphysical Dualism—Descartes' two separate, somehow coexisting substances, namely the physical and the mental dimensions of being human here in space and time. Panpsychism/cosmopsychism has also contributed to viable recent Neodualism theories, as we have seen. [Chalmers 1996, 2015]

The Problem of the Existence of God. Panpsychism is not a brand of theism, and so it entails none of the inherently vexed, age old philosophical theistic conundrums (e.g. the three "proofs" for the existence of God; the Problem of Evil/Suffering, and the rest) that plague theistic belief in an omnipotent, omnipresent, omniscient and perfectly benevolent anthropomorphic Creator God somehow co-existing beside 'His' separate creations in a brutal world of adventitious human and animal suffering. Evil, both natural and human is indeed problematic for any self-respecting all powerful and compassionate Creator God.

The perennial concern and relentless debate about the *existence of God* must first distinguish between such a *theistic,* objective, dualistic concept-belief, even physical anthropomorphic Creator God, and a post-theistic, trans-conceptual, all inclusive, all embracing ontic primordial 'supreme source' or nondual Spirit ground. Such a panpsychic/cosmopsychic, nondual (subject-object unity), noetic (body/mind/spirit) all pervading *Ultimate Truth* dimension (*paramartha satya*) primordial *kosmic* ground state is nothing less than the vast enfolded unbounded whole itself in which, or in whom, all of our spacetime *Relative Truth* dimension (*samvriti satya*) conditional physical and mental spacetime realities unfold and participate. No problematic conceptual monotheism here.

In the absence of such a foundational understanding of metaphysical distinctions between our concepts and beliefs *about* theistic Creator God and trans-conceptual directly experiential non-theistic, nondual all embracing, timeless, selfless godhead—basal unborn, uncreated primordial ground—metaphysical interlocutors face an endless tiresome cognitive and emotional disconnect. Our primary Primordial Wisdom traditions have names and concepts (*namarupa*) for such a nondual primordial ground of being, to wit,

Nirguna Brahman/Parabrahman, dharmakaya/shunyata, dharmadhatu/chöying, Tao, infinite *Ein Sof, Abba,* nondual primordial God the Father of Jesus, and many more.

"What's in a name? A rose by any other name would smell as sweet." [Juliet Capulet]. The primordial indwelling Presence of nondual God is said by those who know to possess a very subtle aroma of roses.

The main problem with endless theism/non-theism debates is that dualistic theistic Creator God and post-theistic nondual primordial godhead are not distinguished, but are invariably conflated. The tragic result has been 5000 years of endless confusion as to the nature of the ultimate primordial ground of appearing reality itself—our soteriological (liberation enlightenment) home. And that means confusion as to the 'supreme identity' of we human beings as we arise in this our ultimate ground, and our choices that are the causes of our relative and ultimate human happiness. Shall it never end?

Approaching Dzogchen Panpsychism/Kosmopsychism

Panpsychism may be broadly construed as the psychophysical nondual ultimate nature of human experience/consciousness of all the arising and dualistic appearing stuff of reality itself, *sub specie aeternitatis.* We understand that through the Buddha's 'mindfulness of breathing'.

Primordial awareness-consciousness in any of its human or other modalities is a fundamental all subsuming unity. Philosophers often refer to this definition of panpsychism as panexperientialism. It does not imply that electrons and rocks have conscious experience. However, our perennial panpsychism view has always seen consciousness as extending far beyond humans, animals, and plants.

Indeed, as we shall soon see, spacetime stuff naturally arises in the all pervading formless, timeless primordial awareness consciousness panpsychic ground. Human consciousness/experience arises in that aboriginal consciousness boundless whole. Such a view is as Jon Schaffer (2010) has told, "Priority Monism", or 'priority monist cosmopsychism'. Here, the one primary unity is the indivisible "implicate order of the vast unbroken whole" (David Bohm)—kosmopsychic primordial awareness ground of everything—*Perfect Sphere of Dzogchen.* This holistic view differs radically from the reductionist view of 'constitutive micropsychism' wherein the vast panpsychic macrocosmic whole is grounded in reductionist microscopic subatomic parts in purely physical electrochemical brain structure and function. 'Scientific reductionism' indeed.

Thus arises the devastating 'constitutive micropsychism' *combination problem* ('subject summing problem'): how do many micropsychic conscious subjects, e.g. the 'experience' of electrons in the brain, combine to somehow create the macropsychic conscious experience in a single macrocosmic human being? The micropsychic cart before the kosmopsychic horse? This problem does not arise for cosmopsychism/kosmopsychism because here we no longer presuppose that brain micro structure and function 'creates' macro structure and function for human experience-consciousness.

Remember here that human consciousness necessarily arises in its nondual timeless primordial awareness-consciousness ground. That is to say, *relative* objective spacetime human consciousness and its physical support—the 'neural correlates' of brain structure and function—are *ultimately* embraced and subsumed by naturally present perfectly subjective formless, selfless basal aboriginal whole that is all pervading consciousness Spirit ground itself in which, or in whom all spacetime appearance arises, participates, and is instantiated. *"Tat Tvam Asi"* That I Am! Beyond all our concepts and beliefs about it.

Buddhist Basics: the Essence of Mindfulness. That primordial love-wisdom mind (*jnana, yeshe,* gnosis) is directly experienced (*yogi pratyaksa*) via contemplative mindful meditation—Buddha's "mindfulness of breathing". That essence of mindfulness is the clear seeing of Gautama the Buddha's "four foundations of mindfulness", the four *satipatthanas,* introduced in his foundational *Satipatthana*

Sutra. These are mindfulness of 1) form/body, 2) feeling states, 3) mind states, and 4) phenomena (*dharmas*) arising in mind.

Mindfulness of breathing (quiescent *shamatha*) establishes the conceptual and the non-conceptual contemplative certainty or 'clear seeing' (*vipashyana*) of these "four objects of mindfulness" as a kosmopsychic *Ultimate Truth* nondual one truth unity, beyond the *Relative Truth* duality of a knower/perceiver separate from the perceived objects of its experience—the primal pernicious subject-object split. Clear seeing of the nondual *samadhi* of the 'penetrating insight' of *vipashyana* is the recognition, then realization of the utter absence or emptiness (*shunyata*) of any iota of *ultimate* intrinsic existence to *relative* interdependently arising (*pratitya samutpada*) phenomena appearing in mind. That primordial emptiness 'groundless ground' is itself empty of inherent existence. It is established by human conceptual designation. It is indeed the ultimate nondual 'Nature of Mind'. And that is the realization of the Buddha's Mahayana foundational Two Truths—relative and ultimate—from which his prodigious Four Noble Truths with his Eightfold Path to liberation from suffering arise for us.

All of that from fundamental mindfulness of breathing. From it we learn to cultivate the means/method of both relative discriminating wisdom (*prajna, sherab*) and nondual ultimate primordial wisdom (*jnana, yeshe*, gnosis) naturally and spontaneously expressed as *bodhicitta*—compassionate engaged ethical action for the benefit of all living beings. And that is the open secret of both our relative and ultimate human happiness.

All of that from the Buddha's Mahayana teaching vehicle with its centrist causal Middle Way Prasangika Madhyamaka, the conceptual foundation of acausal nondual highest *Dzogchen Kosmopsychism*.

Panpsychism: Variations on a Theme of Wholeness. Panpsychism, Buddhist *Dzogchen* or otherwise, is at root trans-physical 'pan-mind', whole mind, all mind. It transcends yet embraces derived both objective physical and subjective mental perspectival, *ontologically relative* phenomena. In other words, we perceptually and conceptually impute then reify our nominal phenomenal physical and mental realities via our deep background sociocultural "global web of belief" (Quine 1969), as we have so often seen in these pages.

Hence, panpsychism in its most cogent non-atomistic, non-micropsychic primary or 'priority monist cosmopsychic' raiment is the proto-idealist view that an all inclusive, all pervading basal grounding awareness consciousness primordial ground is ontologically ultimately prior and fundamental. And that is a "cosmopsychic priority monism" (Schaffer 2010) of the vast mental dimension of basal and omnipresent reflexive (self-revealing) ultimate nature of all arising nominal phenomenal physical and mental reality—vast *buddic* Nature of Mind—'basic space' (*chöying, dharmadhatu*) of trans-rational, post-theistic nondual godhead, unbounded *dharmakaya* whole of reality Being Itself.

Hence, there exists one great fundamental natural timeless reality *process*—all embracing formless *kosmic* whole that transcends and embraces the physical cosmos—perfectly subjective noetic nondual Spirit Itself, the 'grounding relation' of the relative spacetime existence of the holonic participating parts. Mereologically, these holonic parts of the singular unity of the great whole exist only derivatively, relatively and conventionally as parts that instantiate the *ultimate* basal whole of Being Itself.

Priority Monism must be distinguished from an easily dismissed Existence Substance Monism—one great concrete substantially physical real object/entity somehow exists out there absolutely. That is the 'metaphysical extreme' of absolute existence (Absolute Physicalism). The other metaphysical extreme is absolute nonexistence (Absolute Idealism). We seek a centrist middle way that avoids the false dichotomy of *either* existence *or* nonexistence. We shall see that causal Buddhist Middle Way Prasangika Madhyamaka with its acausal nondual *Dzogchen Kosmopsychism* provides such a centrist metaphysical consummation.

Contrary to local realist/physicalist ideological interpretations of monism, primary Priority Monism inures to the primordial whole that is not a physically existing concrete cosmic *object* but an all subsuming metaphysical *kosmic* (body, mind, spirit subject-object unity Pythagorean *kosmos*) reality *process* in which, or in whom relative, nominally real spacetime stuff arises and plays out its time on the phenomenal world stage.

In short, microcosmic phenomena arise in and are mereologically grounded in the primordial *unity* of the all pervading *kosmic* whole. Yes. All *relative* spacetime physical, mental and spiritual reality 'form' arises from, participates in, and is instantiated through this all embracing formless, timeless, selfless nondual *ultimate* all subsuming primordial awareness 'emptiness' ground—"Absent and empty of any whit of intrinsic existence." [Nagarjuna] And yes, Middle Way Prasangika Madhyamaka Buddhists refer to this vast ground/whole as the acausal nondual *Perfect Sphere of Dzogchen.*

That which we seek in this connection is then a syncretic middle way primordial kosmopsychic panpsychism—*Dzogchen Ati, Essence Mahamudra, Saijojo Zen, Advaita (nondual) Vedanta,* mystical *Zohar/ Kabbalah, Christian Hermetic wisdom*—that includes the holistic wisdom of both East and West. Such now respectable holistic panpsychism or kosmopsychism has facilitated a reformation in hitherto 'scientific' materialist/physicalist, micropsychic Western analytic philosophy and religion.

As the theoretical and later empirical evidence of the "spooky" (Einstein's *spukhaft*) metaphysics of subjective quantum entanglement/nonlocality entered academic American analytic and British/French continental philosophy, and through that philosophy of science, panpsychism has found new metaphysical acolytes. Beginning with Niels Bohr's 1928 Principle of Complementarity, Werner Heisenberg's 1927 Principle of Uncertainty, and later John Stewart Bell's prodigious "Bell's Proof" in 1964, metaphysical ontology emerged from its cognitive closet after a half century of imperious, extremist anti-metaphysical Logical Positivist "hidden metaphysics" (Ken Wilber), with its odious "taboo of subjectivity" (Alan Wallace). And none too soon.

This dubious positivist metaphysic, concealed in its linear objectivist empirical cloak, is none other than our old tedious proto-religious fundamentalist realist-materialist/physicalist Scientism. It has hitherto thoroughly controlled the suggestible 20th century scientific and philosophical mind—along with the mass mind of "common sense" Realism and consumerist Materialism that has now colonized the Western mind— grasping at their purely objectivist/physicalist ideological web of belief. To question the idols of orthodox Scientific Materialism/Physicalism is still, sad to say, scientific heresy.

Here we must recall that the prevailing scientific ideology that is monistic Absolute Scientific Local Realism/Physicalism is a non-empirical metaphysical view, just as is monistic Absolute Idealism, that everything is mental or 'mind only, and the physical reality dimension of space and time is somehow only mental. We've seen that ontology—what exists, Being Itself—is a synonym for metaphysics and, rather counter intuitively, admits of no logical, mathematical or even empirical proof! Cosmic irony indeed. Hegel called this discomfiting cognitive situation "the irony of the world". The world exists but we can't prove it! Here we are but our prodigious scientific theories are here mute.

For ontic, theistic, and cosmological acolytes of Absolute Idealism, or of Absolute Physicalism—and you know who you are—any hope for logical, deductive absolute certainty for your favorite metaphysical ontology is now logically *kaput*! [Gödel '1931'] Yes, we need a centrist middle way between these two dubious metaphysical extremes.

Well then, is this all too real world of arising matter and energy—$E=mc^2$—*ultimately* only physical, the belief system of modern physics? Or is it ultimately *avidya maya*—just a metaphysical mental illusion, as is the belief system of Absolute Idealism? Or perhaps an idea in the mind of God? Or a nice amalgam of Metaphysical Cartesian dualism, or of a recent Neodualism (Chalmers) of physical and mental entities or dimensions?

The Intrinsic Nature Argument. Perhaps the dominant view of panpsychism among philosophers of mind (Whitehead, Russell,

Goff, Sprigge) who reject the inadequate purely quantitative view of objective empirical science is the qualitative view that the microcosmic reality dimension of baryonic quarks and leptons and bosonic photons has an *'intrinsic nature'*. The intrinsic nature of matter, if indeed matter can possess an intrinsic nature—Middle Way Buddhists deny that it does—is a cognitive *process* that instantiates its prior basal panpsychic consciousness ground. This is a metaphysical statement. That this is in some sense true clearly cannot be demonstrated by quantum mathematical descriptions of the 'purely physical' structure and behavior of electrons. *Physics informs us as to how matter behaves, not what matter actually is.*

The universe of discourse for physics is entirely quantitative—mathematical and physical. And yes, panpsychism in its 'priority monistic' *Dzogchen* kosmopsychic cognitive array—whether or not matter/energy has an 'intrinsic nature'—has the potential to fill this spacious 'explanatory gap' portal into the Buddha's formless Ultimate Truth dimension with a centrist middle path between the Relative Truth metaphysical extremes of absolute existence and absolute nonexistence. And that pacifies one of the most difficult problems in both Western and Eastern philosophy, namely, how shall we balance the ontic false dichotomy between scientific absolute Realism/Physicalism on the one hand, and Absolute Idealism, whether Objective Idealism, or Subjective Idealism, on the other hand.

The Intrinsic Nature Argument has its foundation in the monistic pararealist panpsychism of Leibniz, Schopenhauer and the German idealists, Arthur Eddington and William James (Neutral Monism). It arose because we need to know the foundational or essential nature of our appearing realities beyond the conceptually empty quantitative limits of physics, logic, and mathematics. Conceptual panpsychism cannot altogether accomplish that result. Its usefulness lies in offering a qualitative, holistic, elegant, parsimonious and intelligible metaphysic for the inherent Nature of Mind and its experience that avoids the essentialism of an absolutely intrinsically existing purely physical 'real world out there'

(RWOT), be it micropsychic or macro cosmopsychic. And as well, panpsychism offers a positive alternative to the nihilism of an utterly unintelligible world where human knowledge and wisdom, and indeed everything else are perforce absolutely nonexistent (*abhava,* solipsism).

The power of *priority monist cosmopsychism,* and the *Dzogchen* kosmopsychic variants of panpsychism lie in their logical *reductio ad absurdum* nature. The burden of rejoinder to such an argument is here correctly placed upon the proponents of other metaphysical explanations as to the nature, origin and aim of physical and mental phenomena; precisely as the centrist Buddhist Middle Way Prasangika Madhyamaka metaphysic of 2nd century Nagarjuna and 7th century Chandrakirti have done.

Once again, these are not scientific questions. These are metaphysical questions. Important ones. Some paraconceptual middle way resolution of these ontic conditions obtains. But alas, we can't prove any of it via the linear two-valued deductive reasoning of classical logic and mathematics; nor can we consistently argue it on empirical grounds. So the indispensible ontic metaphysical conjecture of the philosophy, theology, even the physics and quantum cosmology trades shall go on, and on. Buddhist *Dzogchen* 'logic of the non-conceptual', wherefore art thou? [Klein 2006; Boaz 2023 *Ch. II*]

Toward a Panpsychic Unified Quantum Theory of Consciousness. We have seen that our primary task in this book is to utilize the *Prior Unity of Science and Spirit* to develop a *Dzogchen panpsychic/kosmopsychic Noetic Quantum Ontology.* Let us then explore a noetic theory of primordial consciousness, a quantum consciousness, that is instantiated by our human consciousness-experience arising herein. So we must first conceptually unpack the meaning of Irwin Schrödinger's global superposed 'universal quantum Ψ-wave function'—the prodigious Schrödinger Equation.

The 1926 Schrödinger Equation (1933 Nobel Prize in Physics) is a linear partial differential equation that governs the real time evolution of the Ψ-wave function of all quantum mechanical systems. Ψ is the *quantum* counterpart of Isaac Newton's wondrous Second Law of Thermodynamics of *classical* physics—a mathematical prediction of the evolution of a particular matter state over a given period of classical 'real time (t)'. Just so, the Schrödinger Equation reveals the evolution of a quantum wave function over classical time. That presupposes that the time evolution mathematical operator is 'unitary' (the quantum Hamiltonian).

Other equally consistent theories of quantum mechanics include the 'uncertainty relations' of Werner Heisenberg's 'matrix mechanics', and the later 'path integral formulation' of Richard Feynman. Paul Dirac brilliantly synthesized Heisenberg's matrix mechanics with Schrödinger's Ψ-wave function equation into a single mathematically consistent quantum mechanical theory. Schrödinger's Ψ-wave formulation—especially after Dirac's Quantum Electrodynamics (QED) relativization of Quantum Field Theory (QFT)—proved to be far easier to use and is now, after some initial priority disputes, used almost exclusively.

I shall thus argue herein that the quantum theory, the quantum mechanics of Quantum Field Theory (QFT) and its unification with Einstein's Special Relativity Theory (SRT) via Paul Dirac's Quantum Electrodynamics (QED), is an inherently holistic and primary 'priority monistic' (Schaffer 2010) panpsychic process—the priority of the vast *ultimate* all subsuming boundless primordial awareness-consciousness whole itself in which *relative* QFT arises. Jon Schaffer's cosmopsychism with its primary Priority Monism represents a new vital heterodoxy in contemporary micropsychic panpsychism for present students of philosophy of mind, and for consciousness and religious studies.

Therefore, primordial *kosmos,* which subsumes the *relative* atomistic micropsychic microphysical and macrocosmic cosmos, is inherently a nonlocal quantum entangled ZPE state/system, a fundamentally unified *ultimate* unbounded 'unbroken whole'. "For only the whole is guaranteed to exist. Only the [priority] monist can provide a unified story of the ground of being for every metaphysical possible world." [Schaffer 2010] That is primary 'Priority Monism', the priority of the all embracing primordial whole to its always already included participating instantiating parts. That whole is greater than the sum of its parts.

As Neoplatonist Proclus told, "The monad ['The One', The Whole] is everywhere prior to the plurality... the whole embraces all separate beings in the cosmos." [Proclus 1987 *Commentary on Plato's Parmenides,* Princeton Univ. Press] 'Existence Monism' or Substance Monism is the common physicalist broader view that the physical cosmos is the "one and only actual concrete object [entity] in existence". Schaffer points out that Priority Monism does not entail Existence Monism because the priority monist argues for a multiplicity of the inherently derived parts of an ontic prior monistic unitary all embracing *kosmos.*

Well, should the noble history of Panpsychic Monism be understood as primary Priority Monism, or common Existence Substance Monism, or neither?

Textual criticism is always a sticky wicket. A close reading of the history of ontic monism—Thales, Parmenides, Plato, Plotinus, Proclus, Spinoza, Hegel, Royce, Bradley, Blanshard, Whitehead, the Cambridge Neoplatinists in the West; and in the East Nagarjuna, Chandrakirti, Adi Shankara, Longchenpa, Mipham, and many others—reveal explanatory nuances that preclude a prosaic answer to that question. The relevant texts demonstrate that both Priority Monism, Existence Monism and ontic Pluralism are intimately intertwined within the textual tradition of panpsychic Ontological Monism. That said, "Many of the historical monists directly deny *Existence Monism.*" [Schaffer 2010]

Be all that as it may, following Jon Schaffer we may establish a *prima facie* case for ontic panpsychic cosmopsychic primary Priority Monism, over against common Existence Substance Monism and ontic Pluralism by demonstrating that the *three primary streams* of historical textual monistic view and praxis "presupposes the falsity of *Existence Monism,* while being perfectly compatible with *Priority Monism.*" [Schaffer 2010]

1) The first of these streams or "threads" is "*the [mereological] priority of whole to part.*" "*Existence Monism* denies that there are any parts to the whole ... that there is anything for the whole to be prior to." For proto-monist Gottfried von Leibniz, "In the vast continuum the whole is prior to its parts."
2) The second stream in our noble history of ontic monism is an expression of the first: "*the organic unity of the whole.*" For Plotinus, "The One is that on which all else depends .... All is one universally comprehensive being... Every separate thing is an integral part of this All ... " For Shaffer, "Such a notion of organic unity is incompatible with *Existence Monism*... but is a perfect fit for *Priority Monism.*"
3) The third stream of our great monistic metaphysical tradition is "*the world as an integrated [entangled] system.*" Spinoza told it well, "The whole of nature is one individual whose parts vary in infinite ways." Schaffer explains, "The idea of the cosmos as an integrated system is incompatible with *Existence Monism*, for [it] denies that there is anything other than the whole cosmos... that there are any things to be integrated into the cosmos... But the idea of the cosmos as an integrated [quantum entangled] system is a perfect fit for *Priority Monism.*"

Our conclusion then must be that ontological metaphysical panpsychic/cosmopsychic primary Priority Monism provides a more precise metaphysical understanding; a foundational mereology (part-whole relations); and a better historical textual fit than does Existence [Substance] Monism, or of ontological Pluralism. In any case, perhaps, in the fullness of time, these rather tedious philosophical distinctions will prove to be in a greater view more or less gratuitous.

Now, as to pluralistic 'combination micropsychic panpsychism', in order to avoid an discomfiting infinite regress, physical atoms and their subatomic parts require an all subsuming trans-physical homogeneous *ultimate* 'emptiness' ground in which they arise to instantiate *relative* physical 'form'. As Buddha told so long ago, "Form is empty; emptiness is form." Atomism is *relatively true* in the spacetime dimension of scientific Relative Truth (*samvriti satya*), yet does not complete the metaphysical whole picture that perforce must include the all embracing dimension of Ultimate Truth (*paramartha satya*) in whom basal physical micropsychic atoms and their subatomic parts arise to comprise macropsychic holonic wholes, and ultimately the One singular all inclusive primordial awareness-consciousness ground/whole itself. For acausal nondual *Dzogchen* these Two Truths are a prior *one truth unity* (*dzog*).

There is no form of reductionist atomistic micropsychic pluralistic ontology that can provide a sufficiently holistic grounding relation for primary nonlocal quantum entangled processes.

Yes, the primary priority monistic fundamental foundational Plotinian monadic One exhibits a plurality of relatively, if not ultimately existing stuff. That allows for *anitya*, relative impermance and change (temporal heterogeneity). That is the undeniable truth of ontic pluralism with its multiplicity of many micro and macro parts. *Yet pluralism fails the unitarity requirement of primary Priority Monism.* The holistic formless, timeless universal *kosmos* ground, boundless ultimate whole itself metaphysically provides that. That primordial timeless unity naturally arises from a monistic infinite all subsuming, unified, nonlocal quantum entangled trans-conceptual *kosmic process* of nondual 'unitarity' (Bohr).

The mathematical formalism of the global infinitely 'superposed' 'universal quantum Ψ-wave function' of the prodigious Schrödinger Equation describes this superluminal (exceeding light speed) nonlocal entanglement process that begins at the Big Bang/Big Crunch singularity beginning of any cosmos in the infinite multiverse, and thereafter evolves in accordance with Ψ. As per the theory, the utter subjectivity of the superposed quantum Ψ-wave function with its sea of infinitely superposed reality probabilities 'collapses' into an objective spacetime reality via a conscious observation by the 'consciousness' of an observer-dependant conscious observer/perceiver, or else in a quantum experiment by an experimenter consciously interpreting the result of a pointer, computer camera, or geiger counter. Such a 'wave function collapse' is known to the trade as 'consciousness causes collapse' (von Neumann, Wigner, Stapp).

CCC is one of the 'foundational quantum interpretations' or quantum ontologies of the leading seven such primary theories now on offer (*Ch. II*) None of them address the ontic elephant in the room, to wit, the ultimate ontologically prior primordial awareness-consciousness ground in which, or in whom such relative quantum interpretations arise. [Boaz 2023 *Ch. IV*] I have in Chapter II below developed the fundaments of such a centrist integral Noetic Quantum Ontology. That metaphysic is grounded in Parmenides' monistic homogeneous "perfect sphere of nondual unity" in which "all is one". That ancient primary monistic ontology parallels the Buddhist Middle Way Madhyamaka perfect wisdom equality (*samantajnana*)—all subsuming nondual *Perfect Sphere of Dzogchen.*

Quantum Cosmology

The Cosmological Constant Problem. Now, as post-Bang singularity cosmic quantum evolution proceeds, in due course all of the micro-psychic physical particle-fields in the infinitely vast *kosmos* whole become interdependently, timelessly, nonlocally entangled with every other particle arising via nearly timeless 'borrowed energy' quantum vacuum 'virtual particle' fluctuations of 'empty space' in the paraphysical 'quantum zero point vacuum energy field' (ZPE or ZPF). These diaphanous 'space particles' of not so empty space are said to randomly arise or 'pop' into and out of physical existence of the quantum vacuum, almost instantly, and continuously.

Hence, in every parsec of 'empty space' there exists a quantum qbit of something. There is in nature no such thing as a pure vacuum. Physical stuff is ubiquitous. Heisenberg's quantum Principle of Uncertainty does not permit pure physical nothingness, a void. The universal quantum vacuum is non-zero; it is not entirely empty.

Moreover,the presence of quantum fluctuation 'virtual particles' is quantifiable and measurable (the Casimir Effect). So we must somehow 'weigh' the particles of empty space. Experimental physicist Enrico Calloni of the Italian National Institute of Nuclear Physics is now attempting such a forbidding task with his daunting Archimedes Experiment, designed by Carlo Rovelli and others.

That mysterious quantum vacuum energy may be determined by two different methods. 'Top down' General Relativity Theory (GRT) reveals what measured quantity of the universal vacuum energy (dark energy)—Einstein's Cosmological Constant—is required in order to explain our exponentially expanding universe. That dark energy force acts as a kind of counter gravity to as Einstein told, "Hold back gravity" and prevent large scale cosmic structure from collapsing back into its Big Bang beginning. Dark vacuum energy is believed to comprise 68 percent of the total energy density of this entire physical universe. Dark matter is about 28 percent, and ordinary baryonic matter about four percent of the total.

The other great theoretical pillar of Modern physics, the 'bottom up' Quantum Field Theory (QFT/QED) predicts the value of the universal vacuum energy based upon the masses of all the virtual particles in the physical cosmos that spontaneously arise in that quantum vacuum of empty space. The value of the vacuum energy based upon the calculations of QFT were vastly larger than the value of the Cosmological Constant derived from the GRT measurements of the energy of cosmic accelerating expansion.

In other words, here is the rub, first observed by Wolfgang Pauli in the mid 1920's. These two methods yield two values that differ by an utterly fantasque 120 orders of magnitude! The measured vacuum energy is 120 times smaller than what QFT calculates that it should be. This *'cosmological constant problem'* was called by Steven Weinberg "The worst theoretical prediction in the history of physics." It's especially bad news as to an accurate understanding of the accelerating expansion and therefore the ultimate fate of this universe of ours—Big Chill, Big Rip, or Big Crunch. [Boaz 2023]

So, the problem of the 'missing vacuum energy' is that there is not enough of it. The Cosmological Constant *lambda* Λ should be huge! What to do? Short of modifying Einstein's sacrosanct GRT field equations—for example with Milgrom's MOND or Modified Newtonian Dynamics—cosmology is cosmically vexed.

Perhaps we might modify equally sacrosanct QFT. Is the QFT method of calculating the vacuum energy the problem? Perhaps the QTF equations do not apply to the curved topology of the spacetime of Einstein's GRT. Perhaps we must in the final analysis appeal to the now respectable yet surreal *Anthropic Principle.* If we arise and abide in a multiverse of many universes—if our local cosmos is but a 'bubble' in an infinite primordial *kosmic* sea of universes—then perforce one or more of these must be a universe with the physical constants that rule our present universe permitting the beginning and evolution of life.

The host of other scenarios are highly unlikely to result in a universe that make spacetime cosmic structures such as galaxies stars, planets and the life forms that evolve from the heavy element ejecta of dying stars—the carbon, iron and oxygen that we are.

Superstring theorists like the Anthropic explanation, and so consider the cosmological constant problem a pseudo-problem. In any case it is now clear that all such explanations of the problem require a post-empirical 'new physics' well beyond our present scientific ideologies of classical and quantum physics.

Well, what if the current non-zero Cosmological Constant is actually zero? What if the quantum ZPE vacuum energy is cancelled by God knows what, or was not at all connected to a problematic Cosmological Constant in the first place?

Quintessence (Steinhardt and Caldwell) is a propitious heterodox alternative to the now orthodox accelerating expansion of this local universe of ours. Quintessence is said to be a very subtle form of physical or proto-physical 'phantom energy' that pervades all of space, but with a rather spooky negative pressure. It is not a constant but changes in classical time. Its phantom energy density *increases* as the universe expands in space and time. That results, a few trillion years hence, in an ultimate Big Rip wherein all physical form is torn asunder, right down to its microcosmic baryonic quarks and leptons. Bad news for physically embodied sentient beings, although not necessarily for awareness-consciousness beings in other forms of existence.

So, is the dark energy/vacuum energy of this universe a function of the Cosmological Constant, or is it caused by Quintessence? We must know whether or not the universal ZPE vacuum energy density has changed over classical time.

Recent research (Blanco Telescope in Chile) tends to favor the constancy of the vacuum dark energy. Recent research in gravitational waves (LIGO and VIRGO) shall add theoretical grist to this endless mill of purely metaphysical cosmic speculation. Such research has already demolished several post-GRT theories.

QFT or GRT? These two glorious mathematically incommensurable pillars of Modern physics—these two 'perfect theories'—seem a long way from the unification that results in that consummation devoutly to be wished—a forever incomplete Quantum Gravity Theory (QGT). Our inchoate centrist, integral Noetic Quantum Ontology will further that end. It shall be revealed in Chapter II.

What is certain is that our grail quest to resolve the discomfiting cosmological constant problem shall continue to open new vistas in our providential relationship that is the prior unity of relative objective Science and its ultimate perfectly subjective nondual Spirit ground.

Well, is that it? Is dualistic quantum and dualistic gravitational Science all we desire to know? Shall we "Shut up and just do the calculations." What about Science's relation to nondual Spirit ground in which QFT and GRT both arise?

In the brave new world of cosmopsychic primary Priority Monism it is not. But Jon Schaffer (2010) does not help us here. Should we *choose* to explore the premodern noetic Primordial Wisdom Tradition of our species we discover—both conceptually and direct contemplatively (*yogi pratyaksa*)—that the basal universal quantum ZPE vacuum energy itself continuously arises in its ontologically prior all pervading formless primordial awareness-consciousness ground—*dharmakaya, gzhi rigpa,* Tao, Nirguna Brahman,

Ein Sof)—the prior *kosmic* ground/whole of everything that appears in cosmic space and time to a sentient perceiver/observer. Our *relative* human consciousness-experience arises and plays in that *ultimate* awareness-consciousness "implicate order of the vast enfolded unbroken whole". [David Bohm] *Kosmic* quantum holism indeed.

Such a noetic exploration is a 'post-empirical' trans-conceptual wisdom *choice.* But few will enter in. In that vast Upanishadic 'forest of wisdom' most will choose to remain cloistered in the comfy conceptual cabin built by the prevailing dogma of our ancient Greek legacy that is common Scientific Local Realism/Physicalism, even as quantum physics continues its tireless effort to be free of such bygone classical ideology.

Nonlocal quantum entanglement—both Irwin Schrödinger and John Stewart Bell called it the very essence of quantum mechanics—is *ipso facto* a macropsychic cosmopsychic holistic *ultimate process* that inherently possesses *relative* quantum information states beyond the quantum states of the ontic pluralism of its micropsychic microparticle parts. I shall not here again describe nonlocal superluminal panpsychic quantum entanglement other than to say that it has been decisively confirmed by Irish physicist John Stewart Bell in his 1964 justly famous Bell's Theorem, and further confirmed by some eighteen subsequent 'Bell Tests' through 2021. [Boaz 2023 *Ch. IV*] Nonlocal quantum entanglement—the *relative* modern physics principle of *ultimate* holistic *kosmic* indivisibility (Niels Bohr's "unitarity")—that prior unity of quantum interdependent interconnectedness (Buddhist Dependent Arising or *pratitya samutpada*) is now considered one of the most certain principles of modern physics.

Shan Gao in 2007 argues that when considering the quantum collapse (the 'state vector reduction') of Schrödinger's *Ψ*-wave function:

> Consciousness [human] is not reducible or emergent but a new fundamental property of matter. This may establish a quantum basis for panpsychism, and make it be a promising

> solution to the hard problem of consciousness... the relationship between matter [brain] process and conscious experience (Chalmers 1996).

Emergentism. The perennial debate between the inherent subjectivity of Priority Monism panpsychism/cosmopsychism and objectivist physicalist 'Emergentism' is precisely that distinction. Panpsychism holds that consciousness (human and its primordial awareness-consciousness ground) is a basal fundamental feature of the *kosmos* as a whole. Emergentism holds rather that such consciousness emerges from purely physical brain matter structure and function thereby precluding an integration of the 'spooky' subjectivity of consciousness with the ostensibly pure objectivity of the prevailing scientific ideology that is Scientific Local Realism and its epistemic consort Scientific Materialism/ Physicalism. "Only the great whole is completely objective." [William Earle]

The question now becomes: How it is that human consciousness-experience emerges from mere matter; or is it truly an inclusive non-material fundamental quality of the matter-energy of space and time? Recall here that mereologically speaking, relative derivative human consciousness perforce arises in its ontologically prior panpsychic primordial timeless awareness-consciousness ground (*dharmakaya, gzhi rigpa, Tao, En Sof, Parabrahman, Abba Nondual God the Primordial Father*), David Bohm's "implicate order of the enfolded vast unbroken whole."

Emergentism is the standard realist physicalist 'scientific' resolution of David Chalmer's 'hard problem of consciousness'—how is it that conscious subjective human experience (*qualia*) arises from objective purely physical brain structure and function?

But is physicalist emergentism an intelligible response to the fortunate prodigious 'explanatory gap' that may be seen as the fundamental Basic Bpace (*chöying, dharmadhatu*) that pervades all matter and sentient experience-consciousness? [Gautama the Buddha, Adi Shankara, Chalmers, Stapp, von Neumann, Wigner] Many recent philosophers of physics and philosophers of consciousness now doubt it. Human consciousness as it arises in its trans-conceptual all subsuming noetic nondual aboriginal awareness-consciousness ground, and spacetime matter-energy are clearly two conceptually non-reducible faces or voices of a noetic, ontic prior yet phenomenally present nondual unity—the prior unity of objective Science and its perfectly subjective nondual Spirit ground. And yes, that *ultimate* primordial consciousness ground has secondary derivative objective causal efficacies in the *relative* physical dimension of quantum space and time.

Therefore, on the account of Gao's "Unified Quantum Theory of Matter and Consciousness", "Consciousness [human] must be not reducible or emergent, but a new fundamental property of matter." And the relation of that matter-energy ($E=mc^2$) to relative human consciousness and its ultimate consciousness ground is one of identity. That is required if the consciousness of a human observer-experimenter *causes* the collapse of the subjective infinity of potential probable quantum reality eigenstates of the superposed quantum wave function into an objective locally real spacetime object—an electron. If subjective human consciousness can 'collapse' the quantum Ψ-wave function to reveal objectively 'real' stuff, then that consciousness must be a fundamental property of matter. But is it? If so, how?

Consciousness: Human and Ultimate. Unfortunately, none of the viable panpsychic explanations now on offer address the mereological priority elephant in the room. That is to say, the inherent interdependence and interconnectedness of our *relative* human consciousness to the prior and present indwelling Presence of the *ultimate* primordial ground in which, or in whom it perforce arises is conspicuously absent in these panpsychic explanations. Neither Shan Gao's excellent Unified Quantum Consciousness Theory, nor the acolytes of the closely related Intrinsic Nature Argument manages to unify human consciousness in both its objective and perfectly subjective voices as a "fundamental property of matter", nor as an inherent "intrinsic nature" of matter, with the inherent propitious prior primordial consciousness ground, vast boundless whole in which this all necessarily arises as spacetime matter and energy. Unification gap indeed.

The inherently physicalist explanations—whether 'atomistic micropsychic' or proto-noetic 'constitutive cosmopsychic'—have failed to dodge the conceptual *aporia* (contradictions inherent in purely conceptual cognition) and *ipseity* (selfhood) that result from mere semiotic explanation. We require here both voices of our human *noetic cognitive doublet*—objective conceptual, and subjective contemplative. It is the latter that is the finite conscious cognitive portal or 'gap' that opens into the infinite 'spiritual' direct experience (*yogi pratyaksa*) of the profound nondual panpsychic kosmopsychic primordial ground of Being Itself—vivid, numinous always already present spiritual Presence of That (*tathata*).

And yes, that requires cognitively committing to spooky transconceptual "mindfulness of breathing" as the Buddha called it, *jnanaprana* wisdom meditation practice guided by the very subtle mind of an authentic meditation master. 'Spiritualized' conceptual virtuosity is not an adequate substitute for noetic primordial wisdom (*jnana, yeshe,* gnosis) contemplative practice.

Sadly, this confusing situation has become the 'really hard problem of consciousness' for our 21st century *Noetic Revolution in Matter, Mind and Spirit* (Boaz 2023) that is now abroad in our human cognosphere.

Be all that as it may, the entire endeavor of Metaphysical Scientific Realism/Physicalism has come under attack by the nominalist Antirealism that began with Niels Bohr and his Copenhagen Interpretation of our 20th century quantum orthodoxy. Objective Scientific Local Realism or the inherent subjectivity of Quantum Antirealism? Or perhaps a nice centrist Buddhist Middle Way that unifies this tiresome false dichotomy.

Brief summary. Hence, what we have accomplished thus far through our engagement with primary 'priority monistic' panpsychism/kosmopsychism is a new noetic ontology and a reasonable epistemology that does not nihilistically deny the physical dimension, nor the mental dimension; one that opens an ontological 'middle path' between this false dichotomy of two untenable equivocal ontic extremes—*either* 'eternalist' substantialist physicalist absolute existence *or* 'nihilist', idealist absolute nonexistence. We must develop a centrist integral Noetic Quantum Ontology. Middle Way Prasangika Madhyamaka Buddhism, the causal conceptual foundation of acausal nondual *Dzogchen* view and *Ati Yoga* practice have accomplished that foundation. Let us further develop it.

Panpsychicism and Buddhism

Middle Way Buddhism in India and Tibet. Buddhists have for 26 centuries evolved a consistent centrist panpsychic cosmopsychic epistemology and ontology. It is known as *Dzogchen,* the Great Perfection. On the accord of H.H. Dalai Lama, the conceptual foundation of *Ati Dzogchen* is Mahayana Tibetan Buddhist Prasangika Madhyamaka, the Indian Middle Way (*madhyama pratipad*) Consequence (*prasanga*) School of 2nd century Nagarjuna and 7th century Chandrakirti. Prasangika Madhyamaka is thoroughly embedded in the Indian Mahayana *Prajnaparamita* 'perfection of wisdom' tradition. The other principle school of the Indian Mahayana Vehicle is the Yogachara School with its 'mind only' (*chittamatra*) Metaphysical Idealism. [Boaz 2020 *Ch. V*]

In the 8th century CE both Middle Way Indian Buddhism and its 2nd century *Dzogchen* transmission found their way to the snowy land of Tibet. In the late 10th century, the pre-classical 'valid cognition' (*tshad ma*) Tibetan Buddhist Mahayana/Vajrayana teaching vehicle became firmly established as the centrist 'Middle Way' for the Tibetan Buddhist tradition through the foundational metaphysical principle of Indian acausal Emptiness/*shunyata* with its causal Interdependent Arising (*pratitya samutpada*). The Dependent Arising of spacetime form from its Basic Space (*chöying*) boundless emptiness ground is the utter interdependent interconnectedness of all inherently empty phenomenal reality as it arises from its primordial nondual *dharmakaya* perfectly subjective Spirit ground.

These two, together with the Buddha's noself (*anatman*) rejection of an eternal Hindu Vedic Atman Self, became the essential teaching of all Buddhist schools throughout the premodern world.

This twelve link chain of causal 'conditioned arising' begins with primal 'ignorance' (*avidya, marigpa*), failure of recognition of the Buddha's Two Truths and his Four Noble Truths that cause the cessation of suffering, and finally buddhahood. The 'groundless ground' of the emptiness of interdependently arising form *is* Dependent Arising. The relation is one of identity. Buddha told "Form is empty; emptiness is form." Spacetime form arises from the boundless whole that is its primordial emptiness *dharmakaya* ground.

The teaching of Emptiness and Dependent Arising (Dependent Origination) offer the method to transcend our two metaphysical extremes that are absolute existence of phenomena and its absolute nonexistence. Middle Way founder Nagarjuna told it well:

> Whatever arises interdependently,
> that is explained to be emptiness.
> That being is a dependent designation
> is itself the Middle Way.

Hence, on the accord of the essential Mahayana *Prajnaparamita sutras* all arising appearing spacetime reality is utterly causally interdependent and interconnected in this vast boundless whole that is the primordial emptiness 'groundless ground' of nondual Ultimate Truth reality that is Being Itself, empty ground of the fullness of the Relative Truth spacetime dimension of phenomenal form arising therein. That aboriginal ground is 'groundless' because, along with its interdependent arising of form, emptiness itself possesses no intrinsic absolute existence! It is rather, 'established by conceptual minds'.

Well, what is all that to me? Who Am I? Who is it that I Am? Always present 'innermost' indwelling bright blissful Presence of That (*tathata*) primordial ground is who we actually are, our 'supreme identity' of the aboriginal 'supreme source' condition of that vast boundless whole that is nondual Being Itself. Human happiness, both relative and ultimate, is dependent upon the recognition, then realization of That (*tathata*). What is your mind? That is the very nature of your mind. We all arise, abide, and expire as cognizant participants of that vivid luminous "vast implicate unbroken whole" (Bohm).

Be That as it may, for the Indian Middle Way Prasangika Madhyamaka, its founder Nagarjuna deconstructs these two falsely dichotomous notions of our being here—absolute existence and absolute nonexistence—and then points out that the truth of the matter is actually a natural complementarity of these two seemingly opposite views—this prior and present "unified conjugate pair" (Bohr).

We have seen that quantum pioneer Niels Bohr applied this great wisdom truth in his prodigious 1927 quantum Principle of Complementarity. Werner Heisenberg provided the other foundational quantum physics principle with his 1926 Principle of Uncertainty. In 1928 Irwin Schrödinger then formalized these two foundational quantum principles as the quantum subjectivity of the infinitely 'superposed' universal quantum Ψ-wave function, his amazing Schrödinger Equation that governs the evolution of the wave function of all quantum systems that comprise the whole of appearing spacetime reality.

Hence, appearing reality is empty. Well, empty of what? This arising spacetime reality of ours is decidedly not empty of *relative* spacetime existence! Real stuff is everywhere. Phenomena is empty of any *ultimate* intrinsic existence. Relative spacetime phenomena remain alive and well, and undeniable. Appearing reality is then "Absent and empty of any whit of intrinsic *ultimate* existence (*svabhava*)." [Nagarjuna] That is the truth of 'nihilist' nonexistence or monistic Metaphysical Absolute Idealism. "But we must respect the relative conventional existence in which we live." That is the truth of 'eternalist' substantialist absolute existence or monistic Metaphysical Local Realism/Physicalism.

And because all dualities are in the ultimate view unified in the vast all subsuming whole, these Madhyamaka Two Truths—'concealer *samvriti*' spacetime Relative Truth dimension of form, and its nondual *paramartha* Ultimate Truth emptiness/*shunyata* ground—must be seen as an ontic prior yet phenomenally present *one truth unity* (*dzog*). [Boaz 2020 *Ch. V*] Yes, as Buddha told in his "Fourfold Profundity",

> Form is empty; emptiness is form.
> Form is not other than emptiness;
> Emptiness is not other than form.

Unified love-wisdom centrist 'middle path' indeed. Because form is empty of ultimate existence we need not grasp and covet its manifestations. *Mahasukhaho!* Great joy!

Panpsychism: Theme and Variations. Let us now further consider some recent variations of ancient holistic panpsychism.

To fully understand the profundity of panpsychism's theory of mind we require a "top down" holistic primary 'cosmopshchic' "Priority Monism" (Schaffer 2010). That grounds our conceptual and physical, and our trans-conceptual and trans-physical cosmic participating parts or aspects of a mereologically non-essentialist but inclusive, metaphysically ontic ultimate, all embracing, nondual primordial boundless *kosmic* whole (*dharmadhatu, mahabindu, dharmakaya*). That vast boundless whole is the primordial awareness-consciousness all pervading ground of Being Itself. It subsumes the participating parts arising and instantiated therein. This top down 'grounding relation' is often called 'grounding by subsumption'. The instantiating parts are naturally subsumed in the prior all embracing aboriginal whole itself. Some definitions. *Panpsychism,* broadly construed, is the ontic or metaphysical view that the basal primordial awareness-consciousness ground or whole that includes all of spacetime reality is ultimately fundamental, ubiquitous, all pervading and all subsuming.

Constitutive Panpsychism is the metaphysical view that *relative* 'organic' sentient human and animal consciousness is not fundamental but is derivative of an awareness-consciousness reality dimension that that is more foundational, prior, and original.

Constitutive Panpsychism has two sub-variants: 1) *Atomistic Micropsychism*, the 'bottom up' view that organic human and animal consciousness/experience is grounded, not in the fundamental primordial consciousness whole itself, but in purely physical, conscious, atomistic subatomic microcosmic particle structures of matter-energy. 2) *Cosmopsychism*, including Schaffer's (2010) 'priority monistic cosmopsychism', is the holistic 'top down' view that all spacetime reality—microcosmic and macrocosmic—is grounded in the formless, timeless, selfless all subsuming primordial consciousness panpsychic whole/ground. Yet even here we find a cognitive residue of the prevailing ideology that is Scientific Local Realism/ Physicaliam. We shall see that only *Dzogchen Kosmopsychism* is such local realist/physicalist bias surrendered to the nondual whole.

Hence, Constitutive Panpsychism as primary 'Priority Monism' equals top down Constitutive Cosmopsychism. None of these panpsychism variants sees our ultimate primordial awareness-consciousness ground as perfectly subjective nondual. All of them have bought in to an objectivist proto-physicalist 'intrinsic nature' ontology not only for appearing physical stuff, but as well for the primordial panpsychic ground itself. Because there is present here no trans-conceptual basal nondual contemplative ground, limiting objective conceptual theoretical cognitive artifacts remain. Such explanations are not nondual ultimate explanations. They fail to understand the prior unity of objective Science and its perfectly subjective all embracing nondual Spirit ground.

And yes, for primary priority monistic *Dzogchen Kosmopsychism* dualistic semiotic conceptual imputation and reification is finally undone and transcended in an ultimate noetic nondual panpsychic ontology.

Do not our linear conceptual wisdom seeking strategies finally require a subject-object duality collapse or surrender (*Wu Wei: Tao*

Te Ching, Ch. 48) into the nondual basic space (*chöying, dharmadhatu*) of our foundational grounding *buddic* wakefulness, nondual vivid clarity of *awakened mind* that is ultimately this timeless, formless, selfless awareness-consciousness being itself? That unborn *kosmic* primordial ground is adorned from time to time by self-reflexive human consciousness arising within it. Such is our nondual (nonconceptual prior subject-object unity) *Dzogchen Kosmopsychic* wisdom imperative.

> Objects altogether are a whole, yet separate;
> Being Itself altogether, yet apart;
> In harmony, yet dissonant.
> Of objectivity, there is a great whole;
> And through this all things arise and pass away…
> The Logos speaks: all things are One.
>
> —Heraclitus (Author's translation)

That all said, it seems to me that the many extant variations on this vital panpsychic theme (e.g. 'bottom up', atomistic micropsychism; and 'top down' cosmopsychism) that is our new Western analytic panpsychic adventure into the vexing metaphysics of consciousness (Goff 2017; Goff "Panpsychism" entry in *Stanford Encylopedia of Philosophy* 2017) miss the mark that is this propitious and providential trans-objective, post-empirical, unitary, monistic/holistic—in a word *kosmic*—nondual view. Twentieth century Western analytic ideological habits of mind—Objectivism, Local Realism, Physicalism, Pluralism, Substance Monism, panpsychic atomism-micropsychism, local causal determinism, and the invidious 'scientific' closure principle ("Idols of the Tribe" Boaz 2023)—still haunt this brave new world of East-West panpsychic exploration. Let's take a closer look.

Priority Monistic Cosmopsychism

Let us begin with an all too brief introduction to a vital holistic Western panpsychic 'unified cosmic consciousness' variant known to the initiates of the recent panpsychic cabal by the cloddish epithets "cosmopsychism", or worse, "priority monistic cosmopsychism" that are the panpsychic foundations of an even more obscure noetic nondual *Dzogchen Kosmopsychism.* Let's conceptually unpack this.

With the exception of the nondual *Dzogchen Kosmopsychic* view, any of these are preferable to the fraught, finally unintelligible offerings of a fragmented, physicalist 'atomistic constitutive micropsychism' panpsychism. All suffer from the relentless 'Combination Problem' in one or more of its omnipresent guises: 1) the subject CP; 2) the quality CP; and 3) the structure CP. [Chalmers 2017 "The Combination Problem for Panpsychism"] Broadly construed the CP obstructs coherent explanations of the 'bottom up' arising of panpsychic conscious human macrosubjects from 'concrete' purely physical micropsychic microparticle microsubjects. We still need to know which is ontologically prior—the primordial whole, or its participating microscopic parts.

I shall herein attempt to integrate the promising monistic cosmopsychic view of Jon Schaffer (2010) with the parallel holistic panpsychic wisdom of the East as it has arisen in the Vedic-Hindu *Sanatanadharma* through the Advaita Vedanta of Adi Shankara (8th century); in nondual monistic cosmopsychic Kashmiri Shaivism (9th century); and in the 2nd century Two Truths trope of Nagarjuna, and 7th century Chandrakirti Buddhist Middle Way Prasangika Madhyamaka, the foundation, on the accord of H.H. Dalai Lama (2000), of the quintessential acausal *nondual* teaching that is the Tibetan Vajrayana Buddhist teaching pinnacle *Ati Dzogchen,* the Great Perfection or Great Completion of the causal duality of the Mahayana Causal Vehicle's Two Truths view and praxis. "In one way or another, primordial [*kosmic*] consciousness must ground the reality of evolved sophisticated phenomenologies." [Itay Shani]

In the words of Jon Schaffer (2010),

> The *monist* holds that the whole is prior to its parts, and thus views the cosmos as fundamental, with metaphysical explanation dangling downward from the One. The *pluralist* holds

> that the parts are prior to their whole, and thus tends to consider particles fundamental, with metaphysical explanation snaking upward from the many. Just as the materialist and the idealist debate which properties are fundamental, so the monist and pluralist debate which objects are fundamental... Physically, there is good evidence that the cosmos forms a [quantum] *entangled system* and good reason to treat entangled systems as irreducible wholes.

This perennial debate between ontic monism and ontic pluralism is, as William James told, "The most central of all philosophic problems..." The monists include an illustrious wisdom clan that begins in the East with Gautama the Buddha and Middle Way Buddhists Nagarjuna and Chandrakirti, and in the West with Jesus the Christ, Plato, Plotinus, Proclus and the Neoplatonists, then Spinoza, Leibniz, and Hegel; and ends with the subtle panpsychic minds of Niels Bohr, A.N. Whitehead, and H.H. Dalai Lama.

In the West ontic Monism has, in 20th century philosophy and physics, fallen on hard times. Bertrand Russell naïvely summed it up as "contrary to common sense". Of course modern science quantum physics is also contrary to common sense. It drove the great scientific mind of Albert Einstein to distraction! The logical mathematical mind of Bertrand Russell, co-author of the great *Principia Mathematica,* utterly failed to comprehend the urgent difference between common 'Existence Monism'—that "only one thing exists"—and the 'Priority Monism' of 'constitutive cosmopsychism' that does not deny the common sense pluralism that many things exist *relatively* and derivatively, while revealing that *ultimately* the all subsuming whole is ontologically prior. And that is the view of conceptual cosmopsychism that becomes noetic nondual *Dzogchen Kosmopsychism.*

Is it so hard to understand that, as the centrist Middle Way Madhyamaka Buddhists pointed out two centuries past, spacetime physical and mental 'form' arises relatively and pluralistically as the 'many' from an ultimate monistic 'One'—primordial all embracing emptiness ground—by whatever grand name? Recent 21st century work in 'constitutive cosmopsychism' has again opened the wisdom door to a primary panpsychic 'Priority Monism' beyond easily dismissed common sense 'Existence Monism'. Jon Schaffer again:

> The core tenet of historical monism is not that the whole has no parts, but rather that the whole is *prior* to its parts. As Proclus says, 'The monad is everywhere prior to the plurality... the whole that precedes the parts is the whole that embraces all separate beings in the cosmos.' Such a doctrine presupposes that there are parts, for the whole to be prior to them. The historical debate is not a debate over which objects exist, but rather a debate over which objects are fundamental... The world has parts, but the parts are dependent fragments of an integrated whole... A substantive debate as to whether the whole or its parts is prior can arise only if the whole and its parts both exist... [This] assumption may be understood as a kind of *metaphysical foundationalism,* on analogy with *epistemic foundationalism*... There must be a ground of being... something from which the reality of derivative entities ultimately derives... This is the *question of fundamental mereology*... the question of what is the ground of the mereological hierarchy of whole and part... Armstrong suggests that 'The mereological whole supervenes [depends] upon its parts, but equally the parts supervene upon the whole'.

Ontic idealist F.H. Bradley supports such an ontology, "Everything less than the entire universe is an abstraction from the whole." Hegel opines, "Organic being [is fundamental] ... undivided oneness as a whole." Spinoza also told it well, "There is just one genuine substance, the primordial kosmos itself."

These monistic panpsychic views constitute reasonable parallels to Buddhist causal Interdependent Arising/*pratitya samutpada.*

However, Middle Way Buddhists deny that neither the arising parts nor their all subsuming whole/ground (*shunyata, dharmakaya*) possess any whit of 'concrete', inherent ultimate "intrinsic existence". Rather, the whole 'groundless ground' and its relative derived instantiating parts are all imputed, reified and "established by conceptual minds". [H.H. Dalai Lama]

Note here again that whereas the Middle Way Buddhist Two Truths view permits *relative* conventional really 'real' object-entities, while denying them "any whit" of *ultimate* intrinsically real existence—even Jon Schaffer's prodigious Priority Monist cosmopsychism retains a dualistic cognitive artifact of the classical 'scientific' physics ideology that is the waning knowledge paradigm of the *absolute* existence of real stuff—Metaphysical Scientific Local Realism/Physicalism. Here, spacetime parts arising in their cosmopsychic whole/ground are inherently physical locally real *ultimately* existent 'concrete' object-entities. Worse still, "There can be one and only one basic actual object prior to all of its proper parts, namely the cosmos." For Jon Schaffer our all subsuming cosmos is 'concrete' and ultimately physical—all the way down to minute micropsychic subatomic baryonic quarks and leptons, and bosonic photons. That is his priority monist view. But is this reality *ultimately* 'concrete' and physical?

Russellian Monism. Ontic pluralism denies that ontic Priority Monism is true; the whole of the cosmos is not fundamental. Ontic pluralist Bertrand Russell holds, not the whole itself but multiplicity is fundamental. However, in his 1927 *Analysis of Matter* Russell presents his panpsychic version of 'Russellian Monism' which introduces the now popular Intrinsic Nature Argument as a way of introducing 'consciousness' to the prevailing Scientific Physicalist ideology that has hitherto entirely ignored or denied it. I have criticized that argument above. As the Middle Way Buddhists have demonstrated, matter does not possess an absolute or ultimate *intrinsic nature,* although it is relatively conventionally real—"established by our human conceptual minds." [H.H. Dalai Lama]

Moreover, the thorny 'Combination Problem' here again obtains: how do physical microphenomenal holons combine to produce macrophenomenal properties and human consciousness/experience? How is it that human consciousness is a fundamental participant of the cosmos? How is human consciousness related to physical causality? How is it related to its primordial consciousness ground? Russellian Monism fails to clarify these urgent metaphysical issues. [Goff, "Panpsychism", *Stanford Encyclopedia of Philosophy*] Is Russell finally an ontic Pluralist or a Monist? Hard to tell.

That there arises a plurality of 'real' spacetime stuff is clearly true. Here we are. But neither an ontic Pluralism, nor a naïve Existence Substance Monism—that there is only one physical object/thing that exists—refutes the Priority Monist view that the one vast all subsuming whole, with all of its instantiating parts, is ontologically fundamental.

The pluralist 'common sense' metaphysic that the separate divisible parts are the ontic priority over against the all embracing *kosmic* whole in which these parts arise is indeed a dreary antiholistic metaphysic. And very hard to believe; unless one has entirely bought into the cognitive 'conformation bias' that supports the waning 'constitutive atomistic micropsychic' Scientific Materialism/Physicalism doctrine.

Madhyamaka Buddhist Critique of Monistic Cosmopsychism. Well, what will Middle Way Madhyamaka Buddhists say in response to Jon Shaffer's monistic cosmophychism? They will charge that this priority monist cosmopsychic view does not escape the ontic bias that is 'eternalist' substantialist Metaphysical Scientific Local Realism/Physicalism, the scientific 'metaphysical extreme' that misses the mark of a centrist 'middle path', and errs on the side of an inherent absolute intrinsic prior existence of macropsychic purely physical spacetime matter-energy.

The Buddhist Prasangika Madhyamaka Middle Way has struck a conceptual balance between the 'metaphysical extremes' that is the false dichotomy of *either* absolute existence, *or* absolute nonexistence. And Prasangika is said to be the causal conceptual foundation of acausal trans-conceptual, nondual *Dzogchen Kosmopsychism* view and practice. That nondual view and practice surrenders the ontic local realist/physicalist bias that is Scientific Local Realism/Physicalism into its prior all subsuming ground. [H.H. Dalai Lama 2000; Longchenpa 2001; Boaz 2020 *Ch. V*]

Hence, ontic Monism and ontic Pluralism are exclusive views as to what is mereologically and ontologically fundamentally prior. Thus Schaffer's ontic epithet "Priority Monism" over against naïve 'Existence (substance) Monism'.

In primary *Dzogchen Kosmopsychism* the nondual unified whole of the formless, timeless monistic Pythagorean all embracing *kosmos* is the actual metaphysical primordial ground that transcends and pervades lesser derivative merely physical cosmos, priority monistic cosmopsychism with its 'concrete' physical cosmic object entities.

Hence, in the gloss of the Buddhist *Dzogchen* kosmopsychic view, the *relative* physical and mental particulars of spacetime 'form' arising in its vast *ultimate kosmic* 'emptiness ground'—the *Perfect Sphere of Dzogchen*—necessarily appear, participate, and expire—indeed have never departed—that nondual primordial whole that is non-concrete, insubstantial, immaterial, formless boundless selfless nondual whole that is primordial Being Itself. "Form is empty; emptiness is form." [Buddha]. And that 'groundless ground' of aboriginal *kosmos*, Being Itself, is itself "Absent and empty of any whit of intrinsic existence (*shunyata shunyata*)." [Nagarjuna] That is known as the "emptiness of emptiness" (*shunyata shunyata*). Yet it transcends, embraces and subsumes the entire material cosmos as its nondual ontic prior aboriginal awareness-consciousness, all embracing, numinous Spirit ground itself—primordially awakened *buddic* mind—the very Nature of Mind and all of its relative experience.

Brief Summary. Thus does *Dzogchen Kosmopsychism* provide the foundation for our Panpsychic Noetic Quantum Ontology, as we shall see in Chapter II. Together these two demonstrate the prior unity of objective Science and its perfectly subjective primordial whole, nondual Spirit 'groundless ground'—always present blissful spiritual clear light Presence of That (*tathata*)—great gift (*jinlob*) to all of us guests of the phenomenal world.

Dzogchen Panpsychism/Kosmopsychism

Because it is the metaphysical key to our panpsychic centrist integral Noetic Quantum Ontology, let us explore *Dzogchen Kosmopsychism* a bit more deeply. I have come to call it such because highest *Ati Dzogchen* is a 'Priority Monist' primary panpsychic/kosmopshchic noetic (body-mind-spirit unity), nondual (*nyimed*, 'not two, not one but nondual'), *ontologically relative* and perspectival (phenomenal reality is reified, imputed, and designated via our deep cultural background "global web of belief"), therefore non-essentialist (phenomena are absent any *essential* intrinsic or inherent ultimate nature), prior and present body mind spirit subject/object unity. [That sentence should perhaps be taken out and shot.]

That prior noetic unity of all relative conventional spacetime phenomena abides beyond the odious split of a knowing subject and its perceptual and conceptual objects known—the justly infamous 'subject-object split'. *Maha Ati Dzogchen* View, Practice, Conduct and Fruition/Result is holistic, primary monistic, trans-conceptual and nondual. It embraces the *one truth unity* (*dzog*) of the Buddhist Middle Way Two Truths (relative and ultimate) dominant epistemic trope.

Just so, that *Perfect Sphere of Dzogchen* is all pervading, all subsuming, *ultimate*, fundamental, vast primordial *kosmic* awareness-consciousness ground or base (*gzhi rigpa*)—the all pervading luminous, numinous trans-conceptual unbounded whole itself (*chöying, dharmadhatu*)—nondual Spirit Itself, ultimate *dharmakaya* ground, the vast emptiness/openness in which all *relative*, conditional, utterly selfless post-theistic spacetime physical and mental cosmic phenomenal forms arise, abide and pass away.

Chögyal Namkhai Norbu Rinpoche on *Dzogchen*, the Supreme Source (1999):

> The essence of all the Buddhas exists prior to samsara and nirvana... It transcends the four conceptual limits and is intrinsically pure; this original condition is the uncreated nature of existence that always existed, the ultimate nature of all phenomena... It is utterly free of the defects of dualistic thought which is only capable of referring to an object other than itself... It is the base of primordial purity... Similar to to space it pervades all beings... The inseparability of the two truths, absolute and relative is called 'primordial Buddha'... If at the moment the energy of the base manifests, one does not consider it something other than oneself, it self-liberates [into its ground]... Understanding the essence one finds oneself always in this state... dwelling in the fourth time, beyond past present and future, the infinite space of self-perfection... pure dharmakaya, the essence of the vajra of clear light.

Therefore, this Ultimate Truth of reality itself is generally considered in the Buddhist Mahayana/Vajrayana wisdom vehicle as *Dzogchen.* [*Dzog* means complete or perfect; *chen* means great.] Its nondual contemplative practice is *Ati Yoga,* the ninth and subtlest of the nine vehicles of the Nyingma School *Dzogchen* tradition. [Boaz 2020 *Ch. V*]

The noetic nondual 'innermost secret' view and practice of the accomplished adapt *Ati Dzogchen* yogi is profoundly expressed by the omniscient mind and voice of 14th century Tibetan Buddhist *Dzogchen* master Lonchen Rabjam (Longchenpa 2001) in his sublime *The Precious Treasury of the Basic Space of Phenomena* (*Chöying Dzöd*). The *Chöying Dzöd* is the Great Perfection approach of the *Dzogchen Trekchö* (cutting through conceptual solidity) teaching cycle. Indeed, it is based upon all three *Dzogchen* teaching cycles—*semde* mind series, *longde* space series, and the *mangagde/upadesha* personal pith instruction series. It is considered to be the heart essence of all noetic spiritual teaching by any name. It "establishes the Base, Ground, Path, and Result/Fruition of all aspects of Ati Dzogchen". [Zechen Rabjam]

In the clear words of Longchenpa's *Chöying Dzod* (2001),

> *Trekchö* is the ultimate meaning of the ground of being—for those of the very highest acumen—to effortless freedom… Basic Space (*chöying*) is buddha nature—buddhahood that is spontaneously present [the state of Presence] by its nature… beyond supreme emptiness and sublime knowing, ancestor of all the buddhas, unborn naturally occurring timeless awareness—utterly lucid awakened mind—marvelous and superb, primordially and spontaneously present… It is not existent, for it has no substance. Nor is it nonexistent for it pervades all samsara and nirvana… Thus phenomena are equally existent and nonexistent, equally empty, equally true and equally false. Awakened mind is primordial Basic Space… ultimate truth, fully evident [through] direct experience of it so there is no need to seek it elsewhere. Vividly lucid, it does not entail dualistic perception and is free of conceptual elaboration… and so it is called the 'essence of being'. All arising form and wisdom is the adornment of unborn Basic Space… single mandala of natural timeless awareness… spontaneous equality. Samantabhadra [*dharmakaya* buddha] is that awareness—that innermost buddha nature… The basic view [of Dzogchen] is the view of freedom from extremes… of primordially pure and naked intrinsic awareness, ineffable and unceasing… marvelous and superb.

Dzogchen arises as the trans-conceptual, nondual, non-atomistic, non-reductionist supreme noetic nondual teaching whose View is shared with what contemporary students of panpsychism term holistic *priority monistic cosmopsychism* (Schaffer 2010), a long epithet for a long luminous history of nondual primordial wisdom that arises at the pinnacle of each of our primary wisdom traditions—the Hindu *Sanatanadharma* as Advaita Vedanta; the Buddhadharma as *Dzogchen and Essence Mahamudra*; Taoism as *Tao chia*, Abrahamic Monotheism in which arises nondual mystical *Zohar/Kabbalah*, and in due course Hermetic mystical Christianity. [*Meditations on the Tarot: A Journey into Christian Hermeticism*, Anonymous, Tarcher/Penguin, 1985] *Dzogchen* is often said to be the subtlest of these.

That noetic nondual (body, mind, spirit, subject-object unity) Primordial Wisdom Tradition of humankind has one foundational teaching, namely the monistic basal fundamental *unity* of the arising multiplicity of spacetime phenomena. That essential unity is an ontic and epistemic prior yet present condition for human knowledge, wisdom, and happiness. That prior and present unity grounds all ethical acts of compassion and of wisdom. Spacetime phenomena arise, participate, and are instantiated in that all subsuming unity, by whatever grand name or form (*namarupa*). *Truth*—objective scientific, mathematical, philosophical, and subjective gnostic, religious, mystical and nondual—presupposes such an essential unity of the multiplicity manifesting in this world of ours.

In the absence of that essential unity reality ground one can perforce derive no essential foundational objective knowledge, nor subjective wisdom. Everything is chaotic, separate and disconnected. As to esoteric, trans-conceptual, ultimate 'innermost secret' contemplative direct experience (*yogi pratyaksa*), it is *ipso facto* experience of our nondual unitary 'groundless ground' of arising reality Being Itself, just as Longchenpa revealed in the above quotation.

Now as to these illustrious holistic panpsychic cosmopsychism variants the subtlest or 'highest' *nondual* teaching is as I have said, *Dzogchen* View and Practice. How is this so? We have seen that *Western philosophical panpsychism/cosmopsychism—ancient or recent—is not inherently nondual.* How so? 1) it retains a tenacious grip on conceptual artifacts and subtle realist/physicalist cognitive biases; 2) the requisite nondual contemplative grounding practice or yoga (union, *religio, yogi pratyaksa*), under the guidance of a qualified meditation master is conspicuously absent. In short, even antirealist cosmopsychism still remains, as Longchenpa has told above, a 'conceptual elaboration', albeit an elegant one.

A trans-conceptual "grounding relation" via meditative contemplative mindfulness practice is required in order to transcend these heady dualistic conceptual trappings and actually establish and ground a relative conventional pragmatic, selfless, kind, compassionate practice into the lifeworld moral and political conduct of human beings. A dualistic conceptual, intellectual metaphysical grounding relation without its concomitant grounding in a nonconceptual, even nondual unifying contemplative practice from which spontaneously arises more or less selfless ethical conduct is woefully incomplete.

Thus is the conceptually inscrutable selfless nondual Dzogchen kosmopsychic *buddic* Basic Space 'Wisdom Mind' of our primordial formless 'groundless ground'—all pervading awareness-consciousness itself—grounded in psychophysical spacetime form ($E=mc^2$) as beneficent human love-wisdom conduct. With no such grounding relation in trans-conceptual contemplative meditation practice with its conscious altruistic *bodhicitta* ethical conduct—thought, intention and action for the benefit of living beings—this love-wisdom mind poetry of the selfless nonlocal nondual primordial wisdom ground, while very beautiful to the ear, is little more than prosaic conceptual philosophical self-stimulation. What does all that mean for our ultimate understanding that is nondual *Dzogchen Kosmopsychism*, the foundation of our integral Noetic Quantum Ontology?

Sooner or later there comes a point in the relative time incarnation of personal self-ego-I that this self at least momentarily surrenders itself to all-embracing selfless 'noself' (*anatman*)—our indwelling 'innermost' love-wisdom mind Presence of That (*tat, sat*)—by whatever grand name. Here, "Noself is the true refuge of self." [Buddha] Thus does self take refuge—almost moment to moment—in that fearless perfectly subjective mind state of primordial noself, our innermost Buddha nature of mind. And that changes everything!

So, our human metaphysical ontology, epistemology, phenomenology, and engaged compassionate ethics and political conduct are intimately and practically interdependent and interconnected. In order to know the Nature of Mind—nondual ultimate Spirit—we must know and feel how it is that subjective Spirit continuously manifests and expresses itself through objective Science, and in ethical and political conduct. Science, Spirit, Conduct cannot be effectively split. They are, as we have so often seen, a prior and present unity. Human flourishing requires that we practice then as such.

As Buddha told so long ago, "Let it be as it is and rest your weary mind, all things are perfect exactly as they are." Let our life world practice reflect that subtle love-wisdom mind. No need to suffer the slings and arrows of outrageous self denigration, or compensatory self aggrandizement. No need to try hard to change anything at all. No need to seek some future happiness. Our happiness is always already present *now.* That is Longchenpa's "utterly vivid awakened mind". This very subtle understanding of the selfless perennial Ultimate Truth lifts and heals the chaos and fear of Relative Truth self-ego-I being here in the scary chaos of relative time and space.

Now may we serve our unruly self-ego-I through fearless, courageous engaged action for benefit of other beings—the "two benefits of self and other". That after all is the very secret of our human happiness; is it not? Let that time come sooner than later. Indeed, let it be now. Our personal and therefore our collective happiness depends upon it.

How do we do it? Our conceptual mind is here of little help. It is 'mindfulness of breathing', and pithy mantra prayer—*OM AH HUM*—that instantly connects us to always already present primordial Presence of Spirit ground that we are now. [*Appendix A*]

That is the panpsychic kosmopsychic 'grounding relation' that makes us happy here in this precious present moment *now.* Past is gone, but a present memory. Future has not yet arisen, it is but a present anticipation. Therefore, we cannot *become* happy in some future mind state. But we can *be* happy in this timeless moment here and now. That timeless happiness is always already present now as primordial Presence of the ground. Learning to recognize—"brief moments, many times"—and so realize it is a human choice.

In other words, the *relative* conventional 'spiritual' *practice* of our nondual Primordial Wisdom Mind (*jnana, yeshe,* gnosis) Path grounds and motivates a more profound *ultimate* understanding of the all subsuming primordial ground itself—our 'supreme identity'—bright Presence (*rigpa, vidya, christos*) of that emptiness all embracing Spirit ground. We utilize the Relative Truth dualistic practice of the kosmopsychic *Dzogchen* path to tame the 'wild horse of the mind' and awaken to nondual Ultimate Truth that is our always already present awakened love-wisdom mind—'Big Mind', very *buddic* Basic Space (*dharmadhatu*) Nature of Mind. Bright 'innermost secret' Presence of That! That aboriginal ground always arises and expresses itself in our human life world as spontaneous kind, compassionate Conduct—thought, intention and engaged action for the benefit of living beings. In the Buddhist gloss this altruistic process is known as the '*bodhicitta* of intention' that becomes the spontaneous '*bodhicitta* of engaged human action'. Simple open secret of our human happiness. What do you think Dear Reader?

The Result/Fruition of such a bodhisattva path is the *relative* happiness that is human flourishing (*eudiamonia, felicitas*), and in due course and by grace, the full *bodhi* liberation/enlightenment that is *ultimate* Happiness Itself *(mahasuka, paramanda, beatitudo)*—karma free harmless happiness that cannot be lost. And that 'journey of a thousand miles begins with the first step." [Lao Tzu] Profound *Ati Yoga,* the highest or subtlest ninth stage of the nine *yanas* of Nyingma School's nondual *Ati Dzogchen* view and practice provides such a View, Path and Fruition/ Realization—for those smart enough to stop reading about it, and begin the practice. Or perhaps to vitalize an established but neglected practice.

Dzogchen Kosmopsychic Review. Hence, my intention in this chapter is to complete the best of historical and recent panpsychism, namely holistic yet still incomplete primary 'priority monistic' cosmopsychism in the Great Perfection or Great Completion that we now understand as nondual *Dzogchen* kosmopsychic View and Practice. Let philosophers of mind, scientists, spiritual teachers, and everyone else understand this great teaching. The benefit for living beings, including our precious Mother Earth, is immeasurable.

It bears repeating, our monistic panpsychic/kosmopsychic forebears include such illustrious Love-Wisdom Mind avatars as Gautama the Buddha, Longchenpa, Adi Shankara, Moses, and Jesus the Christ.

In philosophy we have Thales, Parmenides, Plato, Plotinus, Proclus, Spinoza, Leibniz, Hegel, Royce, Bradley, Kant, Fichte, Schopenhauer, F.C.S. Schiller, William James (centrist dual aspect "Neutral Monism"), even a reluctant David Chalmers (1996).

In physics we must include Niels Bohr with his Taoist Copenhagen Interpretation of the quantum theory. In mathematics Albert North Whitehead defends a Leibnizian panpsychism via his "extensive abstraction" Process Philosophy, whose view that the process order of spacetime reality is the very "concretion" or instantiation of the primordial unity nature of nondual godhead, for lack of a better name. Heady wine indeed. Ah, the abstruse genius of Whitehead. Not for the metaphysically squeamish.

In the first half of the 20th century our perennial panpsychic wisdom fell on hard times. The prevailing imperious ideology of Western hyper-objective Metaphysical Scientific Local Realism/Physicalism with its hostility toward subjective metaphysical systems crushed all 'non-scientific' mentalist metaphysics. By the end of the century, due in part to the utter failure of anti-metaphysical extremism of the Logical Positivism/Logical Empiricism movement—and in concert with Bohr's antirealist 1928 Uncertainty Relations, and Gödel's devastating 1931 Incompleteness Theorems with the subsequent decline of the very foundations of logic and mathematics (Boaz 2023 *Ch. V*)—panpsychic metaphysics has now again become mainstream. Indeed, there is now abroad in the noble academic domain of philosophy of mind, philosophy of physics, and newly respectable consciousness studies and Contemplative Science a panpsychism renaissance. It is here that we encounter primary monistic nondual *Dzogchen Kosmopsychism.*

We saw above that priority monistic cosmopsychism is known to the Western analytic philosophy trade as a viable "top down" holistic, "priority monism" (the whole is ontologically prior to and greater than the sum of its participating parts) alternative to "bottom up" realist/physicalist, atomistic micropsychic recent incarnations of our perennial panpsychic wisdom tradition.

In esoteric "top down" holistic primary or 'priority monistic' nondual kosmopsychism the cosmic spacetime located atomistic micropsychic baryonic and bosonic panpsychic parts are grounded (the urgent 'grounding relation') in the vast nondual unbounded primordial all subsuming *kosmic* whole itself. Recall that this grounding relation is one of 'grounding by subsumation', or ontic inclusion.

On exoteric atomistic micropsychic panpsychic accounts the microscopic subatomic purely physical cosmic parts (quarks, leptons, photons) are ontologically prior to the kosmopsychic *kosmic* unbounded whole itself, the nondual primordial ground that embraces of our appearing realities—instead of the other way round. Thus do we avoid, as we have seen, the prickly problem of no micropsychic 'grounding relation', along with the pernicious 'Combination Problem'. Our experience is naturally mereologically (part-whole relations) grounded in the top down prior all subsuming boundless *kosmic* whole in which, or in whom bottom up atomistic micropsychic stuff naturally becomes macrocosmic macropsychic living beings.

So the truth of the matter is that mereologically, the nonlinear nondual boundless all subsuming whole cannot be grounded in its participating micropsychic parts. The whole is greater than the sum of its participating parts. The vast macrocosmic boundless primordial whole in which, or in whom this all arises is necessarily ontologically prior to, yet always already present, as its derived microphysical parts. The linear atomistic micropsychic explanation is ultimately pluralistic and reductionist and so views subatomic particle parts as ideologically fundamental. Our perennial essential *unifying view* is here absent. The micropsychic view is an ominous cognitive relic of the dualistic disunity of our waning cultural classical Western Greek Scientific Local Realism/Physicalism ontology.

Bygone knowledge paradigm indeed. Good for calculating and predicting, not ontology.

Priority monistic cosmopsychic Jon Schaffer (2010) has pointed out: "Just as the materialists and idealists debate which properties are fundamental, so the monists and pluralists debate which objects are fundamental." Just so, *Dzogchen* primary monist kosmopsychic panpsychism is a holistic Priority Monist, centrist middle path between Metaphysical Local Realism, and Metaphysical Idealism—and between ontic Monism and Pluralism. Primary Kosmopsychic Monism is an observer-dependent, ontologically relative, perspectival view that transcends yet embraces dualistic, derivative and reductionist micropsychic views—whether panpsychic or orthodox scientific reductionist.

The entanglement/nonlocality of quantum physics and cosmology, with its requisite observer-dependent "observer consciousness" exoterically parallels such a nondual holistic monistic metaphysical view. David Bohm's "implicate order of the vast enfolded unbroken whole" of that primordial quantum entangled/interconnected interdependent indivisible *kosmic* whole is, as Niels Bohr told, an ultimate "unitarity" whose complementary nonlocal entangled "conjugate pairs" perforce reflexively participate as the interconnected all subsuming boundless awareness-consciousness whole—the *kosmic* Tao of reality Being Itself. That Basic Space ground is our home.

Buddha called such a holistic, monistic centrist ontology "Dependent Arising" (causally interdependent interconnected "interbeing", *pratitya samutpada*), the open empty formless, timeless, selfless unbounded ultimate primordial awareness-consciousness unity, macrocosmic ground/whole itself in which, or in whom this multiplicity of microcosmic relative spacetime form—including all of us—are conscious psychophysical instantiations. Who am I? *Tat Tvam Asi,* That I Am! Without a single exception. We should feel better already.

Dzogchen Panpsychic Kosmopsychic Summary

There is much more to be explored in the dualistic analytics and nondual, contemplative direct *experience* of the metaphysics of consciousness. I have herein very briefly argued that what I have rather obliquely termed *Holistic Primary Dzogchen* K*osmopsychism*) is a promising holistic and inclusive view as to such an ontic monistic metaphysic. I attempt to address the lingering paradox of the Greek Local Realism/Materialism/Physicalism paradigm that has now colonized the Western heart and mind with its dualistic micropsychic panpsychic views. I have utilized the unification of Western monistic panpsychism with Eastern Buddhist panpsychic or *kosmopsychic Dzogchen* View and Praxis. I have further developed this syncretic metaphysic in a 2023 book called *The Noetic Revolution: Toward an Integral Science of Matter, Mind and Spirit.*

As we saw above, the panpsychic 'priority monistic cosmopsychism' of recent Western panpsychic philosophy of mind (Schaffer 2010), while avoiding some of the realist and materialist scientific reductionism of physicalist atomistic micropsychism of 'constitutive panpsychism', still retains subtle dualistic conceptual traces or cognitive biases of the failed ontology of reductionist Scientific Local Realism/Physicalism. The metaphysical materialist-physicalist cognitive bias that matter must be intrinsically only physical substance remains more or less unchanged. And there is in Priority Monist cosmopsychism no trans-conceptual, contemplative imperative. This new and promising panpsychic explanation still remains fixed in the Relative Truth dimension of mere objective conceptual cognition. No real metaphysical progress here.

We require a new holistic metaphysical scientific paradigm ontology and epistemology that integrates our human *buddic* wisdom mind subjectivity with the prodigious objectivity of the scientific quantum paradigm. That is the desideratum devoutly to be wished. It abides in our new *Noetic Revolution in Matter, Mind and Spirit* (Boaz 2023) that is now abroad in our human cognosphere. That nascent Science of Consciousness with its now respectable Contemplative Science has given us a centrist 'post-empirical' *Dzogchen Kosmopsychism.*

Here, in the prevailing Local Realism/Physicalism of Big Science, perceiving subjects and their objects of perception and conception are pre-consciously presumed to be reducible to relative physical substance, observer-independently essentially real stuff in an observer/theory-independent, theory laden absolutely objectively "real world out there (RWOT)". If this be so, even panpsychic 'Priority Monistic Cosmopsychism retains a habitual proto-realist, objectivist physicalist/materialist ontic bias, as we have so often seen in these pages.

Thus have I dared to attempt to integrate Western and Eastern metaphysical ontology by introducing Mahayana Madhyamaka's Tibetan foundational acausal nondual *Ati Dzogchen*, the Great Completion/Perfection in a contemporary panpsychic kosmopsychic context. Admittedly, this does some cognitive damage to the primordial purity of the 'post-empirical' inherently nondual *Dzogchen* View. Without trans-conceptual contemplative practice—under the guidance of the subtle mind of the *Dzogchen* meditation master to balance this heady conceptual cognition with its nondual ground—our process remains, well, all too conceptual.

The Mahayana teaching vehicle with its causal centrist Middle Way Prasangika Madhyamaka view is at root the conceptual foundation of a new non-essentialist ontology, denying that the spacetime stuff of relative physical and mental reality has any inherently existing *ultimate intrinsic nature*, let alone a purely materialist/physicalist intrinsic nature.

Rather, Newton's relative "furniture of reality" is, for the great nondual Buddhist mind of 2nd century Nagarjuna, utterly selfless, empty and absent "any shred of intrinsic existence". Empty of what? Relatively 'real' existing spacetime stuff is empty of any permanent substantial *ultimate* existence (Garfield 1995), yet it remains conventionally really real. After all, here we are, always seeking that happiness that is always already present at the spiritual Heart (*hridaym*).

The good news for those of us who still love our *relative* spacetime appearing stuff—and who are thankful for the all too brief opportunity that time affords for liberating realization of the numinous *ultimate* nature of that reality process—mental and physical phenomena are *relatively*, conventionally really real by virtue of their very appearance in spacetime to a perceiving, designating, reifying consciousness, this often all too real self-ego-I. Still, all this appearing stuff is not essentially, *ultimately* intrinsically real. That *one truth unity* (*dzog*) is the inherent complementarity or unity of Buddha's Two Truths—Relative and Ultimate. That is the Buddhist Middle Way Prasangika Madhyamaka view as to the ultimate primordial ground of the boundless whole that is the *Perfect Sphere of Dzogchen.*

In short, lest I further belabor the point, *ultimately* this view of the great whole that is trans-conceptual nondual reality being itself, formless primordial ground of That, describes the reality limit of all appearing physical and mental phenomena—spacetime instantiations of primordial Presence of the "groundless ground" itself. So yes, this arising stuff is then relatively, observer-dependently real; but not ultimately, observer-independently real.

Indeed, on the nondual *Dzogchen* view the spacetime dimension of Relative Truth (*samvriti satya*) or form, arising as matter-energy—$E=mc^2$—has never departed its formless timeless emptiness dimensional ground that is all embracing Ultimate Truth (*paramartha satya*). Recall that these two truth reality dimensions are an inseparable, indivisible, ontologically prior and epistemologically present one truth unity. Nagarjuna told it well: "There is no ultimate difference between [relative] samsara, and [ultimate] nirvana". This is of course the poetic *kosmic* irony of the causal duality of the Mahayana Buddhist Two Truths epistemic dominant trope that is completed in nondual *Dzogchen* ontology, the Great Completion of the noble Mahayana Causal Vehicle. We come to recognize, then in due course and by grace realize that wondrous natural one truth unity of the Buddha's Two Truths through our human *noetic cognitive doublet*—the mindful practice of objective conceptual study, and subjective contemplative meditation with the *Dzogchen* master.

Thus does *Dzogchen*, through its nondual ultimate view, practice, and fruition/realization transcend and complete not only the Two Truths causal duality of the Mahayana, but as well the implicit, implied or assumed Metaphysical Physicalism and Cartesian

Dualism of recent Western panpsychic and cosmopsychic reality accounts of the all embracing implicate unbounded awareness-consciousness whole (*dharmadhatu, mahabindu*)—noetic (body/mind spirit unity), nondual ultimate Basic Space (*chöying*) of reality itself—pure primordial *dharmakaya kosmic* emptiness (*shunyata*) ground.

Well, what is all that to my own happiness, being here in time? Once again arises the ultimate ontological question: Who am I? *Tat Tvam Asi*; That I Am: vivid, numinous, innermost love-wisdom mind Presence of That (*tathata*). This supreme relationship is one of nondual identity—our 'supreme identity'—harmless human Happiness Itself.

In Buddhist Prasangika Madhyamaka philosophy of mind, causal conceptual foundation of acausal nondual *Ati Dzogchen*, this Two Truths View (Relative and Ultimate) represents a centrist Middle Way between the permanent substantival 'eternalist' material existence of monistic absolute Metaphysical Local Realism/Materialism/Physicalism so beloved of Western physics and philosophy, and the 'nihilist' view of most Eastern and Western monistic Metaphysical Absolute Idealism which sees material spacetime existence as no more than illusory 'mind only' *avidya maya*. Thus does Prasangika strike a conceptual balance between the metaphysical extremes of the false dichotomy of *either* absolute existence, *or* absolute nonexistence. Yes, we require a centrist *Dzogchen Kosmopsychic* Middle Way ontology. The *Perfect Sphere of Dzogchen* provides that trans-conceptual contemplative nondual one truth unity of the matter.

In other words, the great Buddhist Mahayana Prasangika Middle Way acknowledges the reality of the *relative* spacetime dimension of mental and physical *form* or Relative Truth (*samvriti satya*) as it continuously arises from its all embracing formless ultimate awareness-consciousness *emptiness/shunyata* ground of Ultimate Truth (*paramartha satya*), the great all inclusive unbounded whole itself (*dharmadhatu, cittadhatu, dharmakaya, kadag*), the very Buddha Nature of Mind, and its experience of everything arising and instantiated therein. Primordial *Dzogchen* bespeaks luminous primordial Presence (*vidya, gzhi rigpa*) of our always present love-wisdom mind, our indwelling Buddha mind (*samatajnana*) that already knows this great noetic nondual truth—numinous 'clear light luminosity' of the very Nature of Mind and all of its experience.

So, Prasangika Madhyamaka denies arising reality any permanent inherent, intrinsic, absolute or *ultimate* existence. This dualistic Prasangika Two Truths trope is the foundational Mahayana Buddhist philosophy of mind. And yes, its Two Truths ontic and epistemic duality is then completed in panpsychic/kosmopsychic acausal nondual *Dzogchen,* the Great Completion of the Mahayana Causal Vehicle, as we have so often seen.

Engaging Our Panpsychic Wisdom Mind

While the metaphysics of consciousness has been valiantly and relentlessly reexamined through recent explorations of Western panpsychism, still, our noble philosophers need not reinvent the proverbial panpsychic mindwheel. Consciousness studies have been alive and well in our Eastern Wisdom Traditions for at least 35 centuries. Let Western philosophers of consciousness and Contemplative Science now engage this urgent nondual truth of our noetic Primordial Wisdom Tradition.

One wonders how Western philosophy and science has managed to avoid this great nondual Eastern wisdom for so long. Are there not more things in primordial consciousness itself than are dreamt of in canonical Western Philosophy and Modern Science? What are we afraid of? Is it not time to make an individual and collective quantum leap into scary timeless metaphysics?

Our pernicious *taboo of subjectivity* has, for the modern scientific mind, veiled and defended Western dualist and materialist analytic philosophy—with its Physicalism bias in philosophy of physics and cosmology—from a holistic, even nondual metaphysic of *kosmic* consciousness. This is now beginning to lift, due in no small part to the inherent random subjectivity of the quantum theory, and to our recent revealing cognitive adventures in the primary monistic

Antirealism of Western priority cosmopsychism/panpsycism. Perhaps then it's OK to continue our integration of the holistic subjective panpsychic wisdom of both East and West with the prodigious objective science and philosophy of the West. That toward our recognition of the prior and present unity of objective Science and its perfectly subjective all embracing Spirit ground.

Let Western philosophy of mind and philosophy of science—physics, cosmology, biology and an inchoate neuroscience of consciousness (*Appendix B*)—now engage noetic and nondual Buddhist, Taoist, Vedanta wisdom of the East. "O East is East and West is West; and *ever* the twain shall meet". (Apologies to Rudyard Kipling.)

Now that quantum entanglement/nonlocality, along with Buddhist Middle Way Madhyamaka *shunyata*/emptiness/boundlessness has utterly collapsed our uncomfortable comfort zone of a purely objective, physical, observer/theory/model-independent "real world out there" (RWOT); and now that the hitherto despotic culture of science and philosophy has granted us its permission to do the 'post-empirical' metaphysics of the Quantum, and of East-West panpsychism, let's try something completely different already!

Let philosophers of mind—academic and Buddhist—now engage not only the *relative* exoteric/analytic cognitive dimension, but as well an esoteric/contemplative *ultimate* exploration of the Buddhist *Dzogchen* kosmopsychic *metaphysics of consciousness.* Caveat: this shall require—Yikes!—a bit of spooky Buddhist contemplative practice; that is to say mindfulness meditation—Buddha's "mindfulness of breathing—upon our psychophysical spiritual belly buttons. [*Appendix A*]

Or, because meditative-contemplative luminous Presence of the nondual primordial awareness-consciousness whole shebang is always already present at the spiritual Heart (*hridyam*), and renewed upon the wisdom *jnanaprana* wind (life-energy, *ch'i, lung, pneuma*/ Holy Spirit) with every present breath and so cannot be a legitimate future *goal*—mindful, continuous spontaneous *Dzogchen* "undistracted non-meditation" may be the more accurate understanding. [Boaz 2021 *Buddhist Dzogchen: Being Happy Now*]

Recall Dear Reader, that our inherently indwelling, always already present love- wisdom mind, bright numinous Presence of That by whatever grand name or concept, if it is to be more than mere intellectual, conceptual self-stimulation requires the compassionate active *engagement* of both voices of our *noetic cognitive doublet*—both objective conceptual knowledge and subjective contemplative wisdom. That is a coming to meet as it were of the nondual noetic body/mind/spirit dimensions of our being here as honored guests of the phenomenal world.

Wow! What hath God wrought upon the hitherto psychic safety of comfortable Greek local realist/physicalist academic philosophy, physics, and safe and sane intellectual theistic conjecture, and of exoteric but powerful petitionary prayer? Add the prodigious subjectivity of the quantum Ψ-wave function of recent Quantum Field Theory into the ontic mix and we have cause for real cognitive dissonance, as well as for real noetic clarity.

Mindful Noself Help

Be that as it may, there's plenty of scientific evidence based medical and psychological metadata to demonstrate that trans-conceptual contemplative mindful breathing practice, by whatever name, promotes human health and well being. [*Appendix B* "The Neuroscience of Meditation"] And it furthers human evolution toward the conscious discovery and then supra conscious recognition, then 'greater esoteric innermost secret', even nondual liberation/realization of our otherwise spooky human ultimate identity. This selfless 'supreme identity' is our always present indwelling love-wisdom mind Presence of that primordial ground. Is such deep inward knowing-feeling awareness not after all, the real function of the noetic wisdom traditions of our species?

Alas, a mind is a terrible thing to mind. Mindfulness practice, Buddha's "mindfulness of breathing", is blatantly simple; but it's not so easy. Sadly, it requires a bit of self-discipline, patience, and a lot of self-ego-I courage. Yet, there is a veritable Western "mindfulness revolution" now upon us. Check it out for yourself (*Appendix A*).

Please recall here that *both human happiness and unhappiness arise from our present mindstate.* We tame the 'wild horse of the mind' by freeing the narcissistic conscious mind of unruly self-ego-I via the mindful 'placement of attentional awareness' on our already present Presence of The Buddha's *noself* (*anatman*). Thus, breath by mindful breath, we are liberated from the adventitious afflictive negative emotions (fear/anger/hostility, grasping desire, greed, and pride).

That process seems a very sane approach to human *awareness management.* One might even speculate that the real meaning of outer, inner, and "innermost secret" human body-mind-spirit evolution to be precisely that. Such mindful 'placement of attention-awareness' upon the clarity and bliss of the bright noetic Presence of the basal nondual ground, boundless whole itself, is our choice to be happy and at peace here and now.

Still, such dualistic distinctions as outer, inner, innermost, above and below, true and false, samsara and nirvana, relative and ultimate, past and future, and the rest do not obtain in the utterly nondual *Perfect Sphere of Dzogchen.* Such dichotomies are the products of human concept mind and its dualistic cognitive projections. That said, we must show up for work, and balance our checkbooks, and be kind to others. We live in these none too real two worlds at once. That is our happy, if a bit confusing noself human condition.

Dzogchen Kosmopsychism: The Grounding Relation

The *Dzogchen* holistic primary "priority monistic cosmopsychic" and kosmopsychic panpsychism accounts seem to me to be an ontology that clearly dodges not only problematic physicalist, emergentist and the dualist rejoinders, but as well, the presumed "Combination Problem": How is it that panpsychically 'conscious' particle/field micro-subjects (quarks, leptons, photons) combine to constitute the complex consciousness of conscious human macro-subjects? That epistemic problem is here avoided because human macro-subjects are ultimately connected and grounded not in presumed microcosmic particle/field brain micro-subjects—Suzuki Roshi's Small Mind, but in the boundless primordial awareness-consciousness ground that is the vast formless, timeless boundless whole itself—aboriginal Big Mind, our 'supreme identity', indwelling always already present lucent Presence of That (*tathata*).

As to the vexing and mysterious 'grounding relation' of arising conscious macro phenomena, the *relative* unfolding spacetime microcosm is perforce always already grounded in, and arises and participates in the *ultimate* enfolded holistic primary monistic cosmopsychic boundless "implicate order of the vast enfolded unbroken whole" (David Bohm)—by whatever grand name or concept.

Yes. That all subsuming metaphysical ground is ultimately fundamental—formless boundless emptiness 'groundless ground' of all appearing phenomena. Spacetime phenomena are therefore relative, derivative, dependent upon, grounded in, instantiated by, and exist through That all subsuming whole that is most fundamental.

Nice holistic *concept*. Should one desire to directly experience (*yogi pratyaksa*) this truth of the matter—prior to one's concepts and beliefs—that great truth of reality Being Itself, one must perforce engage trans-conceptual contemplative cognition under the gentle guidance of the subtle mindstream of the *Dzogchen* meditation master.

In contradistinction to bottom up "constitutive micropsychism", derived as it is from the materialist metaphysic that everything is grounded in purely physical microcosmic particles and fields, priority monistic cosmopsychism/kosmopsychism holds that the stuff of reality exists relatively, but not ultimately, because it is grounded in the vast macro-*kosmos*, the *kosmic* whole that embraces the physical and mental spacetime cosmos.That may be the view of monistic cosmopsychist Jon Schaffer (2010).

Kosmos here refers to the Pythagorean *kosmos* that embraces not only the physical cosmos, but the whole of reality that transcends the mere physical dimension of space and time. Nondual perfectly subjective Spirit *kosmos* embraces and subsumes the dualistic physical cosmos that is the province of objective Science, as we have so often seen in these pages. Again, that great *kosmic* process is known as 'grounding by subsumption'.

This then is the essential holistic perennial 'one truth unity' (*dzog*) of the subtlest nondual teaching of our noetic Primordial Wisdom Tradition, not the least of which is the all embracing kosmopsychic *Perfect Sphere of Dzogchen*, the Great Perfection.

And yes, "mindfulness of breathing" meditation is the unsurprising, trans-conceptual contemplative technology and methodology for recognizing, then realizing this profound mereological grounding relationship of ostensibly separate microcosmic micropsychic parts to their prior vast nondual macrocosmic panpsychic whole.

Contemplatively merging the unruly wild horse of "conscious mind" with that great quiescent noetic *kosmic* (body-mind-spirit subject object unity) consciousness whole, primordial formless timeless ground itself, is the method of psycho-spiritual practice that reveals not only the conceptual and contemplative voices of the metaphysical understanding of this sublime cosmic process—"the dance of geometry"—but as well the urgent compassionate moral depth of engaged altruistic human relationships with human and nonhuman beings, including our precious Mother Earth.

This 'grounding process' is not essentially an objective physical merging or 'combining' of micro phenomena with conscious macro subjects. Rather, the deeper, subtler perfectly subjective "grounding relation" *process* is trans-physical and trans-conceptual, even spiritual, albeit with analogous physical "neural correlates" in brain structure and function, always in the context of a connecting sociocultural morality that is grounded in our epistemic prior monistic cosmopsychic or kosmopsychic metaphysical objective-subjective understanding.

The presumption that the 'grounding relation' must be somehow a purely physical process of combining purely physical entities is a discomfiting unconscious ideological relic of our prevailing Western (Greek/Hebrew) bias that is Metaphysical Local Realism/

Materialism/Physicalism. Panpsychism is inherently metaphysics, but as Ken Wilber has pointed out, unconscious 'hidden' physicalist metaphysics is "bad metaphysics". Let our inchoate primary 'Priority Monism' *kosmopsychic* metaphysic enter the cognitive light of a bright new day.

Therefore, in our emerging Science of Consciousness, of which Contemplative Science is now a proper participant, the mindfulness panpsychic grounding connection to the trans-conceptual (*nirvikalpa*) primordial 'supreme source'—very ground of being—is the great *kosmic process*, the Way of our inherent perfectly subjective love-wisdom Christ-Buddha mind, our 'supreme identity', innermost clear light Presence of that ground (Basic Space *chöying, dharmakaya*). It changes everything. Beyond concepts, it's like coming home.

Much has been said by philosophers of mind about the nature of this all subsuming grounding relation. Monistic *kosmopsychic Dzogchen* as I have here broadly construed it, employs an acausal/noncausal directly experiential (*yogi pratyaksa*) contemplative grounding relation/connection of dualistic *relative* appearing phenomenal reality to/in our nondual *ultimate* basal ontic panpsychic primordial ground itself—*dharmakaya, gzhi rigpa*, Tao, *Ein Soph*, Nirguna Brahman, *Abba* Nondual God the Primordial Father. And this gnostic love-wisdom process arises in the inherently trans-conceptual, trans-rational, 'post-empirical', even post-metaphysical nondual boundless whole in whom all of our arising and appearing spacetime realities are luminous 'clear light' energetic instantiations.

Our nondual trans-conceptual direct experience of that great noetic process, upon the mindful breath, may then be conceptually, causally, scientifically unpacked. Our inherent *noetic cognitive doublet* is the skillful means for discovering the subjective and objective; inner and outer; physical and mental/spiritual whole. Human cognition includes both at once. These two voices of our human cognition are already a prior and present ontic and epistemic one truth unity. It is that prior yet present nondual *kosmic* unity to which we awaken via the kosmopsychic primordial grounding relation.

As good a soterological (liberation) definition of the human condition and its epistemic human predicament as any. The fruit of that great process is the full *bodhi* of 'awakened mind' "Already accomplished from the very beginning"—deep within us. [Nagarjuna] That indwelling 'awakened mind' is always already awake. Spiritual liberation and enlightenment is the recognition, then realization of that great perfectly subjective nondual truth of Being Itself.

The knowing-feeling contemplative certainty of That is accomplished through objectively and subjectively engaging intrinsic 'open Presence' of the ground of everything, our indwelling always already present *buddic* love-wisdom mind.

Sounds a bit spooky? So how shall we do this? As Buddha told, "mindfulness of breathing". It bears repeating. Paradoxically, we use *relative* dualistic objective and subjective practice to fully awaken to our *ultimate* intrinsic nondual love-wisdom Christ-Buddha mind Presence of the great whole. This then is the prodigious *relative* grounding relation of the microcosmic with the macrocosmic dimensions of the boundless whole that is *ultimate* reality Being Itself—human harmless Happiness Itself.

Our Four Human Cognitive Life Stages

The *Ati Dzogchen* view and practice that is acausal nondual completion of the Mahayana Causal Vehicle with its Buddhist Two Truths philosophy bespeaks our dualistic 'grounding relation' as one of causally interdependent, yet acausal nondual identity—the monistic nondual *one truth unity (dzog)* that is invariant throughout our entire human cognitive consciousness processional, our four human cognitive reference frames, our immediate *mind states* and their corresponding evolutionary *life stages.*

These consciousness mind state life stage cognitive dimensions are: 1) pre-conceptual ordinary direct attention/perception, prior to conceptual naming; 2) outer, exoteric, objective, conceptual-mental, physical embodiment; 3) inner, esoteric, subjective, higher mental, contemplative spiritual; and 4) 'innermost secret' perfectly subjective nondual Spirit ground, nameless direct yogic experience (*yogi pratyaksa*) of That (*tathata*).

These four inherent cognitive reality dimensions are not entirely linear, nor are they reducible one to another. Therefore, one may briefly directly experience the nondual yogic bliss of life stage four via a 'peak experience' *mind state* while living most of the time in *life stage* two or three. Conversely, one may have 'evolved' to life stage three, but suddenly 'backslide' into the primitive destructive afflictive emotions—fear/anger, hatred, random grasping desire, greed and pride—characteristic of the unruly mind states of life stage two.

Cognitive dimension two is where most human beings live most of their lives. It represents the Buddhist dimensional realm of samsara, suffering due to adventitious human primal ignorance (*avidya, marigpa, ajnana, hamartia*/sin) that cause the afflictive emotions. Dimension three represents the contemplative love-wisdom mind opening to indwelling always already present clear light Presence of the primordial ground. It facilitates relative human happiness or human flourishing (*eudiamonia, felicitas*). It functions as a cognitive light bridge into the awakened buddha mind of ultimate harmless Happiness Itself (*paramananda, mahasukha, beatitudo*)—grounded primordial wisdom (*jnana, yeshe,* gnosis) of cognitive life stage dimension four.

These four cognitive realms represent a complementary ontologically prior yet phenomenally present indivisible nondual one truth unity of the Buddha's Two Truths—Relative and Ultimate. Human beings evolve through this wondrous consciousness processional by way of the Buddha's Four Noble Truths with its Eightfold Path to authentic compassionate peace and human relative and even ultimate happiness.

Bold holistic panpsychic *Dzogchen Kosmopsychicism* metaphysic indeed.

Entering In *Dzogchen*

David Bohm's "implicate order of the vast enfolded unbroken whole" of our physical cosmos is subsumed by the even more fundamental, trans-physical, contemplative nondual kosmopsychic *Perfect Sphere of Dzogchen.* This all embracing, all pervading immediate awareness Presence (*vidya, rigpa, christos/*I AM) of *ultimate* formless, timeless, selfless all subsuming nondual primordial awareness-consciousness itself—luminous vivid numinous ground of *relative* conventional human experience, human consciousness. It abides throughout and all about arising cosmic objective material stuff, these myriad spacetime forms that are the contents of the physical, mental, spiritual all embracing formless *kosmic* consciousness whole shebang.

Thus does this *relative* dimension of spacetime form continuously arise in/to our human consciousness mindstream from the *ultimate* formless awareness emptiness 'groundless ground' that is the great unbounded panpsychic whole (*dharmakaya, dharmadhatu, mahabindu*), primordial awareness-consciousness itself—Heidegger's Being Itself; Hegel's nondual Spirit—in whom body and mind are necessarily, luminously already instantiated as always present Presence of that ultimate ground state.

That I Am Now. We have seen that Hindus and Buddhists speak: *Tat Tvam Asi.* That I Am (That Thou Art). Speaking of this primordial "I AM Presence" of Moses and the Prophets (Isaiah 41:10), Jesus told: "That which you seek is already present within you; and it is spread upon the face of the earth, but you do not see it." And from *Dzogchen* founder Garab Dorje, "It is already accomplished from the very beginning; to remain here without seeking [anything more], that is the Meditation". And Buddha told, "Wonder of wonders, all beings are Buddha". Yet, under sway of Metaphysical Scientific Local Realism/Materialism/ Physicalism we miss the mark (ignorance, *hamartia/*sin, *avidya, ajnana, marigpa*)) almost entirely. The antidote to such ignorance is our *noetic cognitive doublet*— relative objective study, and subjective contemplative practice. Practice these two as a nondual unity.

As to the 'innermost esoteric' perfectly subjective life stage four of nondual view and practice of our great noetic Primordial Wisdom Tradition—Hindu, Buddhist, Taoist, and Abrahamic

Hebrew, Christian, Islam—we are taught by these premodern masters and *mahasiddhas* that the numinous I AM Presence of the great all pervading unbounded whole itself that we are, is always already immediately present in this 'eternal moment' *now.* "Just open the door." [H.H. Dalai Lama] Ultimate happiness, enlightenment, liberation is only ever here now. This present moment now. It cannot be elsewhere. So there is nothing to seek elsewhere. As we have seen, past is gone. Future is nowhere. Indeed, everything happens only here and now.

Again, we cannot *become* happy later; but we can *be* happy now. The past is utterly gone beyond, but a present memory. The future is but a present anticipation. So, there is only this present *now.* Yet even this present moment now is to brief to be grasped. It is already past. So, there is nothing solid to which we may cling. There is only this timeless infinite luminosity of the bright Presence of the "feeling of being", all pervading lifeforce love-wisdom *jnanaprana* wind entering in each mindful conscious belly breath. [*Appendix A*] And that's enough. By grace "It is already accomplished from the very beginning." It is that primordial truth to which we awaken through our conceptual philosophical and noetic contemplative practice of this wondrous love-wisdom mind Path—all that upon our mindful *prana* breath in the belly. That great *process* is simpler than we may have thought.

[Philosophical Note. The deeper meaning of philosophy (*philo*/love, *sophia*/wisdom) is the prior unity of compassionate love and noetic wisdom expressed together in human ethical conduct toward the reduction of the suffering (*dukkha*) of living beings. Herein lies the secret of our human happiness (*sukkha*). No *dukkha,* no *sukkha*! Good to know.]

Dzogchen Complementarity. This then is *Dzogchen* primary monistic panpsychic *kosmopsychic* view, practice, conduct, and fruition/realization. Thus do we *choose* to train the unruly mind, through the mindful breath, and under the guidance of a qualified *Dzogchen* meditation master, in the "placement of awareness attention" upon already present luminous numinous Presence of our "innermost secret" love-wisdom Christ-Buddha mind. It happens upon the trans-conceptual (*nirvikalpa*) mindful breath. And yes, it requires a bit of simple—but not so easy—unbiased, Zen Mind/Beginner's Mind contemplative mindfulness practice. And that requires a highly intelligent, fluent, holistically oriented, and courageous self-ego-I.

Facing up to the seemingly bad news: with the exception of a few avatars and *mahasiddhas,* sages and saints, we are scarcely awake to our inherent, indwelling always present wakefulness—radiant 'awakened mind'—nondual ultimate awareness-consciousness-being itself. Now that's a spooky duality!

I have argued here and elsewhere that we might well consider the relation of complementarity (Niels Bohr) in our metaphysical spiritual quest for a panpsychic, *kosmopsychic* 'grounding relation' between the duality of the ultimate consciousness whole, and the pluralistic multiplicity of its participating, instantiating parts. That vast whole/ground perforce already embraces all of us sentient beings as we tread this joyous difficult path to wholeness. Breath, feel, then rest in that Presence now! Please do it now.

Father of recent antirealist quantum theory Niels Bohr, himself a student of Taoism and Vedanta, would have advised that such a relative complementary "conjugate pair" of opposites—relative dualistic objective form and ultimate perfectly subjective Spirit; local particle and nonlocal wave—(subjective yin and objective yang) *ultimately* participates in a complementary all inclusive whole—nameless formless Tao itself. "The Tao that can be named is not the primordial Tao" (Lao Tzu, *Tao de Ching*). Gautama the Buddha told, "Form is empty; emptiness is form." Relative form arises, abides, and passes in its complementary ultimate emptiness 'groundless ground'. Its nature is empty. Empty of what? Spacetime form is empty of intrinsic or absolute existence. Yet, fortunately for all of us form is relatively conventionally really real. That gives us a little time to open and receive this great non-conceptual truth of the matter. It's a Middle Way between the metaphysical extremes of absolute existence and absolute nonexistence. The really good news? Yes. "It is already accomplished from the very beginning". [Nagarjuna]

Just so, Tibetan Buddhist luminary and historically identifiable Vajrayana founding father Padmasambhava (8th century) advised: "The only way to realize the [nondual] wisdom of Ultimate Truth is through Relative Truth dualistic practice of the wisdom path." Therefore, if you want that result, you must practice this cause now—relative cause and effect is the Mahayana Causal Vehicle. As Zen Master Suzuki Roshi told: "To know ultimate Big Mind, work with relative Small Mind." 'One cannot take heaven by storm'.

That nondual monistic *Dzogchen* panpsychic *kosmopsychic* view and praxis of ultimate reality itself parallels, but is not reducible to Jon Shaffer's (2010) panpsychic cosmopsychic primary 'Priority Monism' in that it holds not that the whole has no parts, but that the single ontic contemplatively grounded *ultimate* boundless whole transcends, includes and grounds the *relative* participating parts—indeed, an exemplar of our complementary Buddhist Two Truths *leitmotif.*

We've seen that the primordial boundless whole itself is fundamentally and ontologically prior to, always already embraces, subsumes and is the ontologically *ultimate* trans-conceptual nondual 'groundless ground' for all *relative* spacetime arising herein. Conversely, for pluralistic, dualistic realist/physicalist atomistic micropsychic panpsychic views the micropsychic parts are fundamental and are ontologically prior to and ground the macropsychic whole. The relative epistemic atomistic cart before the ultimate ontic horse of nondual panpsychic/kosmopsychic truth?

In other words, this noetic, nondual, nonlocal *kosmic* whole/ground is ontologically prior to, and transcends yet embraces and includes the local cosmic spacetime located physical atomic baryonic and bosonic parts. In this holistic view of ultimate reality the relative, local physical parts supervene (depend) upon and are grounded in the vast primordial nonlocal unbroken whole itself. Just so, the whole supervenes upon the parts. How so? Mereologically, where there are *relative* parts, there is perforce a greater all subsuming, embracing whole. Where there is an *ultimate* whole, there are participating less fundamental holonic parts. The relative, complementary duality of mereological part-whole are *ultimately* unitary, as Taoist Bohr, Bell, the Buddhists, and indeed our entire premodern noetic Primordial Wisdom Tradition have told for all these many centuries.

Yes, in such an ultimate noetic nondual (subject-object identity) non-conceptual view the primary relation of objective parts to the all inclusive formless, timeless, selfless perfectly subjective awareness-consciousness whole is one of numerical identity—the two *relata* are in the ultimate view one selfsame identity—a relational nondual equality (*samatajnana*). Relative spacetime human consciousness arises from and is instantiated by That (*tat, sat*) ultimate or absolute primordial *kosmic* awareness-consciousness whole ground in which, or in whom all of these concepts about non-conceptuality arise and play. That is the *Dzogchen* complementary kosmopsychic view and practice.

That is the upshot of the objective "grounding relation". The subjective grounding relation is being aware and awake to this great primordial nondual truth—harmless happy Presence of That.

From the dualistic objective 'scientific' relative-conventional view, a whole and its parts are separate. Thus, this conceptual relative/ultimate duality that pervades and permeates all binary discursive semiotic discourse is reflexively (self-revealing) resolved and completed in the *kosmic* nondual all-embracing *Perfect Sphere of Dzogchen*—primary monistic panpsychic *kosmopsychism,* the Great Completion of the Middle Way Buddhist Two Truths Mahayana Causal Vehicle. [H.H. Dalai Lama 2000, 2007; Norbu 1999]

We've seen that the peaceful realization of that great truth cannot be a *goal* for some future happy *kosmic* mind state. *Our happiness and our unhappiness are the result of our present mind state. Human happiness happens only now—indwelling primordial Presence of our always already present love-wisdom mind Presence.* And we have a cognitive *choice* as to our 'placement of awareness-attention' upon that precious love-wisdom mind. That Presence, by whatever happy concept or belief, is always already present this very moment now. That is our great perennial wisdom teaching. Wondrous paradox indeed to our deep background cultural conceptual "global web of belief" (Quine 1969). So don't believe it! Such trans-conceptual (*nirvikalpa*) nondual primordial wisdom (*jnana, yeshe,* gnosis) utterly transcends yet includes our all too human grail quest for comfy objective certainty.

Well, how do we recognize, then realize this truth of human happiness beyond mere concept and belief? Again, we make our *goal,* not happiness, nor liberation, but the *practice of the Path itself,* upon each mindful conscious breath—'brief moments many times'—as it becomes, breath by mindful breath, a radiant continuity of natural present moment to moment awareness. Yes, Zen Master Suzuki Roshi told, "To know Big Mind; work with Small Mind." To personally realize bright 'innermost secret' Presence of our numinous nondual unbounded panpsychic kosmopsychic whole—ultimate trans-conceptual 'groundless ground' of everything—mindfully engage, with compassionate understanding, the whole's primordial consciousness particulars, everyday stuff of our usually distracted unruly concept-belief mind. No dilemma. "No problem at all".

As Gautama Shakyamuni the Buddha told so long ago, "Rest your weary mind and let it be as it is; all things are perfect exactly as they are." That is the ultimate view that transcends yet embraces our destructive cognitive biases and the incredulity of our relative views as to miraculous *Perfect Sphere of Dzogchen.* Awaking to that primordial truth is the ultimate panpsychic kosmopsychic grounding relation that connects us to the perfect ground of all this arising phenomena. [Boaz 2021] It's like coming home.

This then is the body-mind-spirit *noetic imperative* now present in our emerging 21st century Noetic Revolution in matter, mind, and spirit. That is the very complementary unity of objective Science and its perfectly subjective Spirit ground. Entering in *Ati Dzogchen* primary monistic kosmopsychism is a powerful view and practice for its immediate happy realization in human form.

II: Toward a Foundational Noetic Quantum Ontology: Quantum Nonlocality and Buddhist Emptiness

Choosing Reality: Ontological Extremism, a Middle Way, and the Light of the Mind

In Buddhism the early *Pali Canon Abhidharma* of the *Sarvastivada* and *Vaibhashika* Schools, the Greek Democritus and his master Leucippus, along with Western functionalist Material Local Realism (Metaphysical Scientific Local Realism/Scientific Materialism/Physicalism), all hold the realist atomist position wherein ultimate reality consists of indivisible, physical/material atomic baryonic matter particles (Atomism) that have an ultimately physical, objective locally real, observer-independent, substantial, permanent, even absolute and eternal existence. That is the ontological legacy of Greek Materialism/Physicalism that has almost entirely colonized the Western heart and mind.

Here, appearing *relative* spacetime reality is *ultimately* real and purely physical, or reducible to purely physical electrochemical brain structure and function. "It's all just physical." The human being is here reduced to Lewis Carroll's (Alice's) "bag of neurons". Dismal dualistic reductionist 'scientific' metaphysic indeed.

We shall come to see in this chapter how it is that such Modern Big Science with its physical and mathematical quantum explanations of spacetime form perforce arises and is instantiated in its noetic nondual primordial Spirit ground.

We have already seen that this Modernist European Enlightenment (the Age of Reason) rationalist knowledge paradigm, known to the philosophy of science trade as Scientific Local Realism, and Scientific Materialism/Physicalism, is a failed paradigm that not even post-Standard Model quantum physicists and cosmologists take seriously, at least theoretically; although many are still ideologically committed. Scientific and sociocultural knowledge 'paradigm shifts' require a couple of generations to fully establish themselves in the 'global web of belief' of a culture. [Thomas Kuhn 1962]

So some Buddhist schools believe that atoms are eternal; and some particle physicists believe that electrons and protons within these atoms are eternal, that they never decay. As to recent quantum physics the existence of ordinary atomic baryonic matter—our beloved protons and neutrons—is believed to be *observer-independently* real as it emerges from the 'nearly empty space' quantum fluctuations of the proto-physical Unified Quantum Vacuum, the *fantasque* zero point energy field (ZPE), beyond any perceiving, experiencing, experimenting consciousness, or mind. Such realists, whether Abhidharma Buddhists, Hindus, or physicists, are *essentialists,* believing that reality exists essentially and independently of an observer-experiencer—just as it appears from its own side, of its own power, independent of any observer consciousness. A tree in the forest really exists when there is no one about to observe it. This view is known as 'common sense realism'—Bertrand Russell's "Metaphysics of the Stone Age".

On the other hand, Mahayana Prasangika Madhyamaka Middle Way Buddhists view spacetime physical and mental reality as arising *observer-dependently* as a result of an infinite sequence of interdependent prior causes and conditions emerging within a vast interconnected relative, conditional, physical and mental causal 'interdependently arising' (*pratitya samutpada*) spacetime matrix which itself naturally arises within a formless, timeless foundational, ultimate primordial awareness-consciousness ground—basal empty (*shunya*) luminosity/cognizance, beyond or ontologically prior to spontaneous reflexively emerging spacetime light-matter-energy-movement/motion—Albert Einstein's $E=mc^2$.

In our pre-modern wisdom traditions the conceptually uncontrived, unelaborated intrinsic living Presence (*vidya, rigpa, Buddha nature, Parabrahman, Christos/Christ nature*) of That (*tathata*) trans-conceptual, monistic panpsychic/cosmopsychic implicate order of the ground or boundless whole itself dwells always at the spiritual Heart of all human beings.

We have seen in Chapter I that the indwelling Presence of the natural nondual *ultimate* primordial ground of *relative* spacetime reality is accessed via a finite natural conscious awareness portal into that super-conscious infinite 'Basic Space' (*dharmadhatu, chöying*) of ultimate *dharmakaya* ground. That aperture or 'explanatory gap' is known to contemplatives as 'mindfulness of breathing', or trans-conceptual contemplative mindfulness meditation practice.

We also discovered in Chapter I that primary *"Priority Monism Cosmopsychism"* (Schaffer 2010) is a panpsychic top down holistic and expanded variant of bottom up 'atomistic constitutive micropsychic' panpsychism. The view of cosmopsychism is that the ultimate primordial ground of the arising of conscious beings and their experience is perforce the all embracing, all pervading, all subsuming formless, timeless universal awareness-consciousness unbounded whole of *kosmos* itself—'kosmic consciousness' as it were—in which, or in whom all human and animal 'organic' consciousness arises, participates, is physically and mentally instantiated, and ontologically and phenomenally grounded. That *ultimate* relationship is one of mereological (part-whole relations) necessity. The relative parts abide in the great whole; that whole embraces its parts.

Thus does all subsuming nondual *ultimate* primordial *kosmos* ground provide the ontic 'grounding relation' for our perceptual and conceptual experience of the entire *relative* dimension of cosmic spacetime form. This post-classical view is refreshingly and unabashedly antirealist, anti-physicalist, anti-reductionist, and observer-dependent.

I shall continue in this chapter my argument that the holistic Priority Monism cosmopsychic view dodges the prodigious 'combination problem' that plagues panpsychic 'atomistic constitutive micropsychicism' which holds that subatomic micropsychic parts are prior to macropsychic trees, stars, and organic beings; and is thus a more plausible and complete explanation of 'constitutive panpsychism'. [Schaffer 2010]

Moreover, we shall see that Monistic Primary Cosmopsychism that founds *Dzogchen Kosmopsychism* constitutes a phenomenal cognitive bridge from the formalist mathematics of Quantum Field Theory (QFT, QED) with its 'universal quantum Ψ-wave function' to a post-empirical, postformal non-physicalist '*kosmic*' centrist "Middle Way" foundational integral Noetic Quantum Ontology; our 21st century quantum consummation devoutly to be wished. [Schaffer 2010; Goff 2017; Shani 2015]

Just so, while the Middle Way Buddhist view and the quantum view are *observer-dependent* upon a perceiving human consciousness, the 'scientific' local realist, materialist, essentialist view is *observer-independent* wherein this world of spacetime stuff constitutes a separate purely physical 'real world out there' (RWOT), whether or not it's observed by a sentient consciousness. Meanwhile the Middle Way Prasangika Madhyamaka view, and the standard QED quantum view (the universal quantum Ψ-wave function) is *observer-dependent* and 'ontologically relative'—relative to conscious observation and measurement. For that post-materialist quantum view, and for the parallel Middle Way Prasangika Buddhist view, stuff exists not independently, but interdependently 'ontologically relative' to the constituting *consciousness* of an observer/perceiver/experimenter.

It is then urgent to understand that Irwin Schrödinger's 1928 'universal quantum Ψ-wave function' is a *relative* formal mathematical description of a 'post collapse' purely physical, objective empirical reality, not a fundamental metaphysical ontology about the *ultimate* nature of that reality. The quantum Ψ-wave is merely a realist-physicalist mathematical operator that is perforce embraced and grounded in an ontologically prior 'foundational quantum ontology' that ultimately grounds that relative-conventional universal Ψ-wave function in a more basic, ontic prior foundational, and mereologically essential primordial awareness-consciousness ground state in which it arises.

Sadly, we shall soon see that all of the present primary quantum 'foundational interpretations' or ontologies now on offer are observer-independent consciousness-denying defenses of the classical Metaphysical Scientific Local Realism/Physicalism ontic orthodoxy. 'Spooky' post-empirical 'consciousness' is thereby introduced, by both Buddhist and quantum philosophy, into four centuries of settled 'classical' European Enlightenment 'scientific' local realist/materialist observer-independent dogma. Einstein's inner local realist hated it. Antirealist non-essentialist Bohr loved it. They debated the correct ontology for the quantum emergence of spacetime for two decades in the legendary 1935 EPR Einstein-Bohr debates. The debate continues today. Middle Way Buddhism has amply thickened the ontic plot, as we shall soon see.

So let us very briefly once again explore this apparent dichotomy between the 'metaphysical extremes' of the absolute existence posited by Metaphysical Scientific Local Realism, and the absolute nonexistence of Metaphysical Antirealism and Idealism in our ontic grail quest for a centrist 'middle path' that points toward the *ultimate* ontological ground in which *relative* space and time arise and participate. That view shall constitute our new *Dzogchen Kosmopsychic* integral Noetic Quantum Ontology, with an epistemology and phenomenology that integrates and ultimately grounds it into our all too real relative cosmos of arising space and time.

Brief Discourse on Method: Five Modes of Human Inquiry. Let us briefly identify some key terms of our inquiry. *Ontology* may be seen as the objective inquiry into the perfectly subjective metaphysical (literally beyond physics) basal *ultimate* ground of being itself, and how this primordial base arises and appears as our beloved *relative* physical and mental spacetime realities.

Epistemology is then the inquiry into both our objective knowledge, and subjective wisdom, and *how we know* and explain the seeming 'emergence' of all this appearing spacetime cosmic stuff from its formless, timeless, selfless primordial ground—all subsuming *kosmic* whole itself.

Phenomenology is inquiry into how we objectively, subjectively, and contemplatively experience, feel, and act through this wondrous 'intentional' cognitive noetic Presence of being here in our everyday life world of space and time.

Moral Conduct and Governance is inquiry into what and how we 'ought' to express our ontic and epistemic knowledge through our human action and normative compassionate ethical behavior—love guided by wisdom—that avoids harm to living beings, including our Mother Earth, and that results in human happiness, both relative human flourishing (*felicitas, eudiamonia*) and ultimate liberating enlightenment (*paramananda, mahasuka, beatitudo*) that is harmless Happiness Itself.

Science. Modern Science with its 'laws of nature', its physical constants, and its

quantum Field Theory unfolds and evolves in a continuous *process* of discovery. Science is not a grail quest for absolute objective certainty, nor for permanent eternal truths. Science is less about true or false statements of objective 'fact', and more about pragmatic efficacy, useful in a given context. Prevailing scientific theories are not so much correct or incorrect, right or wrong, but subject always to revision and enhanced completeness, furtively awaiting that next more inclusive theory or next 'scientific paradigm'. Scientific theory is relative, conventional, provisional, inherently uncertain, and fallible.

Science unfolds in the causal reality dimension of spacetime Relative Truth which is always already enfolded in its all subsuming Ultimate Truth primordial Spirit ground. That is how it is that objective Science and its perfectly subjective Spirit ground constitute an ontic prior and phenomenal present one truth unity—our prodigious prior unity of Science and Spirit.

These five modes of human inquiry comprise the basis of philosophy, and of philosophical method. Ultimately construed, philosophy may be seen as the unity of love (*philo*) and wisdom (*sophia*) skillfully practiced in the moral dimensional life world of we human beings being here in an often all too real world of relative space and time.

Real Clear Ontology: The Two Truths and the Middle Way Buddhist View

On the essentialist, local realist and reductive materialist/physicalist view with its atomistic constitutive micropsychism reality as it appears to our senses is a perfect 'mirror of nature' (Rorty), a kind of 'immaculate perception' that represents an eternal barrier between inherently unitary human consciousness and an essentially separate Platonic 'real world out there' (RWOT). This observer-independent, theory-independent, realist- physicalist view is opposed by the ontological Idealism of the Hindu *Sanatanadharma*—the hoary Vedas, the Upanishads, and the dualistic Vedanta of Madhva's *Dvaita Vedanta*, but not of Adi Shankara's nondual *Advaita Vedanta* which rather nicely parallels the great nondual Buddhist *Dzogchen* teaching. The essentialist local realist/materialist view is also opposed by Buddhist Idealism, namely the Yogachara Chittamatra or 'Mind Only' *Shentong* school of Asanga and Vasubandhu. [Boaz 2020, Ch. V] It is as well opposed by Western Absolute Idealism—Hegel and Kant, Bradley, Royce, McTaggart—who also construe arising material objective reality as *ultimately* unreal, a subjective apparition or *avidya maya* illusion of a sober, sentient perceiving consciousness.

For Buddhist Yogachara Chittamatra Idealism, appearing relative-conventional physical and mental spacetime reality is relative and illusory (*avidya maya*) as it arises from our mental impressions of its basal nondual ultimate 'groundless ground' (*vidya maya*), which is also *ultimately* illusory, or 'Mind Only'. We shall continue our grail quest for a 'middle path' between these 'metaphysical extremes' of absolute existence and absolute nonexistence. Thus shall we explore the centrist Buddhist Prasangika Madhyamaka Middle Way foundation of ultimate nondual panpsychic *Dzogchen Kosmopsychism* view.

Ironically, for Mahayana Middle Way Prasangika Madhyamaka (*Rangtong*) both emptiness and its arising form are established by human conceptual imputation, designation, and reification. Neither of these Buddhist Two Truth reality dimensions—the Ultimate Truth (*paramartha satya*) of formless emptiness, and the Relative Truth (*samvriti satya*) of spacetime form that arises within and through it—are *ultimately* real, although relative spacetime stuff is most certainly at least *relatively* real. As The Buddha famously told in his nondual *Prajnaparamita Heart of Wisdom Sutra*:

> Form is empty [*stongpa, shunya*]; emptiness
> [*stongpa nyi, shunyata*] is form ... All dharmas
> are emptiness; there are no characteristics.
> There is no birth and no cessation ...
> In emptiness there is no form ... no ignorance,
> no end of ignorance ... no path, no wisdom, no
> enlightenment, and no non-enlightenment ...

Well, ontologically speaking, if "emptiness is form", what is the reality status of emptiness itself? Emptiness too is "absent and empty of any shred of inherent ultimate existence." [Nagarjuna] So, how does emptiness exist? "Emptiness is established by conceptual minds."

As such a non-essentialist view is quite 'beyond belief', Buddha asks us to transcend our beliefs and concepts and 'abide by means of *Prajnaparamita*', numinous indwelling nondual clear light (*prabhasa*) love-wisdom mind clarity of innermost Presence of the basal empty luminosity of our primordial awareness-consciousness ground, beyond even the 'All-ground *Alaya Vijnana* storehouse consciousness' of *Chittamatra School.* [Boaz 2020] Who am I? That I Am! What is my mind? That (*tathata*) is the actual nature of my mind!

Realization of that great truth is harmless Happiness Itself—ultimate human happiness that cannot be lost.

Prajnaparamita, noetic non-conceptual primordial wisdom mother of all the buddhas is our seed of Buddha nature/Buddha mind (*buddhajnana*), immutable and unborn, prior to spacetime mass-energy motion/change ($E=mc^2$), indwelling always already present *buddic* Presence in the ordinary mindstream of the human being. *Prajnaparamita* is nondual (non-conceptual; no subject-object split) wisdom of emptiness/*shunyata,* Ultimate Truth dimension, primordial nature, essence and ground that is Reality Being Itself as it arises and is instantiated in the space and time of our dimension of Relative Truth. It is through assiduous contemplative meditative view and practice of Buddha's Two Truths and his Four Noble Truths—that prodigious love-wisdom Path—that we, as Buddha told, "Fully awaken to unsurpassed, true, complete enlightenment." Meanwhile we may enjoy an abundance of *relative* conventional happiness. Recall that in the ultimate view this double happiness is a prior unity.

And yes, it takes a bit of trans-conceptual 'mindfulness of breathing' practice to *directly* experience (*yogi pratyaksa*) the depth and bliss of this prior and present ontic *one truth unity* (*dzog*) of the Buddha's Two Truths—relative form and its ultimate emptiness ground—as utterly empty of essence; or as Nagarjuna told, without "a shred of inherent existence". This precious gift of spacetime form is then *relatively* really real. We exist as embodied minds. We are alive! But form is *ultimately* absent and empty of 'any shred' of logocentric intrinsic existence—an ontic Middle Way between the two ontological extremes of materialist/physicalist, eternalist substantialist absolute existence, and nihilist idealist absolute nonexistence. That odious false dichotomy finally put to rest.

And that's good news; is it not? While form is *ultimately* empty and absent of intrinsic existence, it is, fortunately for all of us, *relatively,* conventionally really real. Spacetime stuff is not just an idealist illusion. So we have a *qbit* of cosmic karmic cause and effect relative time in which to wake up to the noetic *kosmic* 'invariant *one truth unity*' of the Buddha's prodigious Two Truths—objective, conceptual relative, and perfectly subjective nondual ultimate.

We've seen that for non-essentialist, antirealist Buddhist Idealism school of Yogachara Chittamatra all this appearing phenomenal reality is 'Mind Only'. As with Kant, there can be no objectively knowable absolutely real noumenal things in themselves; only relative phenomenal illusory appearances to mind. For Buddhist realist centrist Middle Way Prasangika spacetime phenomena do indeed exist relatively, conventionally, just not absolutely or ultimately, as we have seen. This then is the great Prasangika Madhyamaka Middle Way, a fine centrist balance between nihilistic nonexistence of Indian Idealism, and of substantialist, eternalist permanence of the absolutely existing stuff of Scientific Local Realism, and of Scientific Materialism/Physicalism. [Boaz 2020]

We saw in Chapter I that causal Prasangika of the Mahayana Causal Vehicle is the conceptual foundation of acausal nondual Buddhist *Dzogchen*, the Great Perfection, the Great Completion of the Mahayana Causal Vehicle; 'highest', subtlest, and most profound nondual teaching and praxis of Gautama Shakyamuni the Buddha of this present age.

Immanuel Kant's 19th century German Transcendental Subjective Idealism—a duality of realist, material objective *phenomena*, and the perfectly subjective and unknowable, utterly transcendent *noumena*, 'the thing in itself'—is a Western (Platonist) version of our Primordial Wisdom Tradition's 'Two Truths' duality—objective relative and subjective ultimate reality dimensions. Kant's ontology parallels the Mahayana Middle Way Buddhist view, and as well the 'Neutral Monism' of William James.

More specifically, Kant's incipient middle way 'Subjective Idealism' also parallels the non-essentialist, yet pragmatically realist centrist Buddhist Middle Way Prasangika Madhyamaka view of Nagarjuna, Chandrakirti, and Tsongkapa. As we have seen, here reality arises and appears interdependently—Buddha's 'Dependent Arising' (*pratitya samutpada*). This ultimate reality is 'ontologically relative' and observer-dependent, that is to say, our realities are dependent upon the relative linguistic semiotic 'global web of belief' (Quine 1969) of the consciousness of a reflexively self-conscious human observer—whether Buddhist yogi or quantum mechanical practitioner.

Is such a 'middle path' between these perennial Two Truths of relative form and formless ultimate emptiness/boundlessness cognitively and behaviorally realizable? Is there a centrist 'middle way' between our seemingly competing paradigms, the metaphysical extremes of the descending, substantialist, objective Metaphysical Local Realism/Physicalism of Science (form), and the ascending, too often nihilistic Metaphysical Idealism of subjective Spirituality (emptiness)? After all, our prodigious awakening to the realization of the prior and present unity of Science and Spirit depends on such a middle path. So again yes! Between these two philosophical extremes—the realist/materialist reification of a permanent, absolute, substantial, eternal and independently existing physical and mental phenomenal reality 'out there', and the idealist nihilistic negation of it—abides the golden mean that is Prasangika Madhyamaka *Rangtong*, the centrist Nalanda Buddhist Middle Way Consequence School.

Yes, Prasangika is the conceptual causal foundation, according to Longchen Rabjam (2007), and His Holiness Dalai Lama (2009) of the utterly trans-conceptual, acausal nondual view and praxis of Buddhist *Nyingma* School's *Dzogchen*, the Great Perfection or Great Completion often seen as the pinnacle of the Buddhist view, even of all metaphysical views. *Dzogchen Ati Yoga* and its panpsychic kosmopsychism is seen as the acausal, non-conceptual, nondual 'correction' or 'completion' of the inherent duality of the Two Truths trope that is the heart of Middle Way Prasangika, and indeed of the entire great Buddhist Mahayana Causal Vehicle. [Boaz 2020 *Ch. V*]

Thus, in Buddhist *Dzogchen* we have not only a centrist Prasangika one truth synthesis of the Two Truths—Relative and Ultimate—that constitutes exoteric Realism/Materialism (matter), and esoteric Idealism (mind/spirit); but an optimistic and freeing soteriology—an 'innermost secret' or greater esoteric 'fruitional' (that happiness you seek is already present) view and praxis for an expedited human liberation/enlightenment, selfless ultimate Happiness Itself—the karma free harmless happiness that cannot be lost.

This concludes our all too brief survey of the Buddhist Mahayana Middle Way Prasangika Madhyamaka philosophy—conceptual foundation of *Ati Dzogchen* view and practice. [Boaz 2020, *Ch. V*]

Human Happiness Secret. Our happiness is present only here and now. The past is gone beyond, but a present memory. The future is but a present anticipation of that which has not yet arisen. But the future never shows up! It's always becoming the present, and then immediately it is past. So, we have nothing to cling to. There is only this breath in the body. What remains of the perennial 'three times', past, present and future? It is this fleeting present moment now! And this present now is far too brief to grasp and cling to. Therefore, we cannot *become* happy and enlightened in the future; we can only *be* happy and enlightened here and now. Nothing happens in the future because it does not yet exist! Human happiness, along with everything else, happens only here and now upon the mindful breath. Yet, we must learn from our past; and show up for work today.

Yes, wonder of wonders, as 1st century *Dzogchen* founder Garab Dorje told, "It is already accomplished from the very beginning. To rest here without seeking more is the Meditation". Our happiness abides deep within us, here and now. Our inherent happiness endures now as this indwelling love-wisdom mind Buddha mind peace/equanimity (*samatajnana*) 'instant Presence' (*vidya, rigpa, christos*) of the primary Priority Monist cosmopsychic (*Dzogchen Kosmopsychic*) primordial awareness ground, nondual vast whole itself, by whatever grand name or concept. That is the *Dzogchen* fruitional view. Our happiness is always already present Presence of that all embracing ground state.

Hence, the fruition or result of our happiness seeking strategies is already present here and now! Seeking the happiness that is already present to release suffering is itself a kind of suffering. So the highest teaching is: there is no need to seek happiness. "That happiness you seek…is already present within you." [Jesus the Christ] "You will not discover happiness until you stop seeking it." [Chuang Tzu] The dualistic 'spiritual' practice of the Path is our trans-conceptual noetic nondual 'wakefulness' to That! Our always present 'awakened mind'. Twenty-five hundred years past Buddha Shakyamuni told: "Let it be as it is and rest your weary mind, all things are perfect exactly as they are." That is the noetic nondual *Ati Dzogchen* fruitional view and practice.

As things are far from perfect in the spacetime dimension of Relative Truth, Buddha was here describing the luminous nondual realm of Ultimate Truth which embraces and pervades the spacetime cause and effect karmic dimension that is Relative Truth. These Two Truths are always an ontic prior, yet phenomenally present invariant (through all cognitive reference frames) noetic *one truth unity*. As Nagarjuna told, "There is absolutely no difference between [relative] samsara and [ultimate] nirvana." Recall, Buddha told of the *Four Profundities* in his noetic nondual *Prajnaparamita Heart of Wisdom Sutra,*

> Form is empty, emptiness is form.
> Form is not other than emptiness,
> Emptiness is not other than form.

Gottfried von Leibniz' view of such a perfect "best of all possible worlds", and recent cosmology's tautological but non-trivial Anthropic Principle (both weak and strong versions), both point out that our unlikely universe with its highly improbable 'super-fine-tuned' physical constants that favor life forms must perforce exist in order that human consciousness arise to reflexively ask such impudent questions. Both Leibniz and the Anthropic Principle suggest that a nondual noetic (no *essential* subject-object separation), non-essentialist view of this ineffable perfect subjectivity is a good metaphysical bet.

On the accord of Tibetan Buddhist Vajrayana epistemology, this perfect understanding is our indwelling Buddha mind (*buddha-jnana*), the Great Perfection of the *Perfect Sphere of Dzogchen*, nondual awareness dimension of Ultimate Truth (*paramartha satya*). Indeed, that may be seen as the very Nature of Mind (*cittata, sem nyid, sugatagarbha*) in whom this all arises. And That (*tathata*) is who we actually are—our 'supreme identity' of that all-pervasive 'supreme source' (*kunjed gyalpo*) ground of arising physical and mental form. Bright indwelling Presence of That. Who am I? Wonder of wonders, *That* numinous aboriginal ground is our home; who we actually are! That is the great teaching.

Heady wine indeed to dualistic concept mind ensnared as it is in the prodigious false quest for Cartesian absolute objective certainty within this dimension of merely realist/materialist 'concealer' Relative Truth (*samvriti satya*).

It is perhaps a bit sobering to remember that all of this heady conjecture is mostly just self-stimulating concepts and beliefs prior to contemplative direct experience (*yogi pratyaksa*) of the monistic cosmopsychic vast all subsuming boundless whole that is nondual primordial awareness ground of everything. Still, there is this unreasonable cognizance or brightness of the Nature of Mind that is always present to our obsessively thinking concept-mind. Luminous Presence of That (*tat, sat*). Presence is at root our penetrating 'direct seeing' (*vipashyana*), a contemplative knowing-feeling 'felt sense' (*samadhi, satori*) that lifts and bestows clarity and bliss to our busy conceptual mind.

Indeed, this nonlocal, nondual noetic awareness Presence penetrates and embraces all four mind states and life stages of our entire human cognitive consciousness processional: 1) pre-conceptual ordinary direct attention/perception, prior to naming; 2) exoteric, dualistic objective conceptual cognition; 3) esoteric emotive, intuitive, and contemplative cognition; and 4) innermost esoteric, perfectly subjective direct nondual cognition (*yogi pratyaksa*). These four mind states embody the lustrous display of the light of the mind that illumines all human cognitive life stages.

Choosing Reality: Universal Ontological Relativity

"Everything that exists lacks an intrinsic nature or identity" asserts Alan Wallace (2003) explicating Nagarjuna's Buddhist selfless (*anatman*) centrist Mahayana Madhyamaka Middle Way ontology. The appearance of objects arising from the basal primordial awareness-consciousness ground (boundless whole, *dharmadhatu, dharmakaya, chittadhatu*) abide in a relation of interconnected interdependence (*pratitya samutpada*). Their reality is 'ontologically relative' or 'universally relative'— established by relative conventional conceptual human minds—and thus observer-dependent upon our relative concepts and beliefs, and other related events and processes in a vast acausal *kosmic* matrix of 'prior causes and conditions'. The 'universal quantum Ψ-wave function' of Quantum Field Theory (QFT/QED) demonstrates a limited relative objective mathematical understanding of this ultimate great process, as we seen so many times in these pages.

In short, human discursive mind conceptually imputes, designates, then reifies these appearances into ostensibly observer-independent, objectively 'locally real' physical/mental/emotional spacetime existent realities in accordance with our limited atavistic, reflexive deep background (pre-conscious) cultural assumptions. Thus arises what W.V.O. Quine (1969) terms our semiotic (logical *syntax* of language; *semantics*/meaning; *pragmatics*/usage) 'global web of belief', which dictates our Western Realist/Materialist worldview with its pursuant predictable beliefs and anti-metaphysical, anti-mystical, anti-subjective cognitive biases. Universal ontological relativity indeed.

Just so, we habitually reify and reduce our bright subjectively real original noetic direct experience to objectified discursive semiotic/linguistic cognitive entities abiding in an emblematic, seemingly separate 'real world out there' (RWOT). With a bit of mindfulness meditation practice—Buddha's "mindfulness of breathing"—we may learn to *choose our reality*; that is, we learn to maintain the initial nondual noetic purity, poetry, and peace of our basal primordial wisdom ground as it arises spontaneously through ordinary direct perception, prior to conceptual intervention and judgment. With a bit more practice, under watchful guidance of a qualified meditation master, we maintain such a happy holistic view simultaneously through all the distractions of our parallel conceptual dualistic relative-conventional dimension—our beautiful precious lives as special guests of this spacetime phenomenal world.

Our Noetic Imperative

We live in two worlds—'self' (*atman*), objective conceptual ego-I; and 'noself' (*anatman*), subjective mental/spiritual at once—whether we are presently cognizant of the prior and present unity of these two dimensions or not. The dualistic objective local world of self-ego-I in time is always already embraced in that nonlocal nondual timeless noself world, bright always present Presence of That. That is our 'human condition'. Great joy! And yes, it's a real balancing act.

Is not our noetic (body, mind, spirit subject-object unity) imperative the recognition, realization and compassionate expression of the primordial unity of these two reality dimensions? To divide or not to divide, that is the noetic question of nondual primordial wisdom (gnosis, *jnana, yeshe*)—this very light of the mind. As we learn *awareness management*, the choice of our 'placement of attention/awareness'—through 'mindfulness of breathing'—upon our nondual love-wisdom mind Presence, recognition and compassionate expression naturally and spontaneously arise. And that is harmless, karma free, reflexive 'self liberating' ultimate Happiness Itself.

Thus do we *choose* a centrist 'middle path' that avoids the ontic extremes of Absolute Local Realism/Physicalism, and the nihilism of Absolute Idealism. We practice skillful, compassionate expression of the prior unity of our always present *noetic cognitive doublet*—relative, objective conceptual, and subjective ultimate contemplative—natural union of objective Science and its Spirit ground. Thus do we freely choose our realities.

Hence, from the metaphysical ontology you choose, arises the phenomenal world you deserve. Karma/action is a human happiness choice—breath by mindful breath; action by action. That is our *noetic imperative.* What do you think Dear Reader?

The Buddhist Two Truths and Dōgen's 'Being-Time'

Dōgen Zenji (Dōgen Kigen 1200-1253), perhaps Japan's greatest Zen master, spoke of this arising, descending dimension of relative time and its phenomenal contents—the spacetime dimension of Relative Truth (*samvriti satya*)—as "a being-time moment flashing into existence" from the vast spacious expanse of the basal, non-logocentric, nonlocal primordial emptiness (*shunyata*) base or ground, boundless awareness whole—nondual reality being itself—the all-embracing dimension of Ultimate Truth (*paramartha satya*).

That Ultimate Truth is nothing less than Dōgen's *Ugi,* or 'Being-Time'. Dōgen's *Ugi* is the here now, always already present prior and present unity of Buddhist Mahayana 'three times'—past, present, and future. So there is no beginning, and no end to this interdependent vast expanse of reality itself. The dimension of spacetime Relative Truth, including all of us, instantiate this primordial awareness emptiness 'groundless ground' of everything that arises and appears to a sentient participating human consciousness.

Yes, we are luminous primordial awareness embodied manifestations of that boundless whole physical/mental/spiritual *kosmic* process. [The many-dimensional *kosmic* whole embraces and subsumes the merely physical cosmos.] Human embodied mind-consciousness arises and participates in, and instantiates nondual primordial *kosmic* consciousness.

For Dōgen Zenji (and for 8th century Tibetan master Padmasambhava), the eternal present moment exists for us only relative to a past and a future. Being-Time/*Ugi* is a simultaneous array of all three. Thus we live in a single vanishing instant now. Yet, this precious moment now derives its meaning from the inter-subjective context of a personal and collective past, and of a possible future. This momentous present moment is so significant because all of our past and future are interdependently, causally enfolded within it, while always unfolding in the timeless continuum of that same moment.

Yes, we live in the moment, but not only in the moment. To live only in the moment, without awareness of our personal and collective past and future is to "make our life meaningless." We must learn from our mistakes. Not to live in the moment now, is "to lose reality itself". Reality happens in the present. As Shakespeare's Antonio told in *The Tempest*, "What's past is prologue." We must learn from our personal and collective past. And we must learn not to fear an uncertain future. [Dōgen 1986]

Philosophers of physics and cosmology, if not always physicists and cosmologists, are now discovering a post-empirical *kosmic* 'presentism'—reality is only now—in Dōgen Zenji's syncretic Being Time/*Ugi*. Such a view unifies the timeless Three Times, past, present, future; and bespeaks the prior unity of our two cognitive voices—objective and subjective—of this inherently reflexive, all-embracing spacious consciousness whole ('Basic Space' of *dharmadhatu, chöying*) of reality being itself, the very Nature of Mind and our human experience-consciousness that arises through it. That is after all who we are, our 'supreme identity' expression of the formless, timeless 'supreme source' ground in whom this all arises.

Dōgen's great insight is this: prior to the superimposition (*vikshepa, distraction*) and intervention of dualistic concept-belief cognition, 'ordinary direct perception' bestows the inherent (*sahaja*), immediate, luminous, liberating, blissful, 'primordially pure' noetic emptiness/*shunyata* Nature of Mind, vast ultimate awareness ground of relative mind and all its relative-conventional experience.

Dzogchen masters agree. Here, in the 'bare attention' of basal 'naked awareness'—ontologically prior to subject/object separation and habitual conceptual imputation and reification—abides trans-rational nondual noetic reality itself! This pristine awareness is the conscious aperture/portal/'explanatory gap' into our all subsuming all pervading primordial awareness wisdom ground (*jnana, yeshe,* gnosis). This vast boundless *dharmakaya* whole whose ultimate nature is emptiness (*shunyata*) manifests as our nondual intrinsic awareness love-wisdom mind, 'Big Mind', Buddha mind that knows and feels this great truth. We awaken to That, breath by mindful breath (*shamatha*/mindfulness meditation) through assiduous practice of the psycho-emotional spiritual path.

Such immediate perception, an instant prior to conception and naming, is pure perception. And we all do this, most of the time, with every perception! Wonder of wonders, we are all 'primordially awakened' (*bodhi, vidya*) to this always 'already accomplished' innate and perfect clear light *bodhi* mind! That is our actual 'supreme identity'. The rub? We must recognize, realize and compassionately express this perfectly subjective truth. We effortlessly and spontaneously accomplish this 'wisdom of kindness' breath by mindful breath in the wake of our 'mindfulness of breathing' and deity practice.

Further, we consult the experts and follow their injunctions, of course. In short, we establish an effective 'awareness management' meditation practice under the guidance of the *Dzogchen* master, and assiduously practice it. We learn the more or less constant 'placement of awareness-attention' upon our always already present Presence of the luminous *kosmic* primordial ground of everything. As H.H. Dalai Lama told (2009), "The clear light mind which lies dormant in human beings is the great hope of humankind."

Lord Buddha, Dōgen Zenji, Guru Padmasambhava, Longchenpa, Ju Mipham, Adi Shankara, Jesus the Christ, and indeed all the sages and saints of our great noetic Primordial Wisdom Tradition have taught this great nondual 'fruitional truth' of human happiness. Our human happiness is already the case. It is that kind love-wisdom mind to which we awaken through the practice of the path.

Hence, there is always, through all of our cognitive mind states and life stages—perceptual, conceptual, emotional, and transpersonal trans-conceptual contemplative—an ontic prior unity of past, present, future, always being here now. We learn to be present to the nondual noetic Presence of That. And yes, it requires a little selfless mindfulness (*shamatha/vipashyana*) contemplative practice. Who am I? *That Tvam Asi!* That I Am! Once again, don't *believe* it. It's utterly beyond belief. Buddha told, "Don't believe what I teach ... come and see (*ehi passika*)."

A Noetic Science of Matter, Mind, and Spirit

Human cognition enjoys objective quantitative and not so objective qualitative dimensions. Physics and cosmology are quantitative. 'The qualitative' (value, volition, ethics, the 'spiritual') is cognitively active in Big Science yet largely suppressed and denied in the common orthodoxy of the physical and social sciences. At long last, 21st century physics and cosmology are now beginning to recognize and strategically develop their inherent qualitative metaphysical dimension. The wondrous quantitative quantum theory is a mathematical bridge to a plausible qualitative integral Noetic Quantum Ontology. A complete Science of matter, mind and spirit requires recognition and praxis in both of these our human cognitive dimensions.

Prelude to a Foundational Noetic Quantum Ontology. What is now urgently required for recognition of the prior and present interdependent unity of Science and Spirit/Spirituality is a settled integral noetic (matter/mind/spirit unity) foundational quantum ontology with a centrist epistemology and methodology that accounts for both faces of our human cognitive experience—objective conceptual, and subjective contemplative/spiritual. We require a 'post-empirical' plausible foundational Noetic Quantum Ontology that grounds the formalist mathematics of Relativistic Quantum Field Theory (QFT/QED) in its prior metaphysical primordial ontic ground or source condition—the formless, timeless, boundless whole in whom this all arises.

To this end we must utilize the methods and noetic technologies of *Contemplative Science* to engage and explore subtle subjective phenomena that are inherently 'something deeply hidden' (Einstein) from our habitual objective conceptual scientific cognition.

Such contemplative praxis gradually reveals the noetic, perfectly subjective, formless, timeless, selfless ultimate reality matrix quantum and Buddhist emptiness base or ground of all arising, evolving physical spacetime form.

In due course this emerging spacetime matter/energy form ($E=mc^2$) evolves self-conscious human beings who desire to know and realize their primary atavistic relationship with that ultimate primordial awareness ground or 'supreme source' in whom their relative spacetime being arises. Human beings have evolved a cognitive life that bestows both quantitative objective conceptual, and qualitative subjective contemplative modes of understanding their experience of that all embracing primordial ground. We must engage and develop both of these, our natural human cognitive capacities.

In the West it is the local realist/physicalist 'scientific', quantitative, objective conceptual 'global web of belief' that has almost entirely colonized the Western heart and mind. The mostly missing qualitative, subjective contemplative technology and practice restores a balance. Without such a balance inherently subjective ontology—the pursuit of untrammeled recognition then realization of the ultimate ground of being—the grounding relation of the quantum theory, remains encaged in mere concepts and beliefs *about* it; if it is considered at all. Sadly, that is the present state/stage of our collective human cognition.

As to the four evolutionary mind state/life stages of our human cognitive life—1) direct attention-perception; 2) objective, conceptual quantitative; 3) noetic subjective contemplative qualitative; and 4) perfectly subjective nondual one truth unity—we remain substantially fixed in the first two. Entering in state/stage three marks the beginning of a our grail quest for a real, clear, complete personal then collective ontological understanding of the whole *process* of being here in time. Stage four realizes it.

Be that as it may, the ontic timeless formless 'groundless ground' of everything—the perfectly subjective, 'implicate', enfolded, ultimate boundless primordial awareness-consciousness whole and 'supreme source' of our noetic wisdom traditions—may be seen as 'Basic Space' (*chöying, dharmadhatu).* Herein arises all unfolding objective, 'explicate' physical relative spacetime particulars—energy, mass, force, charge, particle-waves, the universal quantum Ψ-wave function, and the continuous experience of embodied beings. Herein all of this arising stuff of being participates, interacts, and is instantiated.

Our ordinary human cognizance is not other than that basal primordial ground. Embodied mind arises unbidden and continuously from that vast 'supreme source'. Let us recognize, then realize the interconnected unity of this intrinsic ontic relationship. "No small matter is at stake here. The question [of ontology] concerns the very way that human life is to be lived." [Plato, *The Republic Book* I]

Viewed mereologically (part-whole relations), the panpsychic/primary cosmopsychic prior ontic unity (Schaffer 2010) that is this great awareness whole ground perforce subsumes and embraces its *relative* parts, while the parts necessarily participate in and instantiate the *ultimate* vast nondual primordial whole itself—the prior and present unity of the Buddhist Two Truths. This natural necessary process of *kosmic* subsumation constitutes the 'grounding relation' for all arising physical and mental form/phenomena—including Irwin Schrödinger's global 'quantum universal Ψ-wave function'.

This prodigious quantum Ψ-wave function may be seen as the objective, mathematical conceptual counterpart voice of the transconceptual, post-empirical perfectly subjective nondual whole, primordial awareness-consciousness ground of our quantum being here in space and time.

Recall, our *noetic cognitive doublet*—dualistic, quantitative, objective, conceptual; and qualitative, subjective, contemplative, nondual. We may come to know and experience this doublet as always an ontic prior and phenomenally present complementary interdependent invariant through all cognitive states *one truth unity* of Reality itself.

Clearly, such a *Noetic Revolution in Matter, Mind and Spirit* (Boaz 2023) requires a methodological, 'post-empirical' relaxing of the adventitious limits of obsessively objective positivist 'scientific' view and ideology with its prosaic 'taboo of subjectivity' regarding *a priori* contemplative knowledge. This waning 'old paradigm' Scientific Local Realism and Scientific Materialism/Physicalism dogma still obstructs our emerging inchoate 21st century Noetic Revolution that now arises phoenix-like from the ashes of Greek Metaphysical Realism/Materialism. New Paradigm indeed. Such a Kuhnian 'scientific revolution', is part of the Noetic Revolution that is now upon us.

The basal quantum emptiness ground of the proto-physical 'zero point vacuum energy field' (ZPE)—constant density dark energy, Einstein's cosmological constant (Λ) of Quantum Cosmology's *lambda* Λ-CDM Standard Model—along with the parallel pre-modern wisdom of Buddhist boundless emptiness (*shunyata/dharmakaya/kadag*) is a good beginning for a unified objective/subjective quantum ontology understanding of that diaphanous formless, timeless primordial ground of physical, mental, spiritual *kosmic* reality that is the grounding unity of the whole of merely physical cosmos. The quantum quantitative dimension is embraced and subsumed within the qualitative dimension of the nonlocal, nondual "implicate order of the vast unbroken whole", as David Bohm told it.

The physical, quantitative, cosmic spacetime physical ground that is the superposed 'universal quantum Ψ-wave function' arising from nearly empty ZPE unified quantum vacuum field are thereby ultimately grounded in a subtler, all subsuming, trans-quantitative, trans-rational, post-empirical, formless primordial awareness emptiness *kosmic* 'groundless ground' in which, or in whom it arises and participates. Recognizing then realizing this great truth requires noetic contemplative technologies and research methodologies that utilize both quantitative objective third person data sets, and the qualitative, though still objective data sets of personal, subjective, introspective, even contemplative first person reports of highly experienced meditation practitioners and masters (Wallace 2009; Begley 2007; Boaz 2022) who naturally weave their nondual primordial wisdom mindstream into the splendent fabric of sociocultural space and time.

Thus are the Mahayana Middle Way Madhyamaka Two Truths—spacetime Relative Truth, and post-empirical, all-pervading nondual Ultimate Truth—unified in the Buddhist *Perfect Sphere of Dzogchen*, spacious unbounded whole (Basic Space, *dharmadhatu*), nondual ultimate reality itself (*dharmakaya*) in which QFT, and everything else arises, participates, and is instantiated.

Mereologically speaking, the multiplicity of form—the particular cosmic parts including the quantum Ψ-wave function—are perforce subsumed and grounded by the greater primordial, all-embracing, timeless, formless, selfless, awareness-consciousness *kosmic* womb, Paul Dirac's "zero womb", boundless whole itself. So many words for That that Noetic Science of matter, mind, and spirit that cannot be told in words.

Interpreting the Quantum Theory

Quantum mechanics is proven to be the most successful and predictive scientific theory in the fabulous 400 year history of Modern Science. It gives us our computers, smart phones, microwave ovens, and TV sets. Yet it reveals nothing about actual, even ultimate nature of the physical world it presumes to quantify. The formalist quantum calculus accurately predicts the outcomes of empirical observations yet leaves a view of the cosmos that is quite counterintuitive, paradoxical, anti-empirical, even antirealist. And Irwin Schrödinger's 'universal quantum Ψ-wave function' presumes to rule the entire physical cosmos. O quantum hubris. What does the prodigious quantum of action really reveal as to our being here in time? For that we require a Noetic Quantum Ontology.

Well, how does the quantum calculus actually work? What constitutes a quantum measurement? An observation? What does it mean about the *ultimate* nature of this *relative* objective phenomenal world? Quantum mathematics requires an explanatory interpretation of how this wondrous theory corresponds to our experienced spacetime realities. And what does it imply about the *ultimate* nature of human consciousness and its experienced realities? And about the original *kosmic* ground in whom human consciousness with its

QFT/QED arises and participate. We need to know the relation between quantum Science and nondual Spirit.

Clearly, QFT/QED is incompatible with an objective only physical world. Perhaps there is no logical solution to this 'quantum measurement problem'. To be sure QFT points beyond its objectivist formalist mathematics toward a more inclusive inherently subjective unbounded whole. Meanwhile, quantum physicists busy themselves with highly contrived and convoluted apologetics to reduce inherent quantum subjectivity to classical objective Scientific Local Realism/Physicalism which cannot accommodate such liberating holistic ontology.

We are now beginning to understand that we must seek answers in a more inclusive 'post-empirical' integral Noetic Quantum Ontology.

Of the twenty or so 'foundational interpretations' or ontologies of quantum mechanics that are bandied about in this physics sector, there are seven on offer that receive the most critical attention. All of them are objectivist local realist apologies for the inherent subjectivity in the 'objective random' acausal QFT/QED, and so none are adequate to a satisfactory understanding of human consciousness nor its nondual ultimate awareness-consciousness ground beyond mere objective properties.

Please note that each one of the following 'foundational quantum interpretations' or 'quantum ontologies' is an attempt to escape the 'problem' of a subjective *observer-dependent* consciousness through the metaphysical presupposition that appearing reality is inherently *observer-independently* locally real, physical, or reducible to a purely physical human brain and central nervous system.

In other words classical Metaphysical Scientific Local Realism/Physicalism is the cognitive ontological bias that obtains in and limits all of them. None are metaphysically prepared to reach beyond Scientific Local Realism/Physicalism to engage a centrist middle way foundational *kosmic* monistic kosmopsychic timeless, formless all subsuming primordial awareness-consciousness ground, boundless *order* of the whole in which, or in whom human consciousness with its universal quantum Ψ-wave function arises, collapses, unfolds, interacts and relates, and is *ultimately* instantiated.

In brief, we require more from our quantum 'foundational interpretations' than mathematical explanations. We must at long last engage the 'hidden metaphysics' of quantum physics. So yes, for that we must develop an integral foundational Noetic Quantum Ontology. Let physicists and philosophers of physics read and engage in dialogue with Buddhist Dzogchen scholar-practitioners.

Exploring Quantum Theory Foundations. Let us then take a closer look at the extant quantum 'interpretations' now on offer. The seven primary quantum theory 'foundational interpretations' are:

1) The original antirealist proto-'collapse' *Copenhagen Interpretation* was posited by its founders Niels Bohr and Werner Heisenberg in 1927. On this 'post-empirical' subjectivist antirealist view the realm of quantum mechanics is in principle unknowable to human cognition, nor can it be a description of this cognitive realm. It is inherently indeterministic and probabilities are calculated using the Born Rule, Bohr's foundational *Principle of Complementarity*, and Heisenberg's antinomian *Principle of Uncertainty*.

These seven foundational interpretations of quantum mechanics are grounded in the quantum formalist mathematics for predicting probabilities for microphysical quantum measurement results. And that's all. For Copenhagen no ontic claim for the existence of a 'real world out there' (RWOT) is made nor assumed. For Bohr, metaphysical ontology is for philosophers, not physicists. A developed quantum ontology is ignored. Thus the Copenhagen Interpretation is almost by default an antirealist ontology. Indeed, it has become the default 'collapse' view of present Quantum Field Theory (QFT/QED). It is commonly taught in undergraduate physics as *the* quantum theory.

However, neither Bohr nor Schrödinger were comfortable with the *ad hoc* notion of an objective reality-creating 'collapse' of Schrödinger's universal quantum Ψ-wave function. What then shall we teach in post-graduate physics, and to we inquiring non-physicist minds?

Well, we can eliminate the 'collapse requirement' by rejecting the notion that fundamental objective reality is constituted by a superposed global 'universal quantum Ψ-wave function'. Here our Ψ-wave collapse denier might posit a "hidden parameter" or "hidden variable" at this most fundamental strata of micro physical formation to explain away the need to 'collapse' the subjective superposition of all possible states into an objectively "real" electron with definite *eigenstates* of position and momentum.

2) *Hidden Variables* models, the most satisfying of which is the *nonlocal* 1952 Bohm-deBroglie Pilot Wave Theory. It begins by modifying the standard quantum formalism of Quantum Field Theory (QFT/QED) by positing that the quintessential universal quantum Ψ-wave function should be augmented or even replaced by an as yet to be discovered microcosmic 'hidden variable'. Hidden Variables Theory then should perhaps be seen as incipient proposals for a new more complete quantum theory. These models assume that the Bohr/Schrödinger/Dirac QFT/QED is, as Einstein pointed out, incomplete and in need of an admittedly unlikely "hidden parameter"—local or nonlocal—that describes the true ontic nature of the world it presumes to measure and quantify. For three decades this improbable Scientific Local Realism 'hope for a miracle' remained at least a logical possibility. Then, like an epistemic bolt from the blue—paradigm busting Bell's Theorem.

John Stewart Bell (Bell's Theorem 1964) is supposed to have proven that all Hidden Variable theories must be nonlocal. Whether David Bohm's nonlocal 'pilot wave theory' escapes Bell's Theorem by violating Bell's locality assumption is a matter of recent debate.

In any case Bohm's theory adds *ad hoc* complexity to the elegant simplicity of Irwin Schrödinger's quantum Ψ-wave function non-relativistic quantum mechanics from which evolved Paul Dirac's profoundly successful Relativistic Quantum Electrodynamics

(QED). What is clear is that somehow the vast timeless perfectly subjective *ultimate* primordial ground of physical and mental form perforce descends into *relative* space and time as human experience of objectivity real stuff. The mathematics of Ψ-wave collapse do not explain how. Yes, we need a settled centrist integral Noetic Quantum Ontology.

The prodigious Schrödinger Equation with its wave function collapse is extant in several of our 'foundational interpretations' of quantum mechanics. The nature and process of physical reality is described by the indeterminist quantum Ψ-wave function as it evolves deterministically in classical real time (t). That grand process is explained by the Schrödinger Equation, an elegant and prodigiously predictive explanation.

3) *Spontaneous Collapse* theories presume to complete an ostensibly incomplete quantum mechanics by endorsing Ψ-wave function completeness but they attempt to modify Schrödinger's fundamental dynamic deterministic linear equation into a non-linear indeterministic equation. For example, the 1986 GRW theory (Ghirardi, Rimini, Weber) introduces a probabilistic "collapse" term to the original deterministic Schrödinger dynamics. The dynamics of quantum collapse theories are probabilistic, not deterministic.

4) The *Consciousness Causes Collapse* quantum interpretation of John von Neumann, Eugene Wigner, and Henry Stapp arises from von Neumann's definitive 1932 *Mathematical Foundations of Quantum Mechanics* wherein an embodied 'organic' consciousness or awareness is postulated to be necessary in the process of arriving at a quantum measurement outcome. After all, a pointer on a measurement devise must be consciously read and interpreted by an experimenter! An observing evaluating mind is seen as a *non-physical* measurement process that ultimately 'collapses the quantum wave function' from an infinite number of subjective 'superposed' quantum states into one single objectively real *eigenstate* of an electron's position. The inherent problem of 'causal determinism' is engaged below.

The global 'universal quantum Ψ-wave function' is here 'collapsed' by the observing consciousness of a non-mechanistic organic observer or experimenter somehow producing an objectively 'real' electron from that infinitely 'superposed' field of possibility. David Chalmers (1996) has pointed out that 'collapse' is an almost arbitrary entangled, nonlocal, asymmetric, discontinuous physical process that appears to be anomalous to the Standard Model physics of the continuous, symmetrical, local Schrödinger equation. It seems that we have here at the heart of physical reality two competing dynamical processes—quantum 'collapse', and classical determinist Standard Model physics.

Well, what is the status of physical reality before this cosmos evolved an experiencing sentient observer consciousness to make it all 'real'? Please consider well that sentient embodied consciousness perforce arises from and is not separate from its nondual, all subsuming, all pervading unified primordial awareness-consciousness ground, the formless, timeless unbounded whole itself in whom any spacetime cosmos and its embodied sentient observer-consciousnesses may arise. That natural basal primordial 'emptiness ground' is, mereologically (part-whole relations), always already the case from before any beginning in time. That basal ground *is* an observer-consciousness—physically embodied or not!

Whatever spacetime phenomena that arises from this formless ground then evolves within David Bohm's nondual "implicate order of the unbroken whole" in accord with the psychophysical laws of evolutionary biology. In successive multiple universes (Conformal Cyclic Cosmology) arising in this vast all embracing *kosmos* it has always been thus.

In short, that primordial awareness-consciousness ground itself—by whatever grand name—is the always already present aboriginal 'quantum observer' that bestows 'real' quantum stuff, timeless ontic prior source of any arising evolving embodied 'observer consciousness'.

We've seen that this great trans-physical truth of our human species' nondual noetic Primordial Wisdom Tradition points a way to a non-reductive, non-physicalist Middle Way integral Noetic Quantum Ontology that embraces it. As to such a non-reductive ontology David Chalmers (1965) opines:

> We can give up on the project of trying to explain the-existence of consciousness wholly in terms of something more basic, and instead admit it is [panpsychic] fundamental... similar in kind to the theories that physics gives us of matter, of motion, or of space and time.

5) *Quantum Bayesianism* (*Qbism*). For Qbism a quantum state is not an objective reality but represents 'degrees of belief' or personal judgments of a human agent regarding possible outcomes of any quantum measurement. Bayesianism is then a personalist subjective approach to the treatment of quantum probabilities. Qbism was founded early in the 21st century by Caves, Fuchs, Schack and later adopted by Mermin.

6) *The Many Worlds Interpretation* (*MWI*). In 1957 Hugh Everett composed a now famous doctoral dissertation offering an extremely radical, highly counterintuitive, quite unbelievable 'interpretation of quantum mechanics'. Had it not of late gained the intellectual favor of what one can only term desperate scientific local realist/physicalist ideologues it would not be worth refuting. MWI may be a paradigm case of Greek truth functional two-valued logic gone terribly wrong. ['Paraconsistant Logic Systems' below.]

For the MWI of Hugh Everett, Bryce DeWitt, David Deutch, and Sean Carroll there exist many literal, objectively real worlds or universes which 'parallel' the one that we all inter-objectively experience. For MWI the universal quantum wave function is seen as *objectively* real, and there is no need for a 'quantum Ψ-wave function collapse to obtain 'real' spacetime stuff. Rather, the singular universal quantum Ψ-wave function 'splits' or 'branches' billions of times with every measurement or perception for the mind of each individual perceiver. For every possible quantum outcome, an objectively real new world. Logically possible, yes. Empirically possible, no. *Prima facie* absurd, yes.

So for MWI a quantum wave function must 'divide' into numerous parts, but there is no indication here as to how this 'splitting' occurs. How and why should we identify each part of a single mind state with an infinite ensemble of entirely separate observers in a multitude of empirically impossible parallel universes? Who but a desperate classical physics ideologue mesmerized by inherently spooky quantum subjectivity could buy into such a 'tale told by an idiot'? MWI adds fantasque insult to spooky quantum injury.

For MWI there are 'multiple copies' of each and every personal conscious mind abiding in infinitely many spacetime universes. Thus all possible outcomes of a quantum measurement—countless trillions of them—are somehow physically realized in some parallel universe, somewhere out there, all of which are infinite in number and hidden from one another. How is such an "ontological cloudburst" so? Some have told that such a creation of new worlds violates the inviolable physics Law of the Conservation of Mass-Energy. Can this be a quantum process? Is such a process empirically possible? Is it cognizable at all? And does it not require a viable theory of consciousness to explain such strange behavior of human minds?

John Stewart Bell in 1976 pointed out that in MWI it is not clear under what conditions a splitting or branching event into a parallel world shall occur given the complexities of quantum measurement.

So quantum mechanics for MWI is hyper-realist, purely mechanistic, objectivist, descriptive, deterministic and universal regarding quantum phenomena both microscopic and its macroscopic arising as well. The MWI seems far more *ad hoc*, arbitrary and complex than the various 'wave function collapse' interpretations it presumes to replace. Does not such an objectively extremist quantum theory pay a ruinous price to dodge spooky inherent quantum subjectivity?

OK. MWI gets rid of the unwanted collapse of the Ψ-wave function. But at what price? Despite its fantasque truth claims MWI is logically consistent. As to possibility, MWI is 'logically possible'. But it is not 'empirically possible'. And it certainly violates even physicists' 'common sense' Local Realism notions of the proper local behavior of matter/energy in spacetime. So MWI is not believable. What can truthfully be said of an empirically impossible, unbelievable quantum foundation that has now entered mainstream physics? Perhaps modern physics and cosmology remain stuck in a bygone classical 'scientific' materialism/mechanism dogma that our post-empirical quantum theory has now revealed.

In 1952 Irwin Schrödinger, creator of the universal quantum wave function had an idea nearly identical to MWI in his own search for an alternative to what he considered an ad hoc "absurd" Ψ-wave collapse. After all, there is nothing in the quantum formalist mathematics about any such 'collapse'. Niels Bohr ambivalently added Ψ-wave collapse to his Copenhagen Interpretation as an *ad hoc* attempt to explain how it is that we detect only one solution to the Ψ-wave equation as a result of a measurement. Many solutions may exist in parallel worlds. Irwin Schrödinger's unfortunate cat is dead in one world, alive in another. So it is both dead and alive simultaneously, until an observer opens the cat's box.

In his 1952 paper Schrödinger is incredulous at the notion that a non-objective 'superposition' of many possibilities would collapse merely by an objective conscious observation-measurement. So for him the quantum Ψ-wave collapse is problematic, to say the least! So much for foundational 'quantum collapse theories' from the author of the wondrous quantum wave function. We shall see more of MWI below.

7) *Relational Quantum Mechanics (RQM).* There is now abroad in the quantum cognosphere a recent (1996, 2016) 'soft realist' or antirealist quantum foundational interpretation or 'quantum ontology' developed by theoretical physicist Carlo Rovelli which has gained much attention as a viable observer-dependent,

relational-interactional, relativistic, pararealist epistemology of the occult ontology of post-Schrödinger (1927) Quantum Fielf Theory (QFT/QED). It is usually seen as an enhancement, but not a completion, of the foundational nominalist antirealist, instrumentalist 'textbook' Copenhagen Interpretation (1927) of Niels Bohr and Werner Heisenberg.

As with our other seven 'quantum foundations', RQM cannot suffice as an integral Noetic Quantum Ontology because it fails to engage such ontological considerations as the nature of the nondual, formless, timeless, perfectly subjective primordial Spirit ground in which, or in whom the objective formalisms of quantum Science mereologically arise and interact. [Please ignore the word 'Spirit' should you find it too off-putting or unscientific.] Our urgent ontological or metaphysical mereological concern is here the 'grounding relation' of the entities and theorems of Big Science to that ultimate all embracing all subsuming primordial awareness ground, boundless whole wherein this all perforce arises and is instantiated—by whatever grand name.

For Carlo Rovelli and his RQM the nature of physical reality is not its classical, hitherto intrinsic internal objective physical 'structure' (Metaphysical Scientific Local Realism/Physicalism), but rather the *interdependent relationships* of matter-energy—an interconnected whole network or matrix of causal and acausal relations/interactions among para-physical *'systems'* that constitute the vast boundless whole of event-moments in the totality of the ontic *process* of appearing reality itself. [Middle Way Buddhist Nagarjuna] In RQM and in Middle Way Buddhism any intrinsically existing Kantian 'thing in itself' is denied, or at least considered inherently conceptually unknowable. We can know nothing of 'things in themselves', only relations of interacting systems. Spacetime stuff is ultimately unknowable Kantian *noumena.*

Meanwhile, relatively 'real' relational interactions happen in this all too real regime of spacetime phenomenal existence. So object-things exist not ultimately, but still appear relatively, via their relative conventional interactions. Maps rather nicely onto the

'Two Truths'—ultimate and relative—dominant trope of Buddhist Middle Way Madhyamaka.

Unfortunately Rovelli, as with Bohr, has failed to develop a philosophical ontology, and to make an 'ontological commitment', whether it be ultimate Antirealism or ultimate Realism/Physicalism, or a centrist middle path between these two metaphysical extremes.

For example, Middle Way Buddhism ontologically commits to the ontic view that "appearing reality is ultimately absent and empty of any whit of intrinsic existence", while still acknowledging its nominal 'real' relative-conventional existence. After analyzing the nature of phenomena for 2500 years we Buddhists have come to understand that there is no *ultimate* existing reality! So the ontic question now becomes, "How it is that this cosmos of appearing spacetime phenomena actually does exist? After all, stuff is everywhere, and we all experience it!

The Madhyamaka emptiness/interdependent arising philosophy of Nagarjuna and Chandrakirti threads a centrist Middle Way between the ontic metaphysical extremes of Abhidharma Realism, and Yogachara Cittamatra ('Mind Only School) Idealism. Spacetime stuff is relatively real but not ultimately real. It is both depending on the view, relative or ultimate. So, shall we name it Realism or Antirealism? That centrist 'middle path' is the Madhyamaka Buddhist realist/antirealist 'ontic commitment'. Without such an ontological commitment the participating theorist-scientist remains but a passive observer. What is too often missing in Big Science is a commitment to becoming an *observer-knower.* Buddhist Prasangika with its *reductio ad absurdum* argument has been criticized for a thousand years for missing that ontic boat. Nondual *Ati Dzogchen* corrects that. We can only speculate, as I have done here, as to Rovelli's view of the *ultimate* nature and ground of *relative* arising spacetime phenomena, and how it manifests as this sometimes all too 'real' spacetime stuff that we have come to know and love. Hence, as to quantum ontology we require an *ontological commitment.* Our centrist integral Noetic Quantum Ontology accomplishes that.

Be all that as it may, we do know enough from Rovelli's statements about relations and interactions to apply the epithet 'Antirealism' to RQM—and view it as a metaphysical enhancement of the nominalist 'textbook' Copenhagen Interpretation of the quantum mechanics of Bohr and Heisenberg who have also remained silent as to quantum ontology. Had Rovelli developed a quantum ontology his RQM may well have become the completion for Bohr's nominalist instrumentalist Copenhagen Antirealism.

As to Rovelli's epistemology, we can say with relative certainty that relations/interactions among micro 'systems' are ontologically prior to appearing macro objects. So his para-panpsychism avoids constitutive atomistic micropsychism in favor of a Schafferian priority monistic cosmopsychicism—foundation of *Dzogchen Kosmopsychism.*

For RQM the momentum *or* the position of a physical entity-object, e.g. an electron, as it is in itself, is 'meaningless'. Any object O in a quantum system S is experienced-measured only relative to or in relation to an interaction with another object/system S'. So, all physical variables are inherently 'relational', a necessary relationship between two or more interacting 'systems'. But what pray tell is the actual ontology? Realism, Antirealism, or a middle path? We can only speculate that RQM is a species of Antirealism.

RQM closely parallels Ontic Structural Realism (OSR), but RQM's proto-realistic view is 'softer' and subtler. We presume that Rovelli does see it as an antirealist ontology. It might be said that RQM is an accidental epistemic instantiation of the relational ontology of OSR. RQM also closely parallels the interdependent 'relationality' of the Buddhist friendly incipient quantum ontology of Zeilinger and Bruckner (1999).

[Anton Zeilinger was awarded the Nobel Prize in Physics in 2022—with Alain Aspect and John Clauser—for their paradigm shattering work on nonlocal quantum entanglement. It is Zeilinger's recent work with David Kaiser (2016) that closed the final 'locality loophole' demonstrating that Bell's Proof proves that neither local nor nonlocal hidden variables exist, and that quantum nonlocal entanglement does exist.]

The Quantum Problem of Causal Determinism

Heisenberg's profound acausal Uncertainty Principle undermines our scientific classical and common sense ideology of a local, objective, deterministic real world out there (RWOT). Just so, an electron prior to a quantum measurement by, for example an 'observer's consciousness', is "spread out" in Schrödinger's subjective Ψ-wave 'quantum superposition' of an infinity of uncertain potential positions until it acausally *randomly* interacts with another quantum system—in this experimental case with 1) a conscious observer and 2) a 'decoherent' mechanical measuring device with a pointer revealing a result to the 'consciousness' of that observer. This 'quantum uncertainty', if true, demonstrates that 200 years of the Modernist classical determinism of Pierre-Simon Laplace is false. Few quantum physicists were listening.

'Causal determinism', must not be confused with 'fatalism'—the *belief* that all events in the course of human experience are predetermined and therefore inevitable. Fatalism has justified unspeakable human horror.

Causal determinism is, broadly construed, the philosophical view that all spacetime objects and events are 'determined' to exist as they appear entirely as the 'predetermined' result of prior causes and conditions. A perfect divine intelligence, e.g. an omniscient theistic God, or an 'omniscient oracle', or an 'Ideal Observer' would know the microphysical states of all the particles of the universe and could therefore 'calculate' the result of any future macrocosmic event. Never mind that such an anthropocentric all-knowing intelligence is not empirically nor even logically possible.

We've seen that our past and our future are perforce gone beyond. Past is but a present memory; future is but a present anticipation of what may be. And the present moment is to brief to grasp. It streams through the present moment and is now past. So we have nothing to which we may attach and cling. Yet we do have this timeless, acausal indwelling, always already present luminous primordial *Presence* of the mereologically necessary nondual aboriginal awareness-consciousness ground of it all. That 'clear light' Presence is right here now embedded in our participating human consciousness! That realization (*samadhi, moksha*) is, for our human primordial wisdom traditions, the open secret of ultimate human happiness—compassionate harmless Happiness Itself. Let us not conceptually stray from the timeless Presence of that great love-wisdom mind truth.

Broadly construed, the opposite of determinism is indeterminism, or 'objective randomness'. In physics causal determinism is, as I have said, causality or cause and effect interactions wherein any and all appearing objects and events are completely determined by some prior state or condition.

Determinism is often contrasted with the 'free will' of an acting agent. This involves our notions of karma or human action—beyond the scope of this discussion. Determinism in physics does not require that effects/results be perfectly predictable.

So, it seems that Heisenberg's uncertainty relations of quantum mechanics have shown that 'hard causal determinism' is untenable, yet Schrödinger and many other great quantum physicists hold that the mathematical formalism of quantum mechanical functions, including the Ψ-wave function, are causally deterministic prior to the problematic 'collapse' of that wave function—the mysterious spooky 'state vector reduction'—upon the event of a quantum measurement. That wave function collapse, how or if it collapses, is known as the prodigious 'Quantum Measurement Problem'. But in any case the logical precision of mathematics is necessarily deterministic. Or is it?

We shall see below in our discussion of the limits of classical formal deterministic logic and pure mathematics, that these must be replaced by a less than deterministic 'paraconsistent intuitive' deductive logical system. Classical Greek logic and modern foundational mathematics is riddled with contradiction and logical paradox (e.g. Russell's paradox, and Gödel's Incompleteness) and is therefore inadequate to offer deterministic certainty at either the microcosmic dimension of quantum particles, or at the macrocosmic dimension of compounded spacetime phenomena, like trees, stars and living beings.

Irwin Schrödinger's 'universal quantum Ψ-wave function' (the state vector)—the beautiful Schrödinger Equation—is wholly mathematically causally deterministic. However, its Ψ-wave function does not correspond to any objective natural phenomenal reality. Pragmatically we must then add a causal principle that qualifies what was at the time inherently acausal deterministic quantum mathematics. That causal principal came to be called the 'collapse hypothesis', the collapse of the deterministic quantum Ψ-wave function as it evolves indeterministically in classical real time. And yes, that quickly became the infamous Quantum Measurement Problem.

Well, what in heaven and earth could constitute Ψ-wave function collapse? It seems that some kind of wave function collapse is necessary to bridge between hard determinism and practical indeterminism that are both present in the quantum theory.

We saw above that several 'quantum foundational interpretations', or ontologies have been proffered to explain, or explain away, quantum Ψ-wave function collapse. Schrödinger himself hated it. In any case, notwithstanding the amazing predictive accuracy of the quantum theory (computers, smart phones, TV, laser communications), we require theoretically a hybrid causal deterministic indeterminism to make logical sense of this greatest of all scientific theories.

Perhaps, as several of our quantum pioneers have told, the inherent subjectivity that is acausal nonlocal entangled 'quantum emptiness' is in the final analysis incomprehensible to mere human objective reason.

Quantum Emptiness, Buddhist Emptiness

Individual *microcosmic* quantum events are utterly acausal and 'objectively random', yet there still exists a stochastic/probability pattern. *Objective randomness*—a nonlocal or acausal non-objective reality with no hidden causes—seems to be a primary truth of the inherently subjective quantum nature of our spacetime reality, even a natural law, though we may wish it were not so. It may be contrasted with *subjective randomness* or ignorance of yet to be revealed real but 'hidden variable' causes of phenomena.

Just so, a *relative* objectively 'real world out there' (RWOT) does not *ultimately* exist. This is the Middle Way Buddhist view, as well as the view of Niels Bohr and Carlo Rovelli, the antirealist, some say nihilist Copenhagen view. It may be contrasted with Irwin Schrödinger's realist 'subjective randomness' view as to a 'superposed' quantum reality that reveals itself somehow to be ultimately present to we sentient beings here in space and time. Although Schrödinger was a student of Sanskrit and of our acausal Eastern wisdom traditions, he insisted that his *Ψ*-wave function was inherently a "real" objective cause and effect mathematical function. In a letter to Hendrik Lorentz dated June 6, 1926 Schrödinger stated, "The *Ψ* wave function is surely fundamentally a real function." That is to say, quantum wave functions and their mathematical operators are Cartesian 'natural real numbers' in a 'real number quantum theory'. [Natural numbers are integers that include only positive whole numbers.] He objected to the subjectivity of 'complex' or 'imaginary' numbers.

Middle Way Buddhist centrist ontology views arising cause and effect reality as causally *relatively* real, but *ultimately* acausal and "absent or empty (*shunya*) of intrinsic existence". So it is an error to peg Madhyamaka ontology as a causal determinism.

A centrist integral Noetic Quantum Ontology conceptually reveals for trans-conceptual contemplative/meditative practice such a middle path between these two metaphysical extremes that are absolute existence and absolute nonexistence. Such a mean naturally arises in the causal 'interdependent arising' (*pratitya samutpada*) of Buddhist Middle Way Prasangika relativist epistemology that is, on the accord of H.H. Dalai Lama, the conceptual foundation of the noetic nondual ontology of *Ati Dzogchen* the Great Perfection/Completion, as we have already seen.

Buddhist emptiness (*shunyata*) is the causal 'interdependent arising' of *relative* spacetime phenomena from its acausal *ultimate* formless, timeless, selfless primordial 'groundless ground'. That aboriginal ground is 'groundless' because the vast boundless whole is itself absent and empty of intrinsic existence—Buddha's 'emptiness of emptiness'. It is decidedly a *causal process* of interrelations and interconnections among phenomenal events. That this is so does not mean that such events and relations possess intrinsic, objective, or ultimate existence. Yet this dimension of form, or Relative Truth indeed possesses 'real' relative or conventional existence as it arises in the all subsuming nondual dimension of Ultimate Truth—the Buddha's Two Truths—form and emptiness. Told the Buddha in his 'Fourfold Profundity': "Form is empty [of intrinsic existence]; emptiness is form. Form is not other than emptiness; emptiness is not other than form." These two truths are a prior and always present *one truth unity* (*dzog*) that is invariant through all human cognitive frames of reference.

Indeed, that prior and present unity may be seen as our emptiness, selfless (*anatman, kenosis*) 'view from nowhere'. It represents the view of an always imperfect cognitive bias free perspective of an 'omniscient oracle' or an 'Ideal Observer' that is in the ultimate view, free of an intrinsically biased self-ego-I.

How in heaven and earth shall we accomplish such a mind state? Via the *relative* conscious 'placement of attention-awareness' upon the always present *ultimate* primordial Presence that abides at the indwelling human spiritual Heart (*hridyam*)—upon each mindful breath. That lucent quiescent, often blissful immediate Presence is trans-conceptually nested simultaneously in the love-wisdom *jnanaprana* spirit wind in the belly, the heart center, the forebrain center, and the crown center.

Emptiness? We must understand that for Mahayana Madhyamaka Buddhists the term 'emptiness' refers not to nothingness, or a nihilistic absence of existence, but to the ontic ultimate nature of all that which relatively arises and appears to sentient awareness from the foundational 'Basic Space' (chöying) of its primordial dharmakaya ground.

Quantum physics provides us with profound epistemic scientific evidence for the relative objective existence—yet ultimate absence—of an absolute existing local purely physical spacetime reality. That is quantum emptiness. Buddhism provides the ontology.

Quantum mechanics with its radical nonlocal 'objective randomness' of relative spacetime form points to an ultimate absence or nonlocal entangled quantum emptiness of spacetime reality naturally arising from its primordial emptiness ground. That inherent acausal randomness bespeaks, as His Holiness reasons (Zajonc 2004):

> A meaningful parallel with the Buddhist concept of emptiness ... between the thought of Nagarjuna and Chandrakirti, and the quantum mechanical views on the lack of substantiality of photons ... As far as objective randomness goes ... I feel that they probably will find some [causal] reason for it as research continues.

We all accept the essential formalist mathematics of quantum mechanics, yet we need a settled quantum ontology as to the various ontic 'foundational interpretations' of it.

In short we require a more or less committed centrist (between absolute existence and absolute nonexistence) integral Noetic Quantum Ontology, beyond the minimalist 'objective randomness' explanation—one that makes sense of counterintuitive acausal, superluminal (faster than light), nonlocal quantum entanglement.

David Bohm's 'hidden variables' nonlocal causality attempts such an explanation, yet it still belies our notions of common sense causality. A centrist integral Noetic Quantum Ontology must account for the Relative Truth dimension of nonlocal 'objective randomness' and nonlocal quantum entanglement in its 'grounding relation' to all subsuming nondual Ultimate Truth of the formless, timeless primordial ground in which this all arises, participates, and is instantiated.

How is it that macrocosmic spacetime order arises from such inherently random microscopic states? An ontic tall order indeed. Is macroscopic order observer-dependent, or is it purely objective phenomena independent of a perceiving thinking knowing subject? Who is it that discovers hidden order abiding in quantum data? In a nondeterministic universe is order objective or subjective? If we can harness the subjectivity of nonlocal quantum entanglement, e.g. in a quantum computer, it would seem that indeed the cosmos is actually an objective, even physical structure. But is it?

Quantum 'objective randomness' *ipso facto* precludes causal objective order. This introduces the quantum paradox as to how macroscopic patterned local order—trees, stars, people—*causally* emerges from microscopic disordered randomness, presenting the notorious problem of Einstein's "spooky" stochastic, non-causal, nonlocal quantum entanglement wherein two microscopic particles remain as a single quantum system even when 'space-like separated' by vast intergalactic distances. Einstein hated it. It violated the finite light speed limit of the second postulate of his 1905 Special Relativity Theory. Perhaps even worse, God's creation cannot be based in mere quantum probability. Einstein's reasoning, "God does not play dice with the world." In the 1927 Solvay Conference these two creators of the quantum theory debated 'completeness of quantum mechanics'. Most of the leading physicists present found Bohr the victor. The ultimate nature of spacetime reality is not absolute but probabilistic, and stochastic quantum measurement and Bohr's foundational Principle of Complementarity are vindicated. Einstein and his nemesis and intellectual equal Bohr debated quantum nonlocal entanglement in these EPR (Einstein, Podolsky, Rosen) debates for many years.

Indeed, acausal nonlocal quantum entanglement is the defining essential of the quantum mechanics that questions our classical Scientific Local Realism knowledge paradigm. Is our entire cosmic multiverse inherently internally nonlocally entangled? Can it be said to ultimately exist beyond its relative local appearances? Must we surrender thousands of years of local causality, the very foundation of Science? Can causality be nonlocal as in Bohm's nonlocal 'hidden variables' theory? 'Quantum weirdness' indeed. It would take John Stewart Bell in 1964 to disentangle it all, as we shall soon see.

Our above resulting stochastic quantum measurement is said to now somehow—no one can explain how—"collapse" the system at the instant of observation/measurement from an uncertain subjective superposition into an objectively 'real' electron eigenstate of position with an objective mathematical eigenvalue right here in a problematic classical local time. This seemingly endless 'nonlocal entanglement chain' is ostensibly 'broken' at that very moment of collapse in an objectively real time.

We have seen that Schrödinger himself was quite ambivalent about such an *ad hoc* 'collapse'—and particularly a 'consciousness causes collapse'—of his prodigious universal quantum Ψ-wave function.

As to such wave function collapse, John Stewart Bell disagreed. Surprisingly, even for him, the entire cosmos is inherently acausally nonlocally entangled—forever. [Bell's Theorem below] Rovelli's RQM and Bohr's Copenhagen Interpretation agree with Bell. 'Observation' itself remains unexplained by quantum mechanics. Zeilinger terms 'quantum observation' "a primary unexplained notion". We here encounter the sticky quantum paradoxes of 'Wigner's friend' and 'Schröginger's cat'. For both Rovelli and Bohr a quantum observation cannot be reduced to a classical causal physical model because all the vexing paradoxical quantum 'measurement problems' immediately arise.

Better perhaps to affirm spooky acausal 'objective randomness' and let our spacetime realities be ultimately, or at least relatively nonexistent, or as Madhyamaka Buddhists say, "empty of intrinsic existence". The ontic price paid is 400 years of modernist classical causal Scientific Local Realism, our beloved RWOT. What to do?

Wigner, von Neumann, and the 'consciousness causes collapse' (quantum ontology number 4 above) quantum ontologists attempt to break von Neumann's 'infinite regress' observation chain of the endless web of quantum nonlocal entanglement by appealing to *human consciousness* as the required *non-physical* method to break that infinite regress of physical spooky entanglement to produce a mysteriously 'collapsed' classical objectively real electron with a real location in our beloved 4-D spacetime continuum.

Pragmatically this philosophical consideration is an ontic *fait accompli.* Quantum mathematics always produces our computers and cell phones, regardless of any quantum ontology we might proffer. Yet we must still ask the impudent question: "Who's consciousness breaks the chain of infinite regress when both Wigner and his 'friend' observe Schrödinger's simultaneously dead and alive cat? [Zajonc 2004]

Relational Quantum Mechanics Revisited

Carlo Rovelli: "An electron is nowhere [non-existent] when it is not interacting... things only exist by jumping from one interaction to another." Spacetime stuff does not exist prior to an observer-dependent interaction among 'systems'. On the accord of Rovelli's 'foundational interpretation' of the quantum theory an observer-independent pre-existing 'real world out there' (RWOT) can no longer be the truth of our appearing spacetime realities. Observer-dependent "spooky action at a distance" (Einstein) quantum Antirealism indeed. Einstein hated it. Bohr loved it. We must all face up to it.

So yes, in the RQM antirealist view it is nominally real interacting *systems,* not conceptually imputed and reified 'real' physical object *substances* that constitute our appearing realities. This post-classical, 'post-empirical', nominalist, antirealist quantum physics altogether drops the ideological metaphysical baggage of the presumption of classically 'real' absolutely existing causal physical properties and attributes of interacting hitherto purely physical systems. We no longer need presume a purely physical observer-independent RWOT.

Our classical habitual Scientific Local Realism/Physicalism metaphysic of an *ultimately* real world out there (RWOT) and a separate self-ego-I who perceptually and conceptually imputes, reifies, and defends it is at last surrendered to a nominalist centrist anti-realist theme; just as the Middle Way Buddhists have told for 2000 years. Real physical and mental stuff arise via this relative basal complementary interactional observer-object relationship. This quantum observer-dependent influence—the infamous 'quantum measurement problem' about what arises to our senses—goes far beyond our waning classical physics observer-independent realist knowledge physics paradigm. Niels Bohr with his Principle of Complementarity of opposites (e.g. wave-particle duality) would agree. As would Nagarjuna and Chandrakirti. Through Rovelli's RQM quantum ontology we have taken a big if not yet decisive step toward an integral Noetic Quantum Ontology.

So, for RQM, quantum mechanical 'systems' inherently possess an interdependent relational/interactional character. *Relative* 'real' objective physical cosmic stuff and living beings arise from their mereologically (part/whole relations) necessary *ultimate* noetic nondual, formless, timeless, selfless primordial awareness-consciousness ground—their *kosmic* superposition—when they interact with an observer consciousness. That is to say, we need only experience our physical and mental realities in that way. We need no longer cling to a bygone Greek Apollonian mechanistic physicalist Local Realism, where the mountains and rivers of our realities are truly there, even when we don't look.

Thus is RQM an observer-dependent inchoate ontology over against the observer-independent ideological Greek Western knowledge paradigm that is Modernist Local Realism with its epistemic consort Materialism/Physicalism. This old metaphysic has entirely colonized the Western mind and constitutes our common sense deep background cultural 'global web of belief' (Quine 1969). The 'objective randomness' of inherently subjective post-classical Quantum Field Theory (QFT/QED) has forever changed that view. Yet very few theoretical physicists are listening. They are too busy trying to reduce quantum mechanics and the Ψ-wave function to a bygone classical mechanics.

Just so, RQM as it is left hanging by Carlo Rovelli is a quantum epistemology more than it is a quantum ontology. Its unstated subtle antirealist ontology awaits a proper epistemic unpacking before it appears in its potential ontic glory. Perhaps we cannot conceptually know ultimate Being Itself (ontology), but we can directly experience (*yogi pratyaksa*) both conceptual and non-conceptual contemplative/meditational noetic relative relational *process* of events in this all subsuming, universal, selfless primordial awareness ground of being (Hegel, Nagarjuna) in whom our wondrous quantum realities arise, participate, and are instantiated here in relative space and time.

Well and good. A bit more technically, for RQM any physical system may play as an observer; any physical interaction, conscious or not, qualifies as a quantum measurement! Here the quantum world of interactive relationships ultimately needs no *ad hoc* 'hidden variables', no contrived 'collapse' mechanisms, no fantasque parallel 'Many Worlds Interpretation' (MWI); and it requires no *essential* role for an objective human mind/consciousness to rescue the prodigious quantum theory from subjective oblivion.

Therefore, we need not ask how it is that the stuff of spacetime existed before any conscious beings evolved as observers to give phenomena its objective reality. *Interactions* among matter-energy systems arising in their primordial awareness-consciousness ground will suffice as an observer. For Rovelli, "All physical variables are relational... .Different observers can give different accounts of the same set of events." [Special Relativity] And yes, an 'observer system' need not be a conscious macroscopic sentient observer. Nor is RQM about a subjective quantum superposed Ψ-wave function entity from which our objective world quantum eigenstates emerge at the instant of a conscious observation by a sentient consciousness, or other modality of a quantum measurement.

All seven of our above quantum ontologies—of the 20 or so on offer—habitually interpret Irwin Schrödinger's quantum Ψ-wave function realistically, as did Schrödinger himself, on the ideological ontic accord of classical Metaphysical Scientific Local Realism.

Not so Carlo Rovelli's RQM which is, as we have just seen, a nominalist Antirealism, in the mode of Niels Bohr's Copenhagen Interpretation of quantum mechanics. Rovelli (2016) argues against a Local Realism interpretation for his Relational Quantum Mechanics. Schrödinger's 1927 realistic Ψ is then not required to formulate quantum mechanics! For mathematical Platonist Rovelli, observed measured quantum eigenvalues are necessarily universal. They are objective values of real mathematical operators, not eigenvalues of absolutely pre-existing 'quantum states'. That indeed, is a kind of middle path between the metaphysical extremes of the eternalist substantialism of absolute objective existence (Metaphysical Scientific Realism/Physicalism), and the nihilism of absolute subjective nonexistence (Metaphysical Idealism). Let's settle then for an RQM middle path, in a word, Antirealism. Sounds a lot like Mahayana Middle Way Prasangika Buddhism.

Quantum and Buddhist Causality. Rovelli's RQM causal interdependent 'relational-interactional' cosmic *process*—like the inherently causal 'interdependent arising' (*pratitya samutpada*) of Buddhist Mahayana ontology—may be seen in 21st century 'post-empirical' wisdom as necessary to a pragmatic causal understanding of our naturally arising phenomenal world of relative-conventional spacetime appearing reality without clinging to a metaphysical extreme of either absolute substantial Local Realism, or absolute nihilistic Idealism—the false dichotomy of either only existence or only nonexistence.

We live in a cause and effect phenomenal world. And yes, Big Science depends upon such causality. Both our relative human flourishing, and our ultimate harmless happiness clearly depend upon familiar karmic cause and effect relations/interactions. Clearly causality must obtain at all phenomenal strata of formation—quantum microcosmic to macroscopic human action—physical and karmic ethical. RQM and Middle Way Buddhism each provide a cognitive bridge between such a *relative* objective indeterminist causality, and the *ultimate* determinist acausal subjective randomness of the primordial ground in which it naturally and spontaneously arises.

RQM furthers this natural cognitive evolutionary process, while further weakening our much beloved classical ideology of an ontic 'strong Realism', an absolutely or ultimately existing, observer-independently objective real world out there' (RWOT), irrespective of our countervailing self-reifying mental and emotional states, perspectives, and conceptual-intellectual and spiritual views.

For RQM the physical structure of appearing reality is an interdependent, para-causal network or matrix of continuously interacting, interdependent, inherently interconnected, proto-physical variables as nominally causal interacting 'systems'. That is perhaps the actual nature of 'mutually interdependent' quantum reality, just as Middle Way Buddhist Nagarjuna told some 20 centuries past. Relativistic "relations and interactions" among microcosmic micropsychic 'quantum systems' are then at least nominally causal. Rovelli has salvaged some semblance of causality from the subjectivity of pure quantum 'objective randomness' and saved quantum mechanics from a most discomfiting acausal absurdity!

Perhaps then we might view appearing reality as microscopically acausal while cause and effect clearly obtains in our macroscopic dimension of phenomenal space and time. An absolute 'objective randomness' quantum blanket denial of causal relations at all dimensional strata of spacetime formation is clearly unreasonable and unworkable.

For Mahayana Madhyamaka Middle Way Buddhists the entire dimension of spacetime Relative Truth as it arises in its all embracing acausal, trans-conceptual, timeless nondual Ultimate Truth dimension, is thoroughly causal—linear and nonlinear cause and effect relations. If you desire that result-effect, practice this injunction-cause. The entire buddhadharma, and indeed 'common sense', is grounded in karmic cause and effect 'interdependently arising' relations/interactions. That is after all the way that relative-conventional spacetime reality works!

Thus do we require a 'foundational interpretation'—a metaphysical ontology, of the wondrous quantum theory—what it means for our lives and our happiness as honored guests of this wondrous gift of the phenomenal world. The quantum mathematical formalism alone leaves macroscopic reality out in the formalist cold. Thus have I attempted in this chapter a nascent centrist integral Noetic Quantum Ontology.

Therefore, in RQM, as with Middle Way Madhyamaka Buddhism, there is no observer-independent *ultimately* existing objective RWOT because there exist no intrinsic properties or attributes of physical systems independent of the relational perspectival proto-physical interactions among such systems. The entire *process* of the event-moments of our appearing physical and mental realities are inherently observer-dependent, causal 'interdependently arising', relational/interactional.

These logically paraconsistent relative-conventional relations among causally interacting micro and macro 'systems' are the very 'building blocks' of modern physics' classical spacetime 'presentist block universe'. The new 'Presentism' in physics sees the Three Times—past, present, future—as identical. Here there can be no privileged non-perspectival 'God's eye view', no 'view from nowhere', no absolutely existing 'objectively random' quantum state of this universe of ours because cosmos can be only understood, as Nobel Laureate Anton Zeilinger (1999) told, "from within a given relational perspective." That we know as 'universal ontological relativity'.

Perspectivism and Ontological Relativity

If there is a redeeming benefit to nihilist Postmodernism it is the wisdom of 'perspectivism', or 'perspectivalism', the urgent epistemic principle that human knowledge is dependent upon the conceptual interpretative perspectives and corresponding *beliefs* of an observer or an inter-subjective related cadre of observers. These shared world views are often doctrinaire and dogmatic conceptual fixations within a group of 'insiders'

Our individual and collective human perspectives are grounded in our raw well defended basic belief systems as to how the world actually is to us. These foundational cognitive tenets represent our "primal world beliefs", or "primals". [Clifton and Kerry]

For example, is our lived world essentially dangerous or safe; vibrant, beautiful and alive or grey, grisly and mechanistic; happy or miserable; are others more or less against us, or do they wish to help where possible; are human beings basically good or evil? Is human life much more than 'hell on earth and then we die'? Or a rare gift through which we may learn to be happy? Is there a middle way between such foolish false dichotomies?

Clearly our basic 'primal world beliefs' about the nature of our realities reflect our actual human happiness attainment in this difficult life world. Can we human beings manage the often destructive self-ego-I and 'train the mind' as to our habitual cognitive perspectival "primals" that result in profound felt perceived happiness?

Indeed, that outcome is the soteriological intention of every love-wisdom teaching of the great noetic Primordial Wisdom Tradition of humankind. Yes. We can, and do change even the most entrenched of our perspectives and beliefs. Love guided by wisdom under the guidance of the meditation master informs us as to how we may accomplish it. But don't believe it! As the Buddha told so long ago, "Don't believe what I teach; come and see." How else are we to rise above our doleful unhappy deep cultural background 'global web of belief'. [Quine 1969] Added H.H. Dalai Lama, "Just open the door."

To be sure, no human observer has absolute purely *objective* access to appearing phenomenal reality in itself, Kant's inherently conceptually utterly ineffable "noumenon", always abiding beyond our human perspectives, concepts, and belief systems. As to embodied personal *subjective* direct contemplative yogic experience (*yogi pratyaksa*) of the inherently happy numinous Presence of the nondual boundless primordial ground, perspectivism among 'those who know' organizes and codifies the objective data protocols of such private, subjective, non-conceptual, even nondual 'spiritual' experience.

Perspectivism need not be a form of relativism, but it is decidedly embraced by the Principle of Universal Ontological Relativity—we create, impute, designate, and reify our realities via our cognitive systems of perspectival concept and belief. What we think is what we believe, is who we are. Our *choice* of 'placement of attention-awareness' dictates what we think and so who we are. Great Buddhist *Dzogchen* master Jigme Lingpa has told it well. "By meditating in this way experience whatever you direct your attention to... then rest in that spacious peaceful non-conceptual mind state." Said Buddha, "What you are is what you have been; what you will be is what you do now."

In any case, perspectivism need not be a cognitive assent nor a rejection of any particular objective or subjective reality world view. Nor is it required that epistemic perspectivism itself be absolutely or definitively 'true'. *The principle of perspectivism itself must be reflexively perspectival.* Perspectivism may be seen as a fluent practical non-absolute epistemic and phenomenal tool. [Zeilinger 1999, "A Foundational Principle for Quantum Mechanics", *Foundations of Physics*, 29 (4): 631-643.]

Brief Review. Each of the above seven quantum theory 'foundational interpretations' is an objective conceptual attempt to demystify and explain, or explain away Irwin Schrödinger's 1927 subjectively infinitely 'superposed' indeterminist pre-collapse global 'universal quantum Ψ-wave function' of quantum mechanics, and as well its actual 'state vector reduction' collapse (if indeed it does collapse). That Ψ-wave function followed the 1927 Quantum Theory of Bohr, Heisenberg and Pauli. Quantum mechanics thus need not depend upon Schrödinger's wave function. It was developed before the wave function. [A famous 1926 paper by Born, Heisenberg, and Jordan had previously quantized the electromagnetic field.]

A settled foundational quantitative quantum theory will certainly utilize quantum quantities common to most of these seven, and especially of RQM, but must now at long last engage a qualitative, post-empirical, postformal, holistic integral Noetic Quantum Ontology that is prior to, yet includes and subsumes present and any future revisions of the requisite mathematical quantum formalisms now on offer. A complete quantum ontology metaphysic must perforce refer to the ontological ground of its mathematical formalisms, the aboriginal all subsuming perfectly subjective primordial source condition whence the math and the mathematicians arise and participate.

Thus do we begin to heal and unify objective quantitative cognition (mathematical formalism) with subjective qualitative cognition—consciousness, both human and the formless, timeless primordial ground in which it all is instantiated. These two modalities—objective and subjective—of our unified *noetic cognitive doublet* that our human cognition have been hitherto torn asunder under sway of the prevailing belief systems and local realist/physicalist cognitive biases of Modern Big Science physics. That deep background cultural 'global web of belief' (Quine 1969) includes the dualistic metaphysical 'scientific' dogmas of Scientific Local Realism and Greek Scientific Materialism/Physicalism that have now beset our Modern and Postmodern Western mind.

That paradigmatic perspectival 'mind change' has proven exceedingly difficult after 400 years of local realist Modernist European Enlightenment physics—'the idols of the tribe'—encaged as it is in the mechanistic scientific realist/materialist ideologies that are our Greek metaphysical cognitive legacy.

Quantum Field Theory is Here to Stay

As a predictive theory QFT/QED is nearly perfect. No physics theory has even approximated its accuracy. As to ontology, QFT/QED is physics' still inchoate *relative* epistemic cognitive architecture for accomplishing an *ultimate* ontological understanding of the whole nature of appearing reality, both objective Science and its nondual perfectly subjective primordial Spirit ground. QFT mathematical formalisms must now be integrated with the trans-quantitative qualitative cosmopsychic ultimate whole itself—nondual primordial awareness-consciousness ground whence gravitating spacetime stuff and we who attempt to understand it, and our quantum theories about it emerge.

Dzogchen Kosmopsychism. I shall argue herein that panpsychic (monistic primary cosmopsychism) nondual Buddhist *Dzogchen* Kosmopsychism as it arises from Buddhist Middle Way Prasangika Madhyamaka philosophy constitutes an *ultimate* ontic foundation for the *relative* epistemic 'universal quantum wave function' mathematical formalisms. [Boaz 2022 *Ch. VII*] I have termed such a 'post-empirical' ontology the centrist integral Noetic Quantum Ontology, a new if still inchoate nonlocal 'foundational interpretation' of QFT that dares to tread beyond our seven still classical objectivist local realist quantum ontologies now on offer. Antirealist RQM may well be an important exception to this broad quantum construal, although Rovelli has not developed a quantum ontology.

The immeasurable challenge is this: that greatest of human intellectual achievements, the prodigious Standard Model of particles and forces, with its recent *lambda* Λ-CDM (cold dark matter) Standard Model of Cosmology still clings to the classical orthodox, old paradigm dogmatic materialist metaphysical bias that is extreme objectivist Scientific Local Realism—with its metaphysical Physicalism/Materialism of a bygone classical Galilean-Newtonian cosmos of objectively 'real', substantialist, 'eternalist', purely physical objects existing observer-independently, permanently, and eternally in the absolute, objectively real, purely physical 4-D spacetime manifold of Einstein and his math teacher Hermann Minkowski.

Good news! We have seen that Einstein's classical view of Special and General Relativity is now being integrated with an observer-dependent relativistic quantum mechanics upon the advent of our 21st century Noetic Revolution in Science and Spirit. [Boaz 2023] Such a rapprochement is required if we are to heal the present mathematical incommensurability between Einstein's classical General Relativity Theory and Quantum Field Theory through a yet to be accomplished unified Quantum Gravity Theory.

A purely physicalist objective observer-independent classical spacetime RWOT has now fallen on hard times. Physicists are at last beginning to hear Einstein on time: "Time—past, present, future—is an illusion; albeit a very persuasive one." With new work in quantum cosmology many physicists have now thrown out space as well. That ontic result leaves our beloved spacetime realities obviously relatively real, yet not ultimately real. Sounds like Middle Way Madhyamaka Buddhism. [Boaz 2020, Ch. V]

In any case, the notoriously perverse mathematical incommensurability of QFT/QED with Einstein's General Relativity Theory (GRT)—the formal split between these two great pillars of modern physics—will continue without an ideological softening of Modern Science's hyper-objectivist ontic monistic metaphysical Scientific Local Realism and its epistemic phenomenal consort Metaphysical Physicalism.

That waning classical purely *quantitative* knowledge paradigm view is now considered by most philosophers of physics, and a few theoretical physicists should they bother to think about it, to be a failed ontology. How is this so? 1) The waning old paradigm contradicts the inherently 'objective random' acausal and therefore the inherent subjectivity of quantum theory, to wit, 'always correct' QFT. 2) It fails to engage the *qualitative,* nondual, basal, mereological, perfectly subjective primordial awareness-consciousness ground in which, or in whom our objective, conceptual and mathematical worlds arise. Science must surrender its destructive 'taboo of subjectivity' in order to approach such a unified subject-object noetic understanding.

Yes, we desperately need a unifying Quantum Gravity Theory (QGT) to heal this epistemic split between the minute microcosmic realm of Max Planck's Planck Scale 'quantum of action', and the vast large scale macrocosmic dimension of the cosmic multiverse ruled by Einstein's mysterious entropic gravity.

The Emerging Noetic Revolution in Science and Spirit

Some physicists, and most philosophers of physics, along with Buddhist philosopher practitioners know that there is no innate dimensional separation between our appearing microcosmic and macrocosmic phenomenal regimes. The whole of physical spacetime appearing cosmic reality, with its quantum Ψ-wave function and its 'dehocorent' measurement instruments, is always already unified and subsumed in the formless, timeless, boundless, selfless indivisible, trans-conceptual, nondual primordial awareness-consciousness whole—aboriginal ground in whom this whole shebang arises. It is that holistic metaphysical understanding that must be integrated into the metaphysics of the universal quantum Ψ-wave function with its arcane mathematical formalisms. Such a post-quantum metaphysic lies in the establishment of an integral Noetic Quantum Ontology of our inchoate paradigmatic Noetic Revolution in matter, mind and spirit.

"Aye, that is the rub". Conceptual or even contemplative blissing out in the perfectly subjective ground of being is not enough. We must skillfully engage our *objective* cognitive capacity to conceptually and mathematically explicate that prior unity of objective quantitative quantum form and its subjective nonlocal entangled quantum emptiness ZPE ground while remaining present to the prior qualitative, trans-conceptual, nondual truth of the matter. That is to say, we maintain an awareness of the present state of the nondual *one truth unity* (*dzog*) of our perennial Two Truths—relative dualistic spacetime form or Science, and its formless perfectly subjective nondual emptiness Spirit ground.

We must build our integral Noetic Quantum Ontology upon that prior one truth unity of the Two Truths. That is the rub. Indeed, a bitter cognitive pill for 21st century Standard Model quantum physics clinging as it does to a bygone classical realist/materialist/physicalist reductionist physics paradigm. This scientific and cultural knowledge *'paradigm shift'* is well under way in our incipient 21st century Noetic Revolution in Science and Spirit as we continue to surrender our monistic Metaphysical Scientific Local Realism and Metaphysical Scientific Materialism/Physicalism classical ideologies to the theme of quantum holism. If Thomas Kuhn is right such a paradigmatic 'gestalt shift' shall require two or three generations.

Strange Interlude: A Noetic Gedanken. A thought experiment. Consider for a moment that you suddenly awaken from your 'dogmatic slumber' to a hitherto seeming impossible, but now real cognitive choice in which you behold a now present condition of an utterly unified relative objective form reality dimension abiding in its formless timeless ultimate perfectly subjective emptiness ground. You see clearly that the ultimate relationship of these seemingly separate reality dimensions is one of nondual interconnected, interdependent identity. Your previous cognitive bias for purely objective cosmic form is utterly absent. You trans-conceptually, directly experience (*yogi pratyaksa*) all the mental vivid clarity and emotional peace and bliss that is now present in your new *buddhic* trans-rational *kosmic* love-wisdom mind cognitive dimension. How does that *feel*?

Notice that your choice here is not between a false dichotomy of *either* objective *or* subjective dimensions of a cosmic spacetime reality. Your opportunity now is to recognize, even realize the already present unity/identity of all that appears in this vast *kosmic* whole of ultimate reality itself, both objective conceptual and perfectly subjective nondual, and the all subsuming primordial ground of That, by whatever grand name—trans-theistic nondual God, infinite *Ein Sof*, Samantabhadra, Tao, Parabrahman, and the rest. You now have a *choice* to see and directly experience that nondual clarity, noetic bliss, unity, interconnectedness, and interdependence of what appears to human perception and conception as a dualistic, objective, separate and chaotic 'real world out there' (RWOT).

What would that love-wisdom mind Presence *feel* like? Inter in this dream now for two minutes. Go ahead and do it now. Close your eyes, raise your eyebrows (which generates alpha and 'waking theta' brain activity), and place all of your attentional awareness upon the life force *jnanaprana* breath in your belly, then your heart, then the medial prefrontal cortex just behind your forehead, and then at the crown of your head. Now *feel* your present heart-mind connection to the primordial awareness ground of you and everything else. Rest now in *Presence* of that vast boundless whole. As thoughts naturally arise, say 'distraction', and return to your breath, again and again. How does it *feel* to enter in the actual Nature of Mind? Bright Presence of That that you are now—your 'supreme identity'. What is your mind? *That* is your mind! Who am I? *Tat Tvam Asi!* That I Am! [*Appendix A* below; Boaz 2022, *Mindfulness Meditation: The Complete Guide Ch. VIII*]

Unpacking the Dream. Quantum Field Theory and the history of both Western and Eastern philosophy and religion have told it well: reality is not as it appears! Ordinary perception sees only the parts. Our love-wisdom clear light (*prabhasa*) direct perception (*yogi pratyaksa*) sees the interdependent enfolded parts arising within, and unfolding as that vast acausal *kosmic* whole of nondual Being Itself arising continuously in this primal causal matrix of spacetime phenomena (*pratitya samutpada,* Interdependent Arising).

An open, mostly bias free contemplatively trained mind may enjoy such a conceptually impossible result. Buddhas and *mahasiddhas* live in clarity of that mind state and corresponding life stage most of the time. But don't believe this. It's utterly beyond belief for our thinking concept mind. As Buddha told, "Come and see" (*ehi passika*).

A complete centrist integral Noetic Quantum Ontology that embraces this aboriginal ultimate *kosmic* ground of emerging relative physical spacetime dimension of reality—both microcosmic and macrocosmic—shall greatly facilitate our understanding of the way in which gravity and QFT are already *kosmically,* noetically unified. What is certain is that the task exceeds the conceptual limit of the formalist mathematics of the quantum theory. We require a quantum metaphysical ontology. That task shall engage both of our human cognitive faculties, our *noetic cognitive doublet,* objective conceptual and subjective contemplative.

Thus begins our recognition of the ontic prior yet phenomenally present unity of these two innate awareness dimensions of mind, our human mind as it arises in and through the nondual primordial awareness-consciousness ground or base (*gzhi rigpa*) that is the very Nature of Mind (*sems nyid, cittadhatu, dharmakaya*).

Well, what are these two cognitive worlds that embody our human condition? You guessed it! 1) exoteric, objective, material, dualistic, conceptual, scientific; and 2) esoteric, subjective, trans-conceptual, noetic nondual, contemplative spiritual. In short 'objective' and 'subjective'. May I say it again? These two seemingly separate modes of understanding our arising realities are already a prior yet always present unity, when we choose to enter in, and clearly see, and feel that happy cognitive state of noetic union. That is the noetic reality of our rude awakening from this dreadful dream of a separate reality.

Consciously Shifting Our Global Knowledge Paradigm. Our twentieth century global quantum knowledge paradigm has subsumed three of the four fundamental forces/particle interactions of the wondrous Standard Model of particle physics, namely Electromagnetism, the Strong Nuclear Force that sub-atomically binds the worlds, and the Weak Nuclear Force of radioactive beta decay. Only the Newton-Einstein 'Big G' gravity 'force' or interaction remains to be tamed by the Standard Model's sublime quantum theory with its universal quantum Ψ-wave function, whatever that turns out to be. And yes, we shall require an integral Noetic Quantum Ontology for a Quantum Gravity Theory (QGT) that results in such a providential unity of objective Science and its perfectly subjective Spirit ground.

Quantum theory clearly obtains in both of these dimensional regimes—the three micro forces of the very small, and large macro-scale gravity. This universe of ours is quantum in nature. But we need a propitious new calculus of the *objective relative in the subjective quantum.* The quantum conundrum? How can relativistic gravity be 'quantized' and shown to be quantum in nature resulting in the long sought QGT?

But wait! Is it logically or even empirically possible to mathematically quantize the conceptually unquantifiable qualitative vast whole of objective appearing reality arising in its nondual perfectly subjective primordial ground? Are our reality dimensions of qualitative all-pervading Ultimate Truth and quantitative spacetime Relative Truth arising therein finally quantifiable at all? Is this Big Science ideological insistence a logical 'category mistake'. Can this boundless whole be reduced to a formal quantum mathematical equation? Big Science hubris? What do you think Dear Reader?

Perhaps the perfect mereological subjectivity of noetic primordial wisdom (*jnana, yeshe,* gnosis) of the ultimate primordial ground of everything lies beyond our wondrous scientific knowledge methods of quantification. Perhaps nature possesses inherently subjective *qualities* that preclude objective logical and mathematical quantification.

In any case that has been my argument in these pages. Recall the four ascending mind state/life stage dimensions of our human noetic cognitive consciousness processional—1) ordinary direct perception, prior to concept and naming; 2) objective conceptual cognition (Science); 3) trans-conceptual contemplative cognition; and 4) perfectly subjective nondual cognition (Spirit). We have seen that Noetic contemplative love-wisdom mind state/life stage three and four understand perceptual and conceptual state/stage one and two, but not the other way round.

Toward a Post-Quantum Knowledge Paradigm. Now, as to this spacetime dimension of Relative Truth, we require a new quantum spacetime that reveals how the hitherto smooth non-quantized classical gravity continuum of Aristotle, Newton, and of Einstein's GRT gravity waves may be quantized—'course grained' into discrete quantum bits, Einstein's 'lightquanta', discrete foundational qbits, like photons and gravitons. Indeed, the physics desideratum devoutly to be wished.

But again, it may be asked: is it logically, or empirically possible to grasp the whole of our limited experience of reality by way of mere conceptual mathematical means, no matter how skillful? Is the prodigious quantum gravity dilemma a logical *'category mistake'*? If the nonlocal metaphysical abstractions of the inherently subjective qualities of vast multidimensional nature cannot be *ultimately* reduced to the local mathematical quantities of Quantum Field Theory (QFT/QED), as our present Big Science global knowledge paradigm seems to require, then the 'quantum gravity problem' is a species of category mistake that must be recognized to be so in order to correct and connect.

In any case, it's clear to 'those who know' that modern physics has at last hit its paradigmatic brick wall. Thomas Kuhn (*The Structure of Scientific Revolutions* 1962, 2012) has described the process wherein periods of 'normal science' (e.g. objective classical mechanics) are rudely interrupted by an unbidden 'scientific revolution' (e.g. inherently 'objectively random' subjective quantum mechanics).

Historical cases in point: 1) the heliocentric Copernican Revolution; 2) the Newtonian Revolution that unified the relativity paradigm shift begun by Galileo and Kepler, and continued by Einstein; and 3) the Quantum Revolution—the present scientific and cultural knowledge paradigm shift from objective classical mechanics to the spooky subjectivity of quantum mechanical nonlocal entanglement that I have here referred to as 'quantum emptiness'. Paradigmatic revolutionary Kuhnian 'gestalt shifts' indeed.

Such 'anomalies' or logical or empirical paradoxes and contradictions arising in a previous waning 'old knowledge paradigm' beget a new ascending revolutionary 'scientific paradigm'. This gradual, then sudden sociological process Kuhn termed a scientific 'paradigm shift'. Such a global seismic 'Gestalt shift' in scientific thinking influences not only the scientific culture, but in due course (before that next more inclusive shift descends) pervades our common sense 'global web of belief' (Quine 1969)—the deep background ideology and idiom of the 'global' whole of Western culture.

For Kuhn the old and the new paradigms are so different as to be utterly 'incommensurable'. The ideology and discourse of each one is foreign to the other. Each has its own presuppositions and semiotic gloss that is nearly incomprehensible, and more or less unbelievable to the opposing view.

A radical shift in perceiving and thinking is now upon us as the descending objectivist classical relativistic physics paradigm is ever so gradually replaced with the ascending subjectivity of the relativistic quantum physics and Middle Way Buddhist paradigms. This Quantum Revolution has precipitated our next global revolution in science, culture, and religion/spirituality. I have come to call it *The Noetic Revolution: Toward an Integral Science of Matter, Mind and Spirit*, the title of a 2023 book of mine.

We have seen that a new knowledge paradigm shift requires two or three generations to become a settled dogmatic orthodoxy as the old paradigm tenured acolytes expire, and new paradigmatic blood enters the hallowed halls of academic learning—only to be replaced soon enough by that next more inclusive new knowledge paradigm shift.

Just so, our century old quantum scientific knowledge paradigm is now yielding to a new post-quantum integral noetic knowledge paradigm that recognizes the prior unity of objective Science and its mereologically necessary perfectly subjective nondual Spirit ground. The inherent acausal random subjectivity of the wondrous quantum theory has facilitated this prodigious knowledge paradigm shift. "Past is prologue".

May this rather tedious process be somehow expedited for our ascending post-quantum noetic paradigm? Indeed it may. The random, acausal nonlocal entangled, antirealist subjectivity of the global scientific quantum metaphysic has dethroned the prevailing classical objectivist Scientific Local Realism/Physicalism, and the 'common sense' metaphysic of Newton's and Einstein's observer-independent only physical 'real world out there' (RWOT). Radical 'post-empirical' nonlocal entangled quantum subjectivity has hitherto been vigorously resisted by the old paradigm classical physics orthodoxy—ideological, hyper-objectivist tenacious habitual Scientific Local Realism, Materialism, Physicalism. Quantum physicists are only just beginning to see what random inherently subjective and acausal QFT is pointing out. Perhaps some sunny day they shall surrender their biased 'taboo of subjectivity' and embrace a Noetic Quantum Ontology.

Hence, the next step in this urgent but 'spooky' process of paradigmatic change must be a reasonable, imperfect, non-reductive centrist middle path integral (holistic, essential, fundamental) Noetic Quantum Ontology that unifies the conceptual, objective relative dimension of quantum mathematical formalism with the subjective ultimate dimension of the all inclusive primordial awareness-consciousness ground in which this all arises and manifests. It is this metaphysical 'grounding relation' to which the inherent subjectivity of the universal quantum wave function has always pointed. Such a providential quantum ontology shall facilitate our emerging inchoate scientific and cultural knowledge paradigm shift—the Noetic Revolution in science, religion, and culture. This shall take some sociocultural time.

Albert Einstein himself failed to accomplish the syncretic mathematical consummation that is a universal quantum theory of gravity, although his GRT predicted the massless particle gravitons of his continuous gravity waves. These have now been discovered. [LIGO gravitational wave detectors 2015]

As to a middle way Quantum Gravity Theory, none of the twenty or so ontic 'foundational interpretations' of QFT/QED have done any better. What in heaven and earth could a 'graviton particle of space' possibly be? We might visit premodern Buddhist *Abhidharma* and explore their objectively real 'space particles' ontology. [Boaz 2020 *Ch. V*] Our never ending grail quest for absolute objective certainty is beginning to look like a pipe dream.

What *is* certain is that at the miniscule Planck scale, the classical, smooth spacetime continuity of Aristotle, Newton, and Einstein is now quite problematic, a new scientific quantum paradigm microcosmic discreetness emerges, and the waning objectivist, Greek realist-materialist metaphysical extremist orthodox dogma finally becomes relegated to the proverbial trash bin of history. May it rest in peace, never again to bewitch scientific minds. Meanwhile the 'scientific method' that has evolved from our bygone objectivist/physicalist biases remains as useful as ever in supplying us our smart phones and quantum encryption technologies to thwart espionage by endless political enemies.

Well, what then shall replace the metaphysical *absolute* existence of our 'real' local realities dictated by a failed observer-independent Metaphysical Scientific Local Realism/Physicalism ontology? A centrist Middle Way observer-dependent *relative* existence ontic view, of course. Kuhnian paradigmatic 'scientific revolution' indeed.

The bad news is that at the empirically and even logically impossible theoretical Planck scale, time and distance are utterly immeasurable. How then may we determine which, if any, of several quantum gravity theories are tenable? Here the 'spooky' subjectivist, logic-defying quantum anomaly that is the quantum emptiness or nonlocality/entanglement of a quantum system of two or more quantum qbits (the basal two-state or two-level basic unit of quantum information), will continue to play an important role. *If gravity can be shown to possess the quantum property of entanglement, then it must be quantum in nature.* And that helps. Indeed, we already know that *everything* is quantum in nature. Mathematically 'proving' that infinite order is indeed a tall order—if it is logically possible at all. Well, if that be the truth of the matter, then so much the worse for logic and mathematics which have already fallen from their former grace as candidates for absolute objective certainty. ["Paraconsistent Logics: Gödel and Uncertainty" below]

In the late 1940s Richard Feynman enhanced Paul Dirac's 1927 sublime Dirac Equation that unified QFT with Einstein's Special Relativity to give us Relativistic Quantum Electrodynamics (QED). This refinement suggested that if gravity, the warping of spacetime, is indeed quantizable then a nonlocal entangled particle existing in its logically absurd two 'space-like separated' locations at the same moment in objective time must produce two co-existing entangled spacetime gravitational fields simultaneously.

If this subjective superposed quantum eigenstate does not instantly collapse into an objective reality eigenstate, then it must be nonlocally entangled revealing that gravity is indeed a *quantum phenomenon. Several recent experiments have settled the matter. [Scientific American Extreme Physics,* Spring, 2019*] Gravity is indeed a 'spooky' nonlocal entangled quantum process!*

However, don't we still need a mathematically consistent Quantum Gravity Theory (QGT) to show how it is that we have truly unified GRT and QFT? That formal logic and mathematics are riddled with inconsistency and paradox (e.g. Russell's Paradox) has of late deflated our 'old paradigm' valorization and idealization of each of them. 'Hilbert's Program' is now dead in the water of Gödel's Incompleteness.

We shall see again in this chapter that 'paraconsistent intuitionist' alternative math and formal logical systems have frequently pragmatically replaced the wasteful 'deductively explosive' syllogistic two-valued Greek logic of Aristotle's Three Laws of Thought with its impractical destructive Law of Excluded Middle. Kurt Gödel's Incompleteness and paraconsistent Intuitionist Logic have, as we shall soon see, opened a paradigmatic door to our 21st century Noetic Revolution recognition of the prior unity of objective Science and subjective Spirit.

Now here's a scary thought. What if our beloved, much valorized quantum theory itself is in need of modification? [Boaz 2023 Ch. IV] What if, like gravity, QFT/QED breaks down in the mathematically impossible infinite Planck scale gravitational extremes of neutron stars, and black holes, and at the many primordial Big Bang singularities of the vast timeless *kosmic* cyclic multiverse (Penrose's Conformal Cyclic Cosmology below) of which ours is but a 'big bounce' of some thermodynamically petered out previous universe? What if it is QFT that must be modified and adapted to General Relativity as a few physicists now believe?

What is relatively certain as we consciously shift our global wisdom knowledge paradigm is that in the course of seeking the holy grail of a unified QGT both QFT and GRT will evolve. The psychic pain generated by the two dogmas of incommensurable QFT and GRT have blown open our wisdom doors of noetic perception. Time—if it exists at all—will tell the next result.

Scientific Ideology Becomes Noetic Wisdom. Our understanding of gravity was greatly enhanced by Einstein's GRT gravity field equations. Kuhnian scientific revolution or no, we've seen again and again that what is painfully slow to change is Big Science's cultural zeitgeist, namely the classical bias dogma that is super-objectivist Scientific Local Realism, and Realism's epistemic handmaiden—monistic observer-independent, Greek Metaphysical Scientific Local Realism/Physicalism, to wit, all appearing reality is only 'local' purely physical matter-energy. Notable exceptions to this unwholesome course may be the antirealist, ontologically relative quantum views of Bohr, von Neumann, Wigner, Wheeler, Barbour, and Rovelli. Not to mention Buddhist 'ontologically relative' Prasangika Madhyamaka Middle Way philosophy—conceptual foundation of nondual *Ati Dzogchen.*

Of the many physicists and cosmologists now in recovery from this afflictive obsessive reductive 'scientific' physicalist/materialist' metaphysic, relativistic physicist and cosmologist Stephen Hawking's story is perhaps the most inspiring.

Upon analysis of Kurt Gödel's two 1931 incompleteness theorems (see Gödel below) Hawking became disabused of his grail quest for Cartesian absolute objective certainty through a logically impossible Theory of Everything (TOE) with its realist/materialist metaphysical bias, and at last embraced an antirealist view similar to Bohr's. This epistemic reversal of his hitherto ardent Scientific Local Realism of *A Brief History of Time* (1988) became an ever so reticent antirealist 'Model Dependent Realism' (MDR) ontology revealed in his excellent book, *The Grand Design* (2010). Such rare intellectual openness and honesty in a great mind is indeed a joy to behold. Stephen Hawking, you will be missed.

Well, what might the culture of classical 'old paradigm' modern Standard Model physics and cosmology, with its post-Standard Model physics corrections—Supersymmetry/M-Theory, Multiverse Theory, dark sector ZPE vacuum energy—look like with this methodological enrichment of the ontology, psychology and contemplative science of pre-modern Buddhist Middle Way philosophy? It shall be our path to a 'post-empirical' middle way integral Noetic Quantum Ontology. Stay tuned.

Therefore, let particle physicists, cosmologists, philosophers of physics, neuroscientists and Buddhist scholar-practitioners dialogue in academic symposia. That such symposia are nonexistent demonstrates the tenacious grip of the Scientific Realism/Materialism/Physicalism ideology that so profoundly hinders our emerging paradigmatic Noetic Revolution in Matter, Mind and Spirit. [Boaz 2023]

We've seen that there is now arising in the West a new integral knowledge paradigm. As it merges with Eastern wisdom it becomes a global Noetic Revolution in objective Science and perfectly subjective Spirit in which it arises. That new knowledge paradigm is based in the providential coming to meet of Eastern esoteric Buddhism and Western exoteric physics. The resultant, if inchoate unified integral noetic ontology, epistemology, and methodology, with its Contemplative Science, and its emerging Science of Consciousness, presents a propitious opening for a unified science of matter, mind and spirit; and the healing wisdom that abides therein.

The Unity of Science and Spirit: Quantum Field Theory

Quantum Field Theory: Variations on a Theme of Wholeness. We are now better prepared to once again engage nonlocal entangled 'quantum emptiness' and the emergence of spacetime within our formless, timeless, primordial awareness-consciousness emptiness ground, vast enfolded boundless whole of all unfolding spacetime reality.

"Form is empty; emptiness is form." The Buddha's Two Truth dimensions: *relative* spacetime form ($E=mc^2$), and its *ultimate* timeless boundless emptiness whole. The world of Big Science may be seen as the dualistic dimension of relative, objectively appearing spacetime form, and its matter-energy-motion behavior. The realm of perfectly subjective Spirit then is the nondual primordial dimension of that vast infinite ultimate reality awareness-consciousness ground that embraces spacetime form, and our theories about it, and in whom it arises and is physically and mentally instantiated.

Broadly construed, the province of objective Science with its matter-energy-light form ($E=mc^2$) abides in a relationship of identity with perfectly subjective nondual Spirit that is its formless primordial awareness all-embracing, all subsuming 'all-ground', as we have seen so many times in these pages. The relative dimension of spacetime form/matter and ultimate primordial Spirit are an ontologically prior yet phenomenally present invariant one truth unity. Told the Buddha in his Fourfold Profundity, "Form is not other than emptiness; emptiness is not other than form."

I have argued here and elsewhere that the epistemic evolution of the physics revolution that is Quantum Field Theory (QFT/QED) with its mysterious subjective indeterminist 'superposed' global universal quantum Ψ-wave function expresses this nondual perennial wisdom ontology in its exoteric, relative, conceptual mathematical dimension. The quantum Ψ-wave function is thus a dualistic, conceptual, objective mathematical expression of the emergence of local physical, causal *relative* spacetime within its acausal, subjective, nonlocal, nondual fundamental ontic Spirit ground, *ultimate* primordial awareness-consciousness itself. In short, that timeless ground is a 'post-empirical' foundation for our incipient centrist integral Noetic Quantum Ontology.

However, physicists and philosophers of physics have failed to produce such a qualitative settled quantum ontology in their several reductive philosophical interpretations of QFT—one that serves as a 'grounding relation' for their quantitative quantum math formalisms. This failure is the result of physics' refusal to venture beyond classical local realist-physicalist physics and conceptually and contemplatively engage consciousness, that is, philosophy of consciousness yes, and as well nondual prior primordial consciousness ground itself. Four hundred years of classical modern physics has proven to be an habitual seductive simulacrum of the far deeper truth of the matter.

Ontology, the inquiry into ultimate being itself is by its very nature metaphysics, literally 'beyond physics'. The inherent subjectivity of metaphysical ontology has from the beginning been taboo in objective physics generally, and in quantum physics particularly. The formidable fundamental subjectivity and 'uncertainty' (Heisenberg) of the utter non-causal randomness of quantum mechanics has forced an unbidden confrontation with the inherent subjectivity of human consciousness, and therefore with the primordial awareness-consciousness ground in whom that arises. This ontic and epistemic conundrum has become known as the trans-empirical 'quantum mystery', or the 'quantum enigma', the "lucid mysticism" of quantum pioneer and Nobel laureate Wolfgang Pauli.

The *observer-dependent* freedom of the quantum view is indeed an unforeseen revolution in a hitherto realist/physicalist *observer-independent* 'classical' physics universe of discourse. QFT follows our human wisdom tradition of a spacetime reality that is essentially dependent upon mind, the consciousness of a relative sentient observer arising in and not separate from its ultimate primordial awareness ground—Suzuki Roshi's 'Small Mind' participating in 'Big Mind' in whom it arises and participates. These are the 'Two Truth' dimensions—relative and ultimate—of the vast all embracing aboriginal invariant through all reference frames *one truth unity* (*dzog*) that perforce subsumes and grounds its emerging spacetime reality. It is the mereological (part-whole relations) 'Priority Monist' ontology of this nondual ('not two, not one, but nondual') ultimate source or ground that is fundamental. Such an ontology is conspicuously absent in the metaphysics of the wondrous, 'always correct' quantum theory and its necessarily metaphysical 'fundamental interpretations'—its quantum ontologies. Yes. We require an integral (essential, holistic) centrist integral Noetic Quantum Ontology.

We have seen that if we are to provide a cognitive bridge between the human knowledge dimensions of relative, dualistic, objective Science, and the ultimate nondual perfectly subjective Spirit ground in whom it arises we require a more or less settled metaphysical Noetic Quantum Ontology that reaches beyond the still classical realist-materialist quantum mechanics 'foundational interpretations' now on offer.

Moreover, we've seen that such an ontology may be supported by a mathematically consistent Quantum Gravity Theory (QGT) that quantizes Einstein's gravity—General Relativity Theory (GRT)—thereby unifying these two foundational pillars of physics into a mathematically commensurable GRT and QFT. That is, as Hamlet told, the monumental physics "consummation devoutly to be wished".

In the alternative, instead of quantizing gravity perhaps we should be 'gravitizing' Max Planck's 'quantum of action' itself. Is great gravity sequestered somewhere in the quantum mechanical formalism; 'Einstein's something deeply hidden' in the nonlocal entangled universal quantum Ψ-wave function? Or, are there sunny quantum fields out there, or in here, merrily propagating in the dark recesses of Einstein's gravity field equations? Some philosophers of physics and some theoretical physicists see both as 'post empirical' quantum gravity possibilities.

What is Quantum Mechanics? It is indeed "The dreams that stuff is made of". Quantum mechanics is the very foundation of post-classical modern physics. All disciplines of physics use it. It's the basis of most modern technology. Quantum mechanics ('mechanics' is the sector of applied mathematics that describes physical motion of matter-energy and its forces) represents a seismic knowledge paradigm shift in how we view our relative conventional world of spacetime reality. Quantum mechanics has forced upon us a new, post-realist/physicalist subjectivist worldview. How shall we understand this?

The mechanics of the receding Newtonian 'classical' (pre-quantum) physics paradigm views the world of appearing reality as substantially existing objectively 'real' and independent of any conscious observer. It is thus *observer-independent.* Philosophers of physics know this prevailing metaphysical view as Scientific Local Realism with its epistemic sidekick material Physicalism. That is the view of our everyday common sense worldview of a separate 'real world out there' (RWOT), whether or not it is observed by a conscious observer. The proverbial tree in the forest remains objectively 'real', whether or not it is observed by a sentient being—our much beloved Local Realism 'common sense' ontology abiding at the basis of our deep cultural background 'global web of belief'.

But it is not so. The quantum theory has shown that our appearing realities are indeed dependent upon the consciousness of a sentient observer-experimenter. Appearing emerging spacetime reality is thus *observer-dependent.* Without such observation, stuff cannot be said to truly exist! Now that's a scary quantum thought. Einstein hated it.

It is here that quantum theory engages the mysterious subjectivity of 'consciousness'. Yes, the 'always correct' nonlocal, counterintuitive "lucid mysticism" (Pauli) of quantum mechanics challenges this comfy realist view that is the very foundation of our cognitively biased deep background cultural 'global web of belief' (Quine 1969). Einstein called this nonlocal antirealist view "spooky (*spukhaft*) action at a distance". It flies in the face of common sense Local Realism and Physicalism.

We shall soon see that this 'quantum mystery' of the sentient conscious observer is displayed for all to see as the prodigious '*quantum measurement problem*'—the existence or nonexistence of Schrödinger's wondrous quantum Ψ-wave function. What precisely constitutes a quantum measurement, and what does it reveal about Einstein's 'something deeply hidden' quantum reality, and more, its metaphysical ground that must exist ontologically and mereologically prior to such local measurements?

When we *observe* a quantum system its behavior appears quite differently from the same system when it is not 'disturbed' by a sentient observation. Somehow observation/measurement 'changes' the quantum system being observed. How is it that 'infinitely superposed' quantum systems evolve deterministically in accord with the universal quantum Ψ-wave function before we observe them, then mysteriously 'collapse' into objectively 'real' spacetime objects (electrons) when we observe/measure them? The weird result is that what we see objectively is not what actually exists before we look! That is the 'quantum mystery', the 'quantum enigma': Appearing objective reality is not at all as it appears! Spooky indeed any respectable rational semiotic concept-mind. 'Counter intuitive' writ large.

No real surprise here. The history of human inquiry into the *ultimate* nature of *relative* appearing reality has told this counterintuitive truth almost from the beginning. The quantum enigma may be seen as a modern out-picturing of the ancient ontic engagement between the metaphysical extremes of Local Realism, and nonlocal metaphysical Antirealism/Nominalism, and of even more extreme nihilist metaphysical Absolute Idealism (reality is ultimately only mental or 'mind only').

Still, the radical shift from the Metaphysical Local Realism of classical Greek Apollonian, Galilean, Newtonian, Einsteinian mechanics to the nominalist Antirealism of Niels Bohr in the final 1928 Copenhagen Interpretation of quantum mechanics represents a seismic historical global consciousness and 'Scientific Paradigm Shift'. (Thomas Kuhn *The Structure of Scientific Revolutions*, 1962, 2012)

Interpreting Quantum Mechanics by the Lights of Middle Way Buddhism. There is still no agreement among the many 'foundational interpretations of quantum mechanics' as to a resolution of the 'measurement problem'. Yes, we require a centrist integral Noetic Quantum Ontology that reaches beyond the mere physical reality dimension of quantum formalist mathematics to the trans-conceptual formless timeless noetic primordial awareness-consciousness ground in whom this all arises.

If this all seems a bit fantasque; well, it is. But please don't give up. It's not as complex as it may seem. Our cognitive challenge is to expand our habitual thinking about the nature of the world as it appears to our ordinary sense perception and conception so as to embrace both its real objective, and its noetic subjective dimensions. We have already seen that there exists a metaphysical centrist 'middle path' between the false dichotomy of the metaphysical extremes of *either* common sense absolute observer-independent, causal locally real objective existence (Local Realism), *or* absolute observer-dependent, subjective quantum acausal nonlocal entangled (universally interconnected) nonexistence. These two reality views are complementary as Bohr has told. The truth of the matter requires both.

Meanwhile, let us warm up to this propitious 'middle path' by beginning to think of arising spacetime form as *relatively* objectively truly real as it emerges from its formless timeless *ultimate* perfectly subjective all subsuming primordial awareness-consciousness ground, by whatever grand name. Yet, this arising reality is not *ultimately* intrinsically real. As Buddhist Middle Way founder Nagarjuna told, "We must respect the existence dimension of Relative Truth, while knowing that it is utterly absent of any iota of absolute intrinsic existence." Two Truths indeed. We've seen that this ostensible Mahayana ontic causal dualism is corrected in the acausal nondual view and practice of *Ati Dzogchen*.

Let us then begin to directly see and feel the prior and present indivisible invariant through all human cognitive reference frames *one truth unity* (*dzog*) of these Two Truth dimensions of the nondual boundless whole of reality itself in which, or in whom we all arise and participate. That awakening of our indwelling 'awakened mind' begins by 'mindfulness of breathing' upon the wisdom *jnanaprana* wind in the belly.

After all, the primordially entangled nonlocal, nondual interconnected quantum emptiness of quantum mechanics, and the interdependent (*pratitya samutpada*) emptiness/*shunyata* of Middle Way Madhyamaka Buddhism are two parallel ways of expressing this single providential emergence of primordially enfolded 'form' arising and unfolding in relative space and time within its indivisible timeless formless nondual perfectly subjective Spirit ground. The first represents an epistemic Relative Truth dimensional description; the second an ontic Ultimate Truth dimensional description. We have now learned to view and relate to these two complementary voices of truth as a prior and present indivisible, conceptual *and* trans-conceptual contemplative noetic nondual *one truth unity*, far beyond our mere objectivist realist/physicalist cognitive biases.

Therefore, recent Relativistic Quantum Field Theory (QFT), which includes Paul Dirac's Quantum Electrodynamics or QED, has clearly demonstrated that our appearing realities—both micro-phenomena like electrons and photons, and macro-phenomena like molecules, trees and stars and all of us—are indeed wholly dependent upon observation by a sentient consciousness. Conscious human awareness is thus fundamental to the very nature of our appearing realities. This most unwelcome epistemic imperative has been forced upon us by the nature of conscious human quantum observation/measurement, which has been proven to be 'always correct'. As physicist Sean Carroll has said, "We don't choose quantum mechanics; we only choose to face up to it." [Carroll 2019]

What hath God wrought?! What in heaven and earth is this strange new 20th century quantum mechanics that has subsumed classical physics, and bequeathed to us the mixed bag of blessings that include the computer, the laser, and the bomb? It is said that about a third of the Western economy is due to electronic products that are the direct result of QFT. The practical result of the theory is astounding! Yet few quantum physicists have bothered to consider what it actually means to the *ultimate* nature of our appearing realities. Such an ontic knowledge exploration remains taboo in the mathematical world of physics. "Shut up and calculate" is still the much fraught anti-intellectual norm.

Nonetheless, a few intrepid philosophically minded quantum physicists, along with some peripatetic philosophers of physics have managed to work on more than a few 'interpretations of quantum mechanics', the sector of physics now known as 'Foundations of Quantum Mechanics'.

So, before we pursue further what quantum mechanics means as to the nature of *ultimate* reality, let's see what it teaches us about objective *relative* conventional spacetime reality. Just what is it that is 'waving'? Is stuff particles, or waves, or both, or neither?

The Universal Quantum Wave Function. Quantum mechanics describes the whole of continuously arising emerging spacetime as a global, universal, smooth, vibrating wave—Irwin Schrödinger's 1927 all-embracing, though non-relativistic, determinist *universal quantum Ψ-wave function.* Paul Dirac's prodigious 1928 relativistic Dirac Equation 'relativised' the wave function making it consistent with Einstein's 1905 Special Relativity.

Now if the new superposed 'universal quantum wave function' is to be truly universal it must explain the wave motion of large macro-objects as well micro phenomena, like quarks and leptons. Schrödinger's wholly determinist wave function is a non-relativistic linear differential equation describing *matter waves* in a quantum mechanical system. So the 'universal Ψ-wave function' governs the behavior of microcosmic electrons and atoms, but also indeterminist macrocosmic molecules, cells, living beings, and vast universes. 'Universal' indeed.

Schrödinger's wave function equation was quickly recognized as a physics breakthrough. Indeed the Ψ-wave began the new physics. Einstein called it "true genius". Planck described it as "epoch making". It won for Schrödinger the Nobel Prize in Physics in 1933.

Broadly construed, the wave function is a *relative* mathematical process of *ultimate*, nonlocal, basal universal interdependence and interconnectedness of all arising spacetime physical reality.

Schrödinger's singular universal quantum wave equation (Ψ) rules the evolution of a plurality of infinitely many constituting unrealized 'superposed' quantum wave functions that embody our physical realities, small and large.

If the mighty universal quantum wave function describes the objective *relative* behavior or mechanics of the matter-energy stuff of arising 'emerging spacetime', what then is the subjective *ultimate* nature of that prior more fundamental ontic ground from which this all arises and emerges as the stuff space and time? What is the Noetic Quantum Ontology that describes this wondrous natural all subsuming *process*?

What is the mereological (part-whole relations) relationship of the vast 'holographic' boundless whole to its matter wave parts arising and 'waving' therein? Are not the micro and macro matter waves of living beings participating parts of that vast *kosmic* whole connected to and immersed and subsumed in it, and to one another, through it? How is it that such ontic inquiry and understanding may enhance relative and even ultimate human happiness? Such are the urgent questions of quantum ontology as they naturally arise from their obscure epistemic quantum mathematical formalisms. Wisdom and human happiness lie in our pursuit of such a post-physicalist ultimate centrist integral Noetic Quantum Ontology.

It may be useful to think of questions about and descriptions of the objective behavior of arising ostensibly wavelike physical matter-energy as *relative epistemology*; and the investigation into and engagement with the fundamental subjective nature of the all subsuming ground of emerging spacetime matter and energy as *ultimate ontology*, nondual inquiry into the timeless, formless, selfless ultimate nature of spacetime form—the very Nature of Mind and all of its human experience—that is nondual primordial Being Itself.

Epistemology guides and interprets the *relative* causal scientific, objective, empirical, mathematical observation and investigation of appearing phenomena while ontology is the subjective, metaphysical conceptual and even trans-conceptual direct contemplative exploration of the all subsuming *ultimate* nature and primordial ground of phenomena and of our embodied mind that experiences it. A viable integral Noetic Quantum Ontology must include both relative objective and ultimate subjective variables.

Our human cognition naturally and inherently includes both of these objective and subjective dimensions of our human awareness/consciousness processional—our *cognitive noetic doublet,* and the four inherent dimensions of human cognition—1) ordinary direct perception, prior to concept and naming; 2) objective, conceptual, scientific; 3) subjective, contemplative; 4) perfectly subjective all pervading nondual Spirit ground. Objective Science may describe three of these cognitive dimensions—1, 2, and 3—subjective nondual Spirit grounds and embraces them all.

Be all that as it may, the global universal quantum wave function presumes to describe the epistemic, objective, mathematical, relative-conventional face of spacetime reality as it arises from its ontic perfectly subjective ultimate primordial ground, by whatever grand name. Unfortunately this basal ultimate ground of reality is taboo to most quantum physics practitioners, as we have seen. Is it enough just to build quantum computers, and speculate about quantum gravity? Is not an awareness of our interconnectedness within that vast *ultimate* awareness-consciousness ground of the *relative* physical quantum process also important to human happiness and well-being? That is revealed through an essential foundational integral Noetic Quantum Ontology.

As to our merely physical cosmos as it arises in all embracing whole of *kosmos,* quantum theory describes and engages all of appearing spacetime reality, from quarks and leptons to vast quantum cosmological multiple universes. We may now see that handy quantum electronic products are not the best part of this wondrous quantum narrative. QFT steadfastly, mereologically refers human consciousness to the *kosmic* original ground of all spacetime phenomena, and of all of us. That is the urgent 'grounding relation'.

Irwin Schrödinger was a serious student of Sanskrit and Hindu Vedanta philosophy. He was as well familiar with Buddhist philosophy. His penetrating philosophical commentaries on his wave function equation Ψ (Psi) has demonstrated a conceptual and intuitive understanding of the Buddhist Two Truths—primordial unity of the dimension of relative physical and mental phenomena (Relative Truth), and the formless, timeless, selfless ultimate dimension (Ultimate Truth) in which or in whom it arises, is subsumed, and instantiated as emerging spacetime physical and mental form.

Yes, Gautama Shakyamuni, the Buddha of this present age, told it well, "Form is empty; emptiness is form." QFT seems to see this aboriginal reality relationship of spacetime stuff to its formless ground as: "Form is quantum emptiness; quantum emptiness is form." [I have come to call the physically transcendent distance and time of the basal ZPE Universal Quantum Vacuum and its post-empirical Planck Scale regime 'quantum emptiness'.] 'Quantum emptiness' parallels but does not define Mahayana Buddhist 'emptiness/*shunyata* of intrinsic existence', as we have seen.

Moreover, QFT views the physical 'matter waves' of form as observer-dependent and ontologically relative, that is to say, spacetime form—e.g. an electron—does not exist as a physical reality prior to a sentient, conscious observer's observation, or to an experimenter's measurement which then 'collapses' the superposed subjectivity of the Ψ-wave into really real objective quantum eigenstates of micro-stuff that comprise all of us.

Thus are both quantum emptiness and Buddhist emptiness, as H.H. Dalai Lama has told, "established and designated by conventional human minds". *Universal ontological relativity* writ large. Our appearing realities are subject to, or 'created' by, or 'established' by our human conventional perceptions, concepts, and beliefs. Thus, the basal 'emptiness ground' is full of the potential of spacetime form arising therein. And such an absolute emptiness ground is itself empty of "any whit of absolute existence!" This is known as the 'emptiness of emptiness' (*shunyata shunyata*). Thus the emptiness ground of form is not itself some big objective 'thing' out there, a metaphysical absolute container of space and time, nor is it a theistic godlike Creator of it all. Heady ontic noetic wine to be sure.

Here the 'problem of consciousness' arises as the vexing 'quantum measurement problem'. Does an observer's consciousness, or anything else 'collapse' the subjective linear 'superposed' wave function of an infinity of potential quantum 'eigenvalues' into a single objectively real 'eigenstate of position' of an electron really being here in classical real time (*t*)? And if it does, how? Recall that Schrödinger had a problem with the 'collapse hypothesis' in any of its meanings. Oddly, he finally choose a local realist definition of *Ψ*.

Observation/perception by a sentient human consciousness thus somehow 'creates' our phenomenal realities. Ontological relativity again. How this quantum mystery is so constitutes the prickly 'quantum measurement problem', as we have seen. Such a view is 'antirealist', even proto-idealist in that it violates our unscientific ontic presumption that is the 'scientific' classical dogma of Metaphysical Scientific Local Realism with its epistemic consort Metaphysical Materialism/Physicalism—the metaphysical extreme of an absolutely existing observer-independent, ultimately physical *objective* 'real world out there' (RWOT). That's odd considering the inherent random *subjectivity* of the *Ψ*-wave.

For two decades local realist Albert Einstein vigorously argued the quantum measurement problem with his antirealist antagonist Niels Bohr, discoverer of the prodigious quantum Principle of Complementarity and creator of the Copenhagen Interpretation of quantum mechanics. [The 1936 Einstein-Bohr EPR debates.]

Mahayana Buddhists agree that our appearing realities are 'ontologically relative'—established/created by perceiving conceptual minds—but they don't see a problem here because they do not indulge a local realist ontological cognitive bias in the first place. They see instead a 'middle path' between absolute existence and absolute nonexistence.

Irwin Schrödinger's quantum *Ψ*-wave function, the basis of Quantum Field Theory (QFT/QED), has codified the Bohr-Heisenberg quantum foundational principles—respectively, the Principle of Complementarity and the Principal of Uncertainty—introducing to the physical sciences, and our modern collective cultural consciousness a new post-classical, non-objectivist scientific knowledge paradigm!

QFT/QED thus constitutes a propitious conceptual bridge between conceptual objective Science and its nondual perfectly subjective Spirit ground, as we are beginning to discover. In 1962 Thomas Kuhn designated this not so subtle knowledge 'paradigm shift' a "Scientific Revolution". And indeed it still is.

"Saving the appearances" of the prevailing classical ontology that is Scientific Local Realism from the Antirealism of Bohr and his brilliant student Heisenberg has become a quantum cottage industry fabricating seemingly endless 'foundational interpretations of quantum mechanics'—for those who care to cease calculating long enough to consider the deeper meaning of their formalist mathematics.

Most all the players now agree that the 'classical' non-quantum universe/multiverse is now and has always been quantum in nature. QFT/QED has not replaced classical physics but has transcended and embraced it. That quantum nature of our classical atomic world of spacetime—both microcosmically and macrocosmically—is 'created' or perceptually and conceptually imputed, designated, and reified by sentient conscious observation, as we have seen. Spacetime stuff does not exist observer-independently, 'from its own side'. Spacetime causality somehow arises from this mysterious random/acausal quantum foundation; not from the formalist quantum mathematics, but from its pre-quantum acausal timeless primordial awareness-consciousness ground. Objectively counter-intuitive to say the least.

The point-like particles of classical mechanics reveal "something deeply hidden" (Einstein), something intrinsically fundamental. Wave-like particle-fields are now the 'basic building blocks' of spacetime stuff—our Standard Model 'presentist block universe'. And the universal quantum wave function points directly at their nature, and presumes to objectively describe just what it is that is 'waving' in the quantum wave function.

Add Einstein's classical (non-quantum) universal gravitational field and we have a post-Standard Model 'Core Theory' of the physics of particle-fields and their interactions.

Recall that we desire a Quantum Gravity Theory (QGT) that unifies QFT and GRT. The problem is that the present physics *lambda* Λ-CDM Standard Model Theory doesn't work at intergalactic distances, nor in extreme gravitational fields, like black holes, neutron stars, or at the instant of a 'Big Bang' creation of a new universe participating in Roger Penrose' now respectable CCC conformal cyclic cosmology multiverse.

Moreover, in such extreme gravity classical 'locality'—particles and their fields interacting causally only when in proximity to one another in space—becomes 'nonlocal', known as 'quantum entanglement', the primordial ultimate indivisible interconnectedness and interdependence of all spacetime located form—just as the Buddha told. Thus does quantizing gravity require Einstein's "spooky nonlocal action at a distance." Einstein hated it. To accomplish a quantified gravity by unifying GRT and QFT, these two great pillars of physics, shall be shaken to the core. As Zen Master Dōgen Zenji told, "All that can be shaken shall be shaken." And this of course includes our classical physicalist reductionist biases of rapidly receding metaphysical Scientific Local Realism, Materialism, Physicalism knowledge paradigm—a useful relative-conventional but not ultimate paradigm.

In any case, we do indeed already have a Quantum Gravity Theory (QGT)! And that is expressed in Einstein's gravity equations of his wondrous GRT. It works fine in relatively weak gravitational fields—earth to solar system distances, but not at extreme intergalactic distances, nor in extreme gravity black hole or Big Bang singularities where our notions of classical spacetime utterly collapse into Planck scale 'quantum emptiness'.

The 'old quantum mechanics' (1900 to 1925) of Planck, Einstein (before he turned coat), and Bohr became Quantum Field Theory (QFT) in 1927 with the wondrous work of Irwin Schrödinger, Werner Heisenberg, Wolfgang Pauli, and Max Born (doctoral supervisor of J. Robert Oppenheimer). QFT was then enhanced by the sublime Quantum Electrodynamics (QED) of Paul Dirac in 1928, and further developed by Richard Feynman between 1945 and 1953 to cope with Dirac's prickly 'problem of infinities'.

The Problem of Quantum Ontology. The quantum mathematical formalism of QFT/QED has proven to be an astounding predictive mechanism for developing new and practical technologies for humankind. But, as Sean Carroll has pointed out, "As a fundamental theory of the world it falls woefully short." Physicists have been woefully remiss in addressing its deeper metaphysical meaning and ontological foundations. We've seen that a sociological 'taboo of subjectivity' has cast an ideological pall over the theoretical physics community that almost entirely precludes such ontic inquiry. Any attempt to ontologically illumine the post-classical, nonlocal and random nature of quantum entanglement—the quantum 'measurement problem'—is still taboo. "Shut up and calculate" (David Mermin) became the prevailing cognitive posture for the last half of the 20th century. Let practical electronic quantum inventions suffice. Leave philosophy alone. 'There be dragons'.

That 'scientific' cognitive bias began to shift in the 1980s with the emergence of a new academic discipline known as 'philosophy of physics'. The emerging dialog between philosophers, physicists, and philosophers of physics soon produced over 20 'quantum mechanics interpretations', or ontic 'foundations'. We've briefly surveyed seven of them above. We shall revisit the wacky but now mainstream MWI ontology below.

All of these interpretations of quantum mechanics engage objective *relative* human consciousness, and all fail to engage its 'deeply hidden' *ultimate* ontological primordial awareness-consciousness ground. All are perforce conceptual attempts at grasping an ultimate ontology that embraces the relative math quantum formalism of the universal quantum Ψ-wave function. None have done so. Yes, we need a Noetic Quantum Ontology.

I have elsewhere referred to this 'post-empirical' quantum predicament as "the problem of quantum ontology"—the hitherto largely ignored challenge of addressing the ontic elephant in the room—namely the ultimate essence, nature, and ground of Schrödinger's formalist universal quantum wave function, very conceptual epistemic heart of QFT. Any 'foundational quantum mechanics interpretation' must include a foundational quantum ontology if we wish to explicate "that which is deeply hidden" within the formalist mathematics. Or perhaps we opt for a nominalist default ontology and let the mathematics speak for themselves. Still, ultimately, we desire to know what the formalist math implies and reveals about our world that it presumes to describe; do we not? We need to know how it is that QFT may impact our lives beyond the practical electronic inventions, and the nuclear bombs.

Real clarity in any profound physical theory requires a cognitive amalgam, a middle way that engages both voices of our human noetic (nondual, body, mind, spirit subject-object unity) cognitive doublet—objective conceptual-mathematical cognition, and subjective, intuitive, contemplative, even nondual 'spiritual' cognition.

Therefore, we shall herein attempt to discover how this revolutionary quantum worldview may enable us to bridge the knowledge gap, Heisenberg's *schnitt,* between relative objective experience, and the subjective fundamental ultimate ground of reality itself—bright primordial Presence of That—between the 'dance of geometry' of objective Science and mythopoetic perfectly subjective nondual Spirit, by whatever concept, in which or in whom this all arises and plays.

Unity of Science and Spirit: Objective Physics Meets Subjective Consciousness

Interpreting Reality Itself. Each of the several competing ontological 'foundational interpretations' of quantum mechanics either engages or denies 'consciousness': 1) phenomenal human consciousness-experience, and therefore 2) the ontic all-pervading basal enfolded consciousness ground or 'supreme source' of all our unfolding arising experience. Told quantum physicist Eugene Wigner, "It is not possible to formulate the laws of quantum mechanics without reference to [human] consciousness."

We've seen that each of these mainstream 'interpretations' either engages or denies *relative* human consciousness conceptually, and all fail to engage its relationship to formless *ultimate* primordial awareness-consciousness ground in which, or in whom it arises, participates, and is instantiated in spacetime embodied physical and mental form. For that we require a centrist middle way integral Noetic Quantum Ontology.

We have seen that the universal quantum wave function (*Ψ*, Psi) is a dualistic conceptual mathematical formulation of the inherently uncertain (Heisenberg), conceptually but not contemplatively unknowable primordial awareness-consciousness 'groundless ground' of the nondual boundless all embracing whole of our appearing physical and mental realities.

This "vast implicate order of the enfolded unbroken whole" (David Bohm) is eminently approachable, and may be relatively practiced and thereby ultimately realized via direct human contemplative cognition under the guidance of a qualified meditation master. It is here that human conceptual knowledge (*doxa, kalpana,* our relative conventional 'global web of belief') is embraced by ultimate primordial awareness wisdom (*jnana, yeshe,* gnosis), the direct human knowing feeling awareness-consciousness (*yogi pratyaksa*) of that vast infinite whole.

'Qualified master' cases in point: the Buddha, the Christ, and the countless buddhas, *mahasiddhas,* saints, and sages of our great noetic Primordial Wisdom Tradition. After all, the indwelling love-wisdom mind Presence (*vidya, rigpa, christos*) of that fundamental foundational nondual aboriginal ground (*dharmakaya, mahashunyata, gzhi, Tao, Yahweh, Ein Sof, Abba nondual God the Primordial Father*) in which this all arises is *ipso facto* naturally already present at the 'spiritual' Heart (*hridyam*) of even the most cynical physicist; not to mention all of us peerless, bias-free intellects. Yet it remains ungraspable by the inherent limits of mathematics and the semiotic logical syntax of language. ["Paraconsistent Logics and Incompletness" below] As Hamlet told Horatio, "There are more things in heaven and earth that are dreamt of in your philosophy."

Sadly, under sway of our objectivist, realist classical cognitive biases such contemplative technology is still taboo to most theoretical physicists and even to philosophers of physics, as we have seen. This 'scientific' classical objectivist Scientific Local Realism/ Physicalism belief bias is here quite cognitively paralyzing, even as the inherently subjective 'superposed' Ψ-wave function is a lone quantum voice crying out in a formalist wilderness: "nonlocal nondual"!

Therefore, let these fine minds make a quantum leap into Contemplative Science, the branch of the emerging Science of Consciousness that objectively investigates such spooky subjective spiritual phenomena. For an 'Open Science' no experienced phenomena can remain taboo. The prodigious quantum wave function, whatever it is, has opened that cognitive door. Let us enter in the 'many mansions' of the Nature of Mind and see what abides here. As the Buddha told, "Do not believe what I teach… Come and see (*ehi passika*)."

If quantum mechanics is "lucid mysticism" (Pauli), let us not fear the trans-conceptual, 'post-empirical' foundational noetic ground—that contemplatively knowable nondual Presence which abides beyond the conceptual grasp of even the greatest scientific minds of our humankind.

How shall we accomplish that? Mindfulness of Breathing upon the *jnanaprana* wisdom wind arising and falling with each conscious breath.

What Else is Quantum Mechanics? It's a shiny new conceptual model upgrade of our 400 year old European Enlightenment (The Age of Reason) 'classical mechanics'. It is based in the radical conception of Max Planck and his pal Einstein's assertion about the discontinuous discrete photon particle quantum nature of the wavelike electromagnetic force, in a word, light. Light energy is particle like. Or is it?

Classical physics has mostly understood light as a wave—Aristotle's smooth continuous wave-like flow of energy. Planck and

Einstein forever ended that view. The birth of the quantum revolution that transcended yet included Newton's classical particle ("corpuscular") view was the discovery that light arises from, perhaps, the 'zero point energy field' (ZPE) Unified Quantum Vacuum ground as particle-like discrete energy packets—'quanta of light energy'—now known as massless bosonic 'photon particles'.

By 1930 the inherent nature of this particle nature of quantized light is subsumed in the wave nature of Schrödinger's superposed 'universal quantum Ψ-wave function'. We have seen above, and shall in this chapter further conceptually unpack this quantum conundrum to discover that light is *both* wave and particle! Not as strange as it may seem at first blush because wave and particle are, as Bohr told, complementary modes of being.

The theory of the quantum nature of the whole of spacetime reality was truly a scientific revolution. We proceeded from the 'classical' local realist knowledge paradigm at the end of the 19th century, to the all-subsuming quantum knowledge paradigm in about 30 years! Perhaps we really are as smart as we think we are.

In 1865 Scotsman James Clerk Maxwell had finalized the classical wave theory of light. Light was electromagnetic radiation, waves spread out in space. Utilizing Lorentz' Force Law, Maxwell unified electricity and magnetism in his prodigious Maxwell's Equations. He showed that both electric and magnetic fields move through space as *light waves* traveling at the finite velocity of light. Well, what pray tell is that?

Relativity

Maxwell's relativistic equations proved that the velocity of all electromagnetic signals, indeed all massless particles, including photons of light, and even gravity particles/gravitons in the 'vacuum of space' travel at a finite velocity, and cannot exceed it. This represented a new universal physical constant *c*, with a finite value of 299 792 458 meters per second, or about 300,000 km/s, or 186,000 mph.

Maxwell's Equations, and the mathematical framework of Henri Poincaré, Hendrik Lorentz (Lorentz invariance)—along with the now apparent absence of a 'luminiferous ether' in which light was thought to propagate (the null result of the Michelson-Morley Experiment)—prepared the foundation for Einstein's 1905 Special Relativity Theory (SRT) with its two history altering postulates. SRT then provided the relativistic foundation for Einstein's providential General Relativity Theory (GRT) of 1915.

Einstein's second relativity postulate: the velocity of light in a vacuum is invariant (absolute) for all observers, independent of the motion of the light source or of the 'reference frame' of an observer. Light is not absolute but relative to the perceptual 'frame' of an observer. Electromagnetic particles and or waves travel at the physical constant *c* regardless of the light source or of the 'internal reference frame' of an observer. In short, the velocity of light in a vacuum is identical for any observer, regardless of the motion of the light source or of that observer.

Further, such particle/waves/photons can approach *c*, but cannot exceed it. Physicists call this 'locality'. 'Nonlocality' or 'quantum entanglement' is, among other things, the possible exceedance of this finite locality limit.

The first relativity postulate of Einstein's SRT: The laws of physics (including the constant *c* of his second postulate) are invariant in all inertial (non-accelerating) reference frames. These two postulates represent Einstein's two assumptions of Special Relativity. Quoth the Master: "The insight fundamental for the special theory of relativity and light speed invariance… is contained in the Lorentz transformations." [universal Lorentz covariance] Thus did Maxwell, Poincaré, and Lorentz lay the ground for Einstein's SRT; unlike his prodigious General Relativity Theory (GRT) whence sprang purely and spontaneously from the brilliant master's mind. Or so it is told by those who know.

That said, Einstein was not at all adept at the subtleties of non-Euclidian Riemannian geometry and the tensor calculus required to quantify his GRT field equations. He missed too many classes at the Federal Polytechnic School. Indeed his famous math professor Hermann Minkowski, who later helped Einstein to understand the implications of GRT for a '4-D spacetime continuum', once called him "a lazy dog".

Fortunately, Einstein's erstwhile classmate Marcel Grossmann, a professor of mathematics specializing in non-Euclidian geometry and tensor theory mentored Einstein in the frightening absolute differential calculus. Grossmann is said to have expressed to Einstein his concern that a mere physicist may not have the necessary level of intelligence to grasp such advanced mathematical theory.

Without his old friend Grossmann the sublime GRT field equations may not have happened. After all, Einstein's pal and main GRT competitor David Hilbert had an equally consistent theory at the same time. Hilbert graciously declined to publish it until after Einstein published his GRT in December of 1915. Without his friend and benefactor Marcel Grossman—who got a flat broke Einstein his job at the patent office—General Relativity Theory may well have had the great David Hilbert's name on it.

Einstein's 1905 SRT grounded his 1915 revolutionary GRT which generalizes his first SRT postulate to include non-inertial accelerating reference frames. Here he boldly describes Newton's universal law of gravity, 'Big G', not as a 'force' but as the geometric curvature of a four dimensional spacetime.

Maxwell's Equations of 1865 thus began our universal *kosmic* quest for the nature of light. We arrived in 1930 to find a fully fledged Quantum Field Theory (QFT), complete with Paul Dirac's astounding relativistic Quantum Electrodynamics (QED) which unified SRT with relativistic QFT.

The next step in our knowledge quest for the nature of light in time is the union of QFT with GRT in a most desired Quantum Gravity Theory (QGT). We've seen that presently these two great pillars of physics are mathematically incommensurable.

We must then further unify great gravity with the other three fundamental forces of nature, the non-gravitational physical interactions or forces, namely: the Strong Nuclear Force, the Weak

Nuclear Force, and Maxwell's Electromagnetic Force. The result of such a unification is said to be an empirically if not logically impossible Theory of Everything (TOE). A Grand Unified Theory (GUT) is the unification of the Strong Force and the Weak Force. Such a GUT epoch may have obtained a moment after the proverbial Hot Big Bang singularity beginning of this present universe, just prior to the separation of the Electroweak Force from the Strong Force. [Boaz 2022 Ch. 7]

At extremely high energies the Weak Force and the Electromagnetic Force are unified as a single force/interaction—the Electroweak Interaction/Force. Glashow, Salam, and Weinberg received the Nobel Prize in Physics in 1979 for their mathematical unification of these two fundamental interactions.

The four fundamental forces/interactions of nature are then: 1) *Great Gravity* ('Big G') that binds together the worlds. All bodies with mass/energy are 'attracted' to or gravitate toward one another. Einstein demonstrated Sir Isaac Newton's gravity force 'G' is not an 'attractive force' (the inverse square law), but is the physical result of masses of differing sizes following geodesic lines in curved 4-D spacetime in accordance with the volume of their masses. 2) *The Strong Nuclear Force* which binds quarks (mediated by massless gluon particles) into hadron particles, like protons and neutrons to create the atomic nuclei (the nuclear force) of atoms of which matter and energy is comprised. 3) *The Weak Nuclear Force* is the subatomic interaction that regulates nuclear fission and radioactive beta decay of atoms. 4) *The Electromagnetic Force* is carried by electromagnetic fields which produce electromagnetic radiation, like light. It attracts atomic nuclei to orbital electrons binding atoms together. This causes chemical bonds between microcosmic atoms producing macrocosmic molecules and intermolecular forces. The Electromagnetic Force is the foundational theory for electronics and digital technology.

Relativistic Quantum Field Theory

In 1927 English prodigy Paul Dirac quantized Maxwell's electromagnetic field. His QFT/QED unifies Maxwell's classical field theory of 1865, Einstein's Special Relativity Theory (SRT) and the non-relativistic 'old quantum mechanics' of Planck and Einstein circa 1900 to 1925.

As to Dirac's 1928 relativistic wave equation—the justly famous 'Dirac Equation'—he used the term Relativistic Quantum Electrodynamics (QED) to describe this work. It unified the 'old' non-relativistic quantum theory, and Schrödinger's non-relativistic quantum wave function equation with Einstein's Special Relativity Theory, and so became the foundation of present 'relativistic quantum mechanics'. QED was henceforth an integral part of QFT. Dirac's QED describes not just atomic spectra (electrons), but light quanta/photons as well. The Dirac Equation also predicted the existence of antimatter, and so of the antiparticle of the electron, to wit, the positron which has given us medical PET imaging. If that's not enough, the astounding Dirac Equation unified the Schrödinger Equation with Heisenberg's competing matrix mechanics. Paul Dirac was 24 years old.

In 1929 a 27 year old Werner Heisenberg (matrix mechanics) and Wolfgang Pauli then established the foundational structure of QFT with the first general theory of quantum fields and the method of their 'canonical quantization'. Quantization is the process of limiting energy to discrete particle-like 'wave packet' values rather than continuous wave-like values. The individual discrete 'light quanta' energy packets we call photons are thus seen as particle-like 'quantized' light waves.

Schrödinger's wave mechanics with its revolutionary 1926 Schrödinger Equation (Ψ) was similar to Heisenberg's matrix mechanics. The two geniuses at first disparaged one another's theories. Later both Schrödinger and Dirac demonstrated that they are mathematically equivalent. Schrödinger's version is more visual and mathematically friendly and is now used exclusively. Dirac's unification of Heisenberg's matrix mechanics with the Schrödinger Equation into a single equation—his profound QED Dirac Equation—thus became an integral part of present Quantum Field Theory (QFT).

Quantized light is particle-like photons. But matter moves as a series of physical 'matter waves', as we have seen. An electron's motion is wave-like; yet at the instant of observation/measurement it is particle-like. Schrödinger's foundational global 'universal quantum Ψ-wave function' equation rules the motion of microscopic electrons within atoms, and atoms and molecules, cats, people, buddhas, and the entire macrocosmic physical cyclic universe/multiverse. Universal indeed.

It was not until 1953 that Richard Feynman tamed the unruly QFT/QED 'problem of infinities' with his 'path integral formulation' and a quite uncomfortable 'renormalization' strategy. Dirac called it "bogus". QFT now bloomed as a near complete and hugely successful probabilistic theory with astounding predictive capacity. And yes, it has given us the mixed bag of the laser, the integrated circuit (microchip), and the nuclear bomb.

Will QFT/QED as it evolves toward completeness, in the fullness of time, quantize even the eternal mystery of great gravity itself resulting in physics' grail quest for a consistent Quantum Gravity Theory (QGT) unifying at long last QFT and General Relativity Theory (GRT)? And then the green grass grows all around, all around; and all's right with the quantum world! Or is it?

Is QFT 1) but another of our perennial objective cosmic desire strategies to penetrate and enter in the *kosmic* depths of the perfectly subjective ultimate primordial ground of reality itself and 'know the mind of God'? 2) Is QFT a relative proto-objective quantum exercise in search of a perfectly subjective ultimate Noetic Quantum Ontology? 3) Can any objective merely conceptual cognitive strategy conceptually grasp such nondual perfect subjectivity? Will such impudent questions never end?

The probable answer to the first two presumptive questions is yes. To the third, no. QFT has woefully neglected a Noetic Quantum Ontology; and perfectly certain relative conceptual knowledge does not exist.

How is this so? We've seen that the *ultimate* nondual primordial Nature of Mind and its arising experience is *ipso facto* beyond the *relative* dualistic grasp of even the highest virtuosity of conceptual thinking. That mere concepts can grasp the inherently non-conceptual Ultimate Truth dimension is known to those in the philosophy trade as a logical 'category mistake'. QFT itself has revealed this perennial wisdom truth. For QFT, truth is relative and stochastic. Its all embracing Ultimate Truth ground is ignored, or denied, or explained away. Let philosophers of physics reveal a settled centrist integral Noetic Quantum Ontology that theoretical and experimental physics may move forward.

In other words, the great noetic Primordial Wisdom Tradition of humankind has told again and again that our vast, nondual, aboriginal, perfectly subjective awareness-consciousness boundless whole that is the ultimate 'groundless ground' of all being cannot be grasped by inherently dualistic relative conceptual human mind. Not even QFT. Try as we may. This is the 'wisdom of uncertainty'. If one desires certainty, it will perforce transcend the relative conventional limit of human objective conceptual cognition.

However, our perennial desire to unify our hearts and minds with the indwelling love-wisdom mind Presence of this originary 'supreme source' of all reality has already lead us to more suitable ontic trans-conceptual contemplative modes of knowing that ultimate ground. Mahayana Buddhism has accomplished such a centrist 'Middle Way' ontology. Perhaps we might consider integrating such premodern contemplative wisdom to establish a post-empirical Noetic Quantum Ontology. Indeed, that is my intention in this chapter.

Be all that as it may, quantum mechanics is Quantum Field Theory which is a global inherently subjective superposed 'universal quantum Ψ-wave function' of objective 'quantum measurement'. But we still require a Noetic Quantum Ontology that grounds the mathematical formalism in its prior metaphysical foundation, and reveals what the theory means for our understanding of appearing phenomenal reality. Thus we proceed.

Wave-Particle Duality, Complementarity, Dechoherence, Uncertainty

What does it actually mean to measure a quantum system of dual complementary wave-particle fields? What we find is that for 400 years of scientific Modernity the stuff of our emerging spacetime dimension is not at all as it objectively appears. As to quantum theory, what we see in the quantum measurement of an electron's position *or* momentum but not both at once, the foundational particle of physical matter, is not at all the absolute objectively real 'point-like' particle of classical physics. Quantum measurement reveals that the precise location/position and the precise velocity/momentum (momentum is mass x velocity) of a given measured electron cannot be accurately predicted; an insult to the desired empirical predictive certainty of the noble endeavor of Newton's and Einstein's classical mechanics. What random quantum mechanics measurement accomplishes is to predict an extremely accurate *probability* of finding our diaphanous electron's position/location, or its velocity/momentum, but not both simultaneously, in the same measurement. To be sure a huge advantage over classical mechanics. And yes, wondrous quantum probability has bestowed upon us our computers and smart phones.

Stranger still, the more accurate the measurement of an electron's position, the less accurate the measurement of its momentum! If we pin down its absolute position we can know nothing about its momentum. And vice versa. This curious situation is not simply that we don't know the electron's exact position or momentum. On the accord of Werner Heisenberg's foundational *Principle of Uncertainty*, these two simply *do not exist* simultaneously! Their quantum state is intrinsically uncertain. But it's not that 'everything is uncertain'. There can be absolute certainty as to electron position, and absolute certainty as to electron momentum, but not in the same measurement. The more we know of position the more uncertain the momentum, and vice versa. 'Quantum weirdness' indeed.

Therefore, a particle's Ψ-wave function may present as particle position, or as particle momentum, but not both at once. Just so, particles (electrons) present as particle-like, or as wave-like, in dependence upon whether the experiment is designed to detect a particle or a wave; but not both simultaneously. Niels Bohr's foundational *Principle of Complementarity* clarifies this relation as one of complementary. Quantum entities have pairs of complementary properties which cannot be observed or measured simultaneously. We can set up an experiment to measure the position of an electron, and that excludes the possibility of measuring its momentum. A measurement of position disturbs/changes the value of the particle's momentum. Therein lies the 'quantum enigma' of our new knowledge paradigm Quantum Revolution.

Quantum Decoherence. Furthermore, in Niels Bohr's view, subatomic objects cannot be separated from the measuring instruments. The indivisibility of the 'quantum of action' precludes a definitive separation between the behavior of a quantum system and its 'coherent interaction' with the measuring instruments, and even with the 'consciousness' of the observer of the instrument's pointer. Quantum coherence is coupled to the measurement environment and instruments, and its information is ultimately irreversibly lost, a process termed quantum decoherence, a loss of quantum information in a measured system to the measurement equipment and environment.

Decoherence does not offer a mechanism for 'wave function collapse', but provides a framework for it as information "leaks" into the measuring device and the environment, including the consciousness of an observer-experiementer. Wave function contents are thus decoupled from the coherent quantum system.

Decoherence is required to explain how it is that an indeterminist acausal quantum system can become compatible with determinist causal classical rules of probability upon an interaction with an environmental measuring instrument in classical time. Here, the 'superposition' of many quantum states of the universal Ψ-wave function may still exist, but its ultimate fate is subject to a subjective 'quantum interpretation' of the process by an observing consciousness interpreting a pointer.

Bohr's Copenhagen Interpretation of quantum mechanics assumed wave function collapse to be an *a priori* fundamental quantum process. Decoherence as an explanation for collapse was first proposed by David Bohm in his 1952 Bohm-deBroglie Pilot Wave Theory. His Bohmian mechanics begat the 'nonlocal hidden variables' interpretation of quantum mechanics in order to evade spooky quantum nonlocal entanglement. We shall see that John Stewart Bell in 1964 proved this to be the impossible dream.

Sadly, decoherence was not revisited for almost 50 years when Carlo Rovelli implied it in his RQM Relational Quantum Mechanics, a profound but incomplete contribution to quantum ontology, as we have seen.

The management of decoherence presents a challenge in quantum computing. Moreover, decoherence offers a tentative explanation for the prodigious 'quantum measurement problem'. Elements that decohere due to environmental interactions are said to be quantum entangled with the environment, that is, with a measuring device and an observing measuring 'consciousness'.

Bohr's complementarity propitiously engages this epistemic quantum conundrum. The 'quantum uncertainty' of which Heisenberg spoke—his revealing quantum Principle of Uncertainty—was for Bohr a manifestation of the more fundamental reality of complementarity, his lapidary quantum Principle of Complementarity.

Physicist, cosmologist and popular author Sean Carroll (Carroll 2019):

> The lesson we learned [from the uncertainty principle] was that 'position' and 'momentum' aren't properties that

> an electron has; they are just things we can measure about it. In particular, no particle can have a definite value of both simultaneously. The same is true for 'vertical spin' and 'horizontal spin'. These are not separate properties an electron can have; they are just different quantities we can measure ... The uncertainty principle expresses the fact that there are different incompatible measurements we can make on any particular quantum state.

Both the Uncertainty Principle and the Complementarity Principle apply to the nature of a quantum 'eigenstate', not to the physical act of measurement itself.

Quantum Reality: The Schrödinger Wave Function and Wave-Particle Duality

A quantum Ψ-wave function is a mathematical operator that yields a probability distribution, by way of the Born Rule, for each possible 'superposed' measurement outcome of a quantum system of particle-fields. Therefore, ultimately there is only the prodigious singular global universal quantum Ψ-wave function of the Schrödinger Equation. It describes the motion of matter as wave-like, as 'matter waves'. Because Dirac's QED rules Quantum Field Theory (QFT), the unity of quantum mechanics and Special Relativity is established. QED works with both relativistic and non-relativistic equations.

The wondrous wholly deterministic Schrödinger Equation provides a means to calculate the wave function of a quantum system and how it evolves dynamically in time. The spacetime particles and fields of appearing physical reality are an oscillating wave function evolving in a classical real time (t). And the propitious 'double slit experiment' demonstrates objectively that the wave function is not just a theoretical conjecture. There's something 'real' in it that accurately describes the motion of objective micro-physical and therefore macro-physical spacetime stuff as our quantum realities emerge from the ineffable formless, timeless, selfless nondual primordial ground of being itself.

Quantum Ontology Again. How shall we interpret this mysterious superposed universal Ψ-wave function? Does it represent an objective observer-independent 'real world out there' (RWOT), whether or not an observer is observing or measuring it; or the subjective observer-dependent conscious experience of the 'consciousness' of an observer/experimenter; or neither; or something else entirely?

Niels Bohr and the Copenhagen Interpretation view Ψ as a mathematical *probability* of the wave function 'collapse' into an objective definitive 'real' quantum 'eigenstate of position'. There is here no presumption of an observer-independent existing 'real world out there' (RWOT) at all. This nominalist, 'antirealist', 'instrumentalist' even 'operationalist' view opposes Einstein's classical (non-quantum) 'local realist' view. For Einstein and the local realists nature cannot be utterly random. "God does not play dice with the world".

Einstein begged this question of Metaphysical Local Realism with Bohr for twenty years. Einstein argued that quantum mechanics demonstrates the "spooky" 'quantum entanglement' of a nonlocal quantum system and therefore fails to provide a believable (local realist) complete physical description of appearing reality. Therefore quantum mechanics is not dead wrong, just incomplete. There's something missing, namely, a microcosmic "hidden variable" or "hidden parameter" that explains the quantum wave function in a way that is compatible with respectable classical 'scientific' Local Realism.

Reality, so this local realist argument goes, cannot be only the quantum wave function. There can be no one-to-one correspondence between quantum Ψ-wave and the locally 'real' nature of spacetime reality. Thus for Einstein and his pal David Bohm, and the acolytes of Scientific Local Realism, quantum mechanics is incomplete as a description of appearing spacetime reality. Copenhagen antirealists Bohr and Heisenberg vigorously disagreed with this local realist view.

Thus the deeper meaning of the Schrödinger Wave Equation and the actual relation of its mathematical entities to our physical spacetime realities is *ultimately* a question of which of the many incomplete 'quantum mechanics interpretations' we choose. And none of these venture out of the realist/physicalist orthodox belief bias. What to do?

Yes. We need a post-empirical non-reductionist foundational centrist integral Noetic Quantum Ontology. Bohr's antirealist 'textbook' Copenhagen Interpretation was the original default view. We have engaged it and some of the others above. None of them address the ontic elephant in the room, namely, in what, or in whom does the universal quantum Ψ-wave function with its description of relative spacetime stuff arise? What is the timeless, formless, selfless ontological ground of this wondrous epistemology that is the universal quantum Ψ-wave function of spacetime form? Recall, mereologically, parts are perforce embraced and subsumed in their greater wholes. The quantum wave function is thus grounded in that prior ontic whole.

As physics and Buddhism continues its felicitous dialogue we shall see a new Middle Way foundational quantum interpretation arise in which Ψ (*psi*) is seen as a causal mathematical description of an observer-independent, *relative* locally really real existing spacetime reality emerging within an observer-dependent, formless, acausal, nonlocal, *ultimate* perfectly subjective primordial awareness-consciousness 'groundless ground'. We have seen the need for such an integral Noetic Quantum Ontology. I have suggested the rudiments of such a process below.

The Mechanics of the Quantum of Action. A quantum wave function may be an 'eigenvector' of an 'observable'—a particle's position, momentum, energy, and spin—in which case it is called a quantum 'eigenstate'. The eigenvalue represents the value of the observable of a given eigenstate. A linear combination of such eigenstates constitutes the 'quantum superposition' of an infinity of all possible objective eigenstates. When an observable is subjected to an objective quantum measurement by the 'consciousness' of an observer/experimenter the subjective wave function superposition somehow 'collapses' into a causal objectively 'real' eigenstate. The measurement result will be one of its eigenvalues with a *probability* expressed in accordance with the Born Rule. Although the post-measurement wave function cannot be known prior to measurement, the probabilities can be accurately calculated and predicted using the Born Rule.

From this quantum process we receive the handy gifts of our computers, smart phones, TV, and quantum algorithms for determining the pricing of the economic derivatives which we all so urgently require. And nuclear bombs to keep us safe.

Well then, what else has this spooky mind-boggling quantum wave function ever done for us? How about the internet, laser communications, GPS, *f*MRI image scans, atomic clocks, microwave ovens, and guided missles. A mixed bag of quantum benefits, to be sure.

The Born Rule (1926) named after Einstein's philosophy mentor Max Born, is an integral postulate of quantum mechanics. It gives us the *probability* that the measurement of a given quantum system will yield a result that is proportional to the square of the magnitude of that particle's wave function at that point in a classical time (t). That the Born Rule can be derived from the radical Many Worlds Interpretation (MWI) of quantum mechanics remains controversial. The Born Rule applies to all our primary foundational 'quantum mechanics interpretations' except perhaps MWI which we shall further explore.

Hence, particle position and momentum are the classical physics expression of Schrödinger's all-inclusive universal quantum wave function (Ψ)—the quantized unity of the point-like particles and their wave-like fields of a bygone classical mechanics.

The Quantum and its Discontents. Local Realist Einstein hated it. He was, as we've seen, intuitively certain that quantum mechanics was 'incomplete'; that the mysterious random subjectivity of the universal quantum wave function merely demonstrated that something *objective* was missing in the probabilistic quantum theory, some overlooked, highly improbable "hidden variable" 'hope for a miracle' that would rescue our ontologically comfy old deterministic, non-random, empirical, cause and effect, non-stochastic classical view. Dice games are random. "God does play dice with the world."

Einstein hated the Uncertainty Principle of Heisenberg, that nature was random and God would 'play dice with the world'. But mostly he hated quantum 'spooky (*spukhaft*) action at a distance'—quantum nonlocality/entanglement, which ostensibly violated his local realist Special Relativity Theory (SRT) prohibition of the infinite speed of light. How in heaven and earth can a quantum system of two entangled particles 'space-like separated' by many light years 'know' its partners' spin state instantaneously? After all, the speed of light c is finite. And it would require infinite superluminal transmission of an electromagnetic signal over intergalactic distances to acquire such information.

Einstein and his philosophical nemesis Niels Bohr argued this point in the legendary 1936 Einstein-Bohr EPR Debates for two decades. For Herr Professor Einstein the quantum Ψ-wave function was an expression of our present ignorance of that "something deeply hidden" but ultimately objective nature of appearing spacetime reality. Quantum subjectivity drove him to distraction. He believed this to the end of his life.

Wave Particle Duality. Therefore, the emergence of spacetime reality is fundamentally a wave, the quantum wave function. But when we consciously observe it through a quantum measurement apparatus it presents, appears, and is measured as a particle. Particles move through space as waves. When we measure them they display their particle nature. Thus, every quantum entity may be described as both a point-like particle, and a wave spread out in space. Matter wave and particle are a complementary unitary *process*. Told Einstein,

> We have two contradictory pictures of reality; separately neither of them fully explains the phenomena of light [electromagnetic radiation], but together they do.

So physicists understand Bohr's complementary nature of light as the *wave-particle duality*. It expresses the inability of the concepts of 'particle' and of 'wave' in classical physics to describe the diaphanous probabilistic nature of quantum entities and states.

The facts are: all elementary micro-particles display a wave-like nature, and all waves display a particle-like nature. As Prince Louie de Broglie demonstrated in his 1924 PhD dissertation (1929 Nobel Prize in Physics), this is true as well for macroscopic atoms and molecules, trees and stars. All matter has wave properties.

This 'duality paradox' is fundamental to the very nature of matter-energy ($E=mc^2$) spacetime reality; and it is the foundation of the mathematical formalism of Quantum Field Theory (QFT) with its Quantum Electrodynamics (QED). Elementary spacetime particles interact through their wave-like fields. These nonlocal fields—magnetic, electric, and gravitational—pervade all of space and time. Indeed they *are* spacetime. Weird wave-particle duality. Get used to it.

Thus does classical scientific certainty about point-like particles and field-like waves become a single universal quantum wave function. For QFT our universe of space and time are simply that (Ψ). The nature of matter-energy is a spooky cosmic quantum *'cloud of probability'*. This formless indefinite mass is not other than the monumental global indeterminist 'universal quantum Ψ-wave function' itself, a 'probability wave' that evolves in a classical time (*t*). Spacetime form is quantum empty; quantum emptiness is form. We have seen that this view closely parallels Mahayana Middle Way Buddhism where, as the Buddha told, "Form is empty; emptiness is form". Relative form and its ultimate emptiness comprise the Buddhist Two Truths dominate ontic trope, as we saw above.

Irwin Schrödinger assigned a 'wave amplitude' number for every possible measurement result of a particle's position and/or momentum. Before a quantum measurement all of these possibilities 'exist' in a radically subjective infinitely 'superposed' potential state of what I have come to call 'quantum emptiness'. In 'collapse theories' at the instant of an observation-measurement by the 'consciousness' of a sentient human observer the wave function instantly 'collapses' into a 'locally real' objective causal spacetime 'real' electron. This is known as the 'collapse postulate' of QFT. Note that 'collapse' is *caused* by the human consciousness of an observer-experimenter interpreting an experimental result. John von Neumann and Eugene Wigner told, "Consciousness causes collapse" which has become perhaps the primary quantum ontology now on offer. No one explains just how collapse happens.

Quantum Measurement and Quantum Ontology. The five essential questions for 21st century quantum mechanics (QFT) involve the 'quantum measurement problem': 1) What constitutes a microscopic quantum measurement? 2) Does the quantum wave function collapse? If so what constitutes collapse? In short , how does a quantum 'superposition' of infinitely many possible eigenstate values evolve into a single eigenstate *probable* value when it collapses or interacts with the classical external world via a quantum measurement? How does the subjective quantum become objective classical? 3) How is it that such a measurement appears to 'disturb' or change the data being measured? 4) What is the role of human observation/consciousness in this quantum measurement process? 5) How are quantum micro-measurements of electrons related to macro-phenomena, like trees, stars, and conscious beings?

Our quantum ontology question is this: How will a settled centrist integral Noetic Quantum Ontology that describes the interdependent interconnectedness of human consciousness to its primordial awareness-consciousness ground answer the hitherto unanswered ontic questions that present 'foundational quantum interpretations' have failed to answer?

These five persistent and as yet unanswered questions arise as the '*quantum measurement problem*' which has, as we have seen, created an entire quantum physics sector—the ontic 'Foundations of Quantum Mechanics' or 'the interpretation of quantum mechanics'—which include the 'consciousness causes collapse' (of the quantum wave function) theories, the 'hidden variables' gambit that denies quantum wave function collapse, and the wacky 'Many Worlds Interpretation'. We shall require such a metaphysical Noetic Quantum Ontology in order to resolve the thorny post-empirical 'quantum measurement problems'.

In other words, we need a metaphysical post-physicalist/reductionist centrist, integral Noetic Quantum Ontology that grounds the objective formalist mathematics of the superposed quantum Ψ-wave function, and transcends yet includes the waning classical metaphysic that is the objective Scientific Local Realism/Physicalism paradigm. A cognitive bridge too far?

The history of philosophy and religion has clearly demonstrated that reality is not as it appears. Quantum mechanics is further demonstration of this truth. QFT shows that what we observe is not at all what actually is. Relative 'classical' human perception and conception of phenomenal reality is *prima facie* illusory, obscured in a dark cloud of uncertainty, just as European, Hindu, and Buddhist philosophical Idealism, and the Buddhist Middle Way para-realism have told for 25 centuries. Taoist Niels Bohr would have agreed.

Relative conceptual quantum mechanics of QFT/QED has yet to reveal, or even consider the *ultimate* underlying ontological ground of this emerging spacetime. As I have said relentlessly, we require a metaphysically robust view of the presence, essence, and nature of the *ultimate* original ground of our appearing *relative* quantum spacetime realities, including the mathematical formalism of the quantum wave function. Such formalist mathematical *relative* objective cognition does not constitute an *ultimate* quantum ontology. In what, or in whom does the wondrous universal quantum Ψ-wave function arise and participate? If Ψ is 'universal' a reasonable ontology is required to illumine and to ground it in its greater perfectly subjective whole or ground.

So, let quantum theory continue with its formalist relative calculations. Let us enjoy the wondrous products that result. But let us not fail to observe the ontic elephant in the room, namely, the post-classical, post-empirical *ultimate* essential nature of the reality that its wave function calculations presume to describe. Establishing that urgent 'grounding relation' is our main goal for this chapter.

We need to know how a theory of panpsychic micropsychic micro-phenomena maps onto our relative conventional macro-psychic cosmopsychic life-world, and the primary monistic prior ultimate ontic foundation of these relative dimensions. Let quantum mechanics make a 'quantum leap' out of our cognitive classical realist-physicalist biases and classical textbook quantum mechanics—our deep cultural background local realist 'global web of belief'; and out of its mathematical quantum formalism—into a postformal, 'post-empirical' ultimate foundational Noetic Quantum Ontology. All the while let the calculations continue.

We have seen that until quantum physics produces such a settled post-Local Realism quantum ontology there shall be no Quantum Gravity Theory (QGT) to unify postmodern physics—to quantify 'that which lies deeply hidden' in the great mystery of gravity—the Bhagavad-Gita's "creator and destroyer of worlds". Somehow the secret of the nature and essence of gravity lies 'deeply hidden' in the dark uncertain recesses of QFT mathematics and ontology. 'Cloud of uncertainty' indeed.

Ultimate Ontology: Realism, Idealism, and an Antirealist Middle Way. Essentially antirealist QFT, along with Middle Way Buddhism, has indicated that micro and macro spacetime stuff does not truly exist until it is observed by a consciousness, whether an individual awareness, or by natural extension, before the presence of sentient beings to observe the universe, by the all-pervading perfectly subjective primordial awareness-consciousness ground itself, vast cognizant *kosmic* awareness whole in which sooner or later all objective appearing spacetime reality arises, manifests and is instantiated.

Broadly construed, the QFT ontic view is known to the quantum philosophy trade as 'Antirealism'; although few quantum physics practitioners have made this subjective 'leap in fear and trembling' (Kierkegaard) from the objective consciousness of an individual observer to the *nondual* aboriginal primordial awareness-consciousness ground itself in which, or in whom the whole process of an observing subject and its myriad objects observed arise and participate. 'Nondual' is the prior and present invariant one truth unity (*dzog*) of that odious but adventitious primal 'subject-object split'.

The metaphysical ontology of Antirealism aligns itself but is not identical with perennial philosophical Metaphysical Idealism in its many vestments, the metaphysical view that arising and appearing reality is founded and grounded in the ontological priority of mind, or universal awareness-consciousness over against mere physical matter. In short, 'mind over matter'. Both Metaphysical Idealism and Antirealism deny that local spacetime matter is ontologically prior to all embracing nonlocal mind or consciousness.

The Buddhist Mahayana *Yogachara Madhyamaka* school with its 'Mind Only' (*chittamatra*) view represents Buddhist Idealism in the broad 26 century historical spectrum of Buddhist philosophy. The Buddhist Prasangika Madhyamaka Middle Way view is said to constitute a centrist 'middle path' between the metaphysical extremes of Absolute Idealism (nihilist nonexistence) and Absolute Realism/Physicalism (eternalist substantialist existence). Western Idealism is perhaps best represented by German Idealism—Hegel, Kant, Fichte, and Schelling. Our quest herein has been such a proto-realist, even antirealist 'middle path' between the metaphysical extremes of absolute Idealism and absolute Realism/Physicalism.

Both Buddhist and quantum emptiness point the way to such a propitious middle way ontology.

A bit ironically, Bohr's Copenhagen Antirealism of what was to become Relativistic Quantum Field Theory (QFT/QED) is a nominalist proto-Idealism that opposes the prevailing Scientific Local Realism and Physicalism of Modern Newtonian 'classical' physics. That habitually realist and classical materialist mind—only a recent reticent resident of the quantum abode—is still trying to cram the nonlocal antirealist freedom of the quantum world into an old procrustean bed of classical Scientific Local Realism. The sad result is that the grail quest of modern physics for a Quantum Gravity Theory that unifies the two hitherto mathematically incommensurable pillars of physics—the QFT/QED of Dirac and Feynman, with Einstein's General Relativity Theory (GRT)—has become utterly stalled. Noetic Quantum Ontology wherefore art thou?

Caveat Lector: From the metaphysical ontology you choose arises the phenomenal karmic reality you deserve. What you think is what you believe, is what you do, is what you get. Well, that's inexorable karma for you. Noetic wisdom ontology is always active in our habitual, defensive conceptual and belief biases. See it as a subtle monitor; a still small voice whispering noetic truth.

Bell's Theorem: 'Spooky' Nonlocal Quantum Entanglement Exists

John Stewart Bell's 1964 Bell's Theorem is arguably "The most profound discovery in science in the last half of the 20th century." [Steven Weinberg] Broadly construed, Bell's Theorem allegedly 'proves' spooky quantum superluminal (faster than light) universal nonlocality, entanglement, and interconnectedness. Quantum Theory predicts that all entangled quantum systems violate Metaphysical Local Realism which holds that all particle-fields have definite 'real' properties and spacetime locations for any possible measurement. Further, interactions between particle-fields cannot exceed Einstein's finite speed of light. Thus, it is argued, that

Scientific Local Realism stands refuted as an *ultimate* ontology for spacetime existence. Bell and the many recent 'Bell Test' experimental confirmations of his theorem (18 as of 2022) have demonstrated that Niels Bohr was correct about 'spooky' nonlocality/entanglement and Einstein's classical Local Realism was incorrect. Quantum nonlocal entanglement? Early quantum pioneer Einstein hated it.

Yes. Our relative spacetime reality is pervaded by an ultimate primordial universal interdependent quantum nonlocal entangled connectedness! Objects and beings are altogether intrinsically related in this infinitely vast all embracing primordial boundless unbroken whole—name it as you will.

Moreover, spacetime physical reality is universally 'ontologically relative'. It is established by observer-dependent sentient perception, observation, and quantum measurement. Physical and mental objects exist by way of the sentient consciousness of a sentient observer; and there 'exists' a basal acausal matrix—natural formless, timeless, selfless ultimate ground—that subsumes and in which, or in whom everything abides in a relationship of interdependent interconnectedness. *Kosmos* (physical cosmos and its timeless *kosmic* ground) is observer-dependent, and ontologically nonlocally entangled and interconnected! As H.H. Dalai Lama told, "Emptiness [Buddhist or quantum] is established by conceptual minds." 'Universal ontological relativity' indeed.

The Two Truths: Objective Science (form) and Perfectly Subjective Spirit (emptiness). We are now approaching an incipient *ultimate* centrist integral Noetic Quantum Ontology that transcends yet embraces its *relative* quantum mathematical formalisms. Yet we still await philosophy of physics quantum interpreters to derive a settled postformal quantum ontology that unifies the conceptual math in a transconceptual metaphysical quantum theory ground, as we are doing here. That primordial 'empty of essence groundless ground' is for our concept-minds mostly conceptual! However, we may recognize, then realize it though contemplative non-conceptual nondual direct yogic experience (*yogi pratyaksa*). Our 'middle path' integral Noetic Quantum Ontology merely points the way.

John Bell has provided the objective mathematical data on which to construct such an ontic noetic metaphysical foundation. Such an ontology requires that we surrender, or at least cognitively bracket our 'scientific' local realist and physicalist biases in order to discover what reality lies beyond. Thus do we enter our emerging 'scientific revolution' that is nothing less than a paradigmatic Noetic Revolution in Science, Spirit, and Culture.

We have seen above that nonlocal entangled quantum emptiness parallels Buddhist Middle Way Prasangika Madhyamaka emptiness/*shunyata*. Yes, as Buddha told, "Form is empty; emptiness is form." Relative quantum local form may be viewed ultimately as perfectly subjective nonlocal entangled 'quantum emptiness'. Though things appear separate, there is no *ultimate* separation in this vast infinite enfolded all subsuming interdependent whole! Form and its emptiness ground are an ontic prior yet always present invariant 'one truth' Spirit unity. Quantum Field Theory admits of 'no boundary' between the microscopic dimension of 'matter wave' particle-fields and the macroscopic dimension of the physical reality of all of us bundles of wavelike particle-fields abiding for a few brief years (*anitya*) in our uncreated beginningless, endless cyclic *kosmic* multiverse.

Thus does post-Bell's Theorem Quantum Field Theory offer a bridge between relative objective Science and its ultimate perfectly subjective primordial awareness-consciousness Spirit ground. And yes, these aboriginal Two Truths—relative form and its ultimate formless awareness 'supreme source' ground—are always already a prior ontic and phenomenally present indivisible *one truth unity (dzog)*. The dualistic relative parts perforce participate indivisibly in their nondual ultimate ground, the *Perfect Sphere of Dzogchen*. There is no inherent separation between Science and Spirit; try as we may to split them. That is the emerging integral ultimate scientific and soteriological (enlightenment, salvation) view. All that little more than self-stimulating concepts and beliefs before we submit to self-revealing trans-conceptual noetic 'spiritual' practice—prior unity of objective Science and its perfectly subjective nondual Spirit ground.

However, this quantum "lucid mysticism" (Pauli)—universal nonlocal quantum interconnectedness—does not mean that quantum mechanics 'proves' the mystical holism view of our noetic Primordial Wisdom Tradition, much less the retro-romantic pop quantum mysticism of popular culture. Our non-conceptual primordial awareness wisdom (*jnana, yeshe, gnosis*) cannot be derived from, or verified by conceptual/mathematical quantum electrodynamics (QED). These are parallel views. The concepts of the dimension of *relative* Science cannot conceptually 'prove' the trans-conceptual direct contemplative experience of the dimension of *ultimate* Spirit in whom it arises. Perfectly subjective *ultimate* nondual Spirit is ontologically prior, yet embraces and subsumes the dualistic objective *relative* domain of Science. Thus the prior and present nondual unity of objective Science and ultimate Spirit. Praise be!

Therefore, recent relativistic quantum theory—QFT/QED with its natural nonlocal entanglement connectedness—is based upon the now nearly universally accepted 1964 'Bell's Proofs' of the great Irish physicist John Stewart Bell. 'Bell's Proofs' have been verified by eighteen 'Bell test' laboratory experiments—from 1972 to 2022. All of the 'Bell test loopholes' have now been closed. All have found that the hypothesis of possible local or nonlocal hidden variables' developed by Einstein, David Bohm, and others to save classical Metaphysical Local Realism from the dragon of antirealist quantum mechanics cannot be correct. [Boaz 2022, *Ch. VII*]

Hence, the counterintuitive quantum nonlocality/entanglement of QFT is now 'proven' by Bell's Theorem to be inherently correct. That is the now mainstream view of orthodox post-classical quantum theory—still resisted by too many physicists and cosmologists who desire to reduce QFT/QED to safe and sane classical realist/physicalist physics. Nonlocal antirealist quantum mechanics, in some formulation, is here to stay.

Moreover, Bell's Theorem has shown that any future quantum theory theme variation—and there will be several as the theory evolves toward nonduality—must necessarily describe our world of space and time as universally nonlocal and interdependently entangled. In short, the prodigious quantum theory shall continue to represent a physically real *relative* reality that is intrinsically interconnected, interdependent, inseparable, and indivisible; in short, an *ultimate* noetic nondual whole.

On the accord of both David Chalmers (1965) and Lee Smolin this holistic view is subject to debate depending upon the interpretation of Bell's view of locality in his justly famous 1964 "Bell's Theorem" paper. Can Einstein's and Bohm's *nonlocal* hidden variables theory survive Bell's inequalities? It cannot, as we shall now see.

Well, what was it that inspired John Bell's Theorem in the first place?

Bell's Theorem and Einstein's Hidden Variables. Theoretical physicist David Bohm attempted to prove that his friend Einstein's unlikely 'hidden variables' represented a realist, physicalist, deterministic viable alternative to the indeterminist utterly random 'spooky action at a distance' (quantum nonlocal entanglement) of quantum mechanics universal wave function global quantum connectedness.

Bohm denied wave function 'collapse', as does the bizarre 'branching' or splitting of worlds of the Many Worlds Interpretation (MWI) of QFT. Yet realist Bohm's holism affirms the universal connectedness, interdependence, and indivisibility of both the subatomic world, and the macro world of trees and stars and of self-conscious human beings. It was this brilliant effort, along with the Einstein-Bohr EPR debates, that inspired John Bell to derive his now famous Bell's Theorem. He began by attempting to prove the possible existence of a 'deeply hidden' variable that would save the classical Local Realism of Einstein and Bohm. What he discovered changed his and our worldview forever.

We've seen that Bell's Theorem has demonstrated that quantum nonlocal (faster than light speed) entanglement—Einstein's "spooky action at a distance"—actually obtains in nature. A 'hidden parameter' to refute quantum nonlocal entanglement with its global interconnectedness cannot mathematically exist! Classical physics' local realist 'separability' of physical objects stands refuted. An event/cause occurring anywhere on earth can instantaneously produce an event/effect elsewhere on earth, or in deep space, without any physical electromagnetic force/signal to connect them. Is Einstein's proscription on faster than light signal connections refuted? Theoretical attempts to 'explain away' this disturbing violation of Einstein's second postulate of his Special Relativity have proven unsatisfactory. And more 'idols of the tribe' are sure to fall.

For example, in a single two particle entangled quantum system a measurement on one electron's spin may instantaneously reveal the spin state of its partner in a distant galaxy, violating Einstein's second postulate of his SRT as to the finite velocity of a 'local' light signal. Quantum universal nonlocal, indivisible, interdependent interconnectedness indeed. Middle Way Buddhists agree. Einstein hated it.

Clearly, such a nonlocal view casts a pall over the essential scientific principle of causality—local cause and effect—the unimpeachable foundation of the methods of Science, and of Scientific Local Realism. A scientific sticky wicket indeed. You can see the cause for concern by Einstein and Bohm, and the entire assembly of classical scientific local realists, not to mention 'common sense Local Realism' (Bertrand Russell's metaphysics of the Stone Age). So, the desperately needed 'hidden parameter' to restore real local sanity to Big Science's sacrosanct Principle of Local Causality is ultimately *kaput*! Nonetheless, local causality must obtain in the conventional domain of spacetime Relative Truth.

In other words, Bell's Theorem and its 18 subsequent 'Bell test' experiments demonstrate that quantum nonlocal entanglement is superluminally instantaneous! A photon in an entangled quantum system of two photons 'knows' its 'space-like' separated partners' spin state on Alpha Centauri directly and immediately without having to wait 4.37 light years for verification. Spooky quantum nonlocality indeed.

However, two conscious human observers attempting to measure such superluminal photon behavior, one on the earth and one on Alpha Centauri, may only become aware of the photon signal result at the speed of light. They must wait 4.37 light years for the physical result; a prime exemplar of the quantum 'measurement problem'. Quantum weirdness just gets, as Lewis Carroll might have said, "curiouser and curiouser".

There is no logical or empirical proscription on embodied human beings with 'highly advanced' subjective contemplative minds—'omniscient' Buddhas for example—to spontaneously 'know and feel' many superposed quantum consciousness states simultaneously. Such an inclusive all-pervading nondual Christ mind (*christos*) Buddha mind (*buddhajnana*) would remain free of any dualistic quantum measurement problem. On the accord of the great noetic Primordial Wisdom Tradition of humankind such omniscient embodied *buddic* minds have existed in this and other worlds almost forever.

But don't believe this! It's clearly beyond belief; transcending the limit of mere human concept and belief. Gautama the Buddha of this present age told it well: "Do not believe what I teach ... come and see (*ehi passika*)." In short, personally enter in such transpersonal cognitive spaces via trans-conceptual direct (*yogi pratyaksa*) 'mindfulness of breathing' practice. Reading books about it won't work. Find a meditation master.

Subjective contemplative cognition is such a post-semiotic, 'post-empirical' noetic wisdom technology. We prepare for such a cognitive quantum leap by training dualistic conceptual mind in the relevant contemplative, even nondual philosophical foundation, as we are doing here. I have here referred to that very pragmatic process as *awareness management.* Indeed, for scientists of all stripes, engaging the far out spooky metaphysics of the prodigious universal quantum wave function with its natural but conceptually ineffable nonlocal quantum entanglement is the most reasonable preparation for such super-spooky post-quantum nondual wisdom. Meanwhile, we 'bracket' our still classical realist cognitive biases and beliefs and proceed to post-rational contemplative mindfulness meditation under the guidance of a qualified meditation master. [*Appendix A* below; Boaz 2022 *Ch. 8*]

This wisdom adventure is definitely not for the metaphysically squeamish! Understandably, most 'scientific minds' have chosen to remain in the comfy cognitive cabin that robust reason built in a dense 'Upanishadic forest' of cognitive quantum uncertainty. After all, 400 years of European Enlightenment (the Age of Reason) rationality can be habitually quite persuasive, even for the subtlest of conceptual minds.

Be that as it may, this is all a *fait accompli* before actual trans-conceptual practice is mindfully established. Course minds have a contemplative 'hidden parameter' already present within the Nature of Mind itself. Yes, told Buddha, "Come and see." Check it out.

Moreover, from an Everettian MWI view, an all-pervading omniscient *buddic* mind abides in, and is always already in a state of nondual unity with all possible wave functions in all possible worlds; all of this embraced and included in the *ultimate* dimension of the primordial awareness-consciousness ground itself, Bohm's "vast implicate order of the unbroken whole" in which or in whom this *relative* spacetime dimension arises and appears to the 'many worlds' of ordinary dualistic mind.

Please recall that 18 subsequent 'Bell test' experiments since 1964 have removed all of the possible 'locality loopholes'. [*Scientific American* Dec. 2018; Special Spring issue 2019] Bell's Proofs have now become quantum orthodoxy. How it is that this perplexed orthodoxy integrates such post-empirical, even nondual global quantum connectedness with 'old paradigm' Scientific Local Realism and Scientific Materialism/Physicalism remains to be seen. We need a new 'interpretation' or 'foundation', a Noetic Quantum Ontology of the prodigious quantum theory that dares to tread the path of our noetic primordial wisdom to which it has been pointing from the beginning. It's a big step.

Bell's Theorem Fundamentals. Bell's Theorem begins with the three metaphysical assumptions of modern classical Newtonian physics: 1) *causality* is local, effects follow their causes and must be proximate in space and time to their causes; 2) *determinism*: all physical events are determined by an external physical force; 3) *locality*: no measurement may have more than one result in any given spacetime location in which the measurement is made—the 'one world' stipulation. Causal results happen in the locally connected world in which they are made, and not elsewhere. If Bell's Theorem is true, then QFT is correct and at least two of these three classical metaphysical assumptions must be false. How shall we understand this?

We've seen that quantum entanglement/nonlocality demonstrates that in a quantum system consisting of a pair of measured 'entangled' particles, each particle seemingly *causally* effects the other, even when the two particles are 'space-like separated'—not connected by a causal electromagnetic (light) signal—by hundreds of light years. Such counterintuitive, inscrutable 'spooky' noncausal, nonlocal entangled quantum behavior clearly violates both 1) and 3) above—causality and locality—for it shows that entangled particles can effect one another's measured states (e.g. spin-up or spin-down) under *nonlocal, acausal* superluminal—faster than light—conditions. And that violates Einstein's sacrosanct first principle of relativity, namely, the velocity of light in the vacuum of space is finite—no electromagnetic signal can exceed light speed *c*—a denial of nonlocal quantum entanglement. John Stewart Bell proved it wrong.

Clearly, we need a bit of post-quantum therapy to assuage our anxious cognitive dissonance, and our self-ego-I ontopathic fear of nonexistence. Let us then in this spirit of fearless exploration revisit the least viable, least comforting, but sadly perhaps most popular 'scientific' MWI quantum ontology. This may be painful.

Critique of the Many Worlds MWI Quantum Ontology

Why must I belabor such a preposterous ontology? My own cognitive biases being as they are, far too many quantum physicists and cosmologists now believe that the fundamental local causality and Scientific Local Realism that physicists have come to know and love can only be salvaged by ontological extremism, to wit, by adopting an even spookier, absurd and more surreal, radically mechanistic, hyper-objectivist, super-deterministic Many Worlds Interpretation (MWI) of Quantum Field Theory (QFT/QED) and its universal quantum Ψ-wave function. Let us then revisit MWI by the lights of our above consideration of John Stewart Bell's definitive 1964 Bell's Theorem.

We have seen that MWI was developed by Hugh Everett in a 1957 PhD dissertation under the supervision of great gravitational physicist Archibald Wheeler, and popularized by physicist Bryce DeWitt in the 1960s, and later by David Mermin and Sean Carroll. MWI rejects (3) above—locality the 'one world' stipulation of classical physics.

MWI begins with the classical local realist/physicalist mechanistic paradigm—the ontological/metaphysical mechanistic assumption or presupposition that the forms of emerging spacetime existence are "entirely physical", or reducible to purely physical brain structure and function. "Many Worlds quantum mechanics is a quintessentially mechanistic theory." [Carroll 2019] MWI, like the rest of our foundational quantum ontologies, including our Noetic Quantum Ontology, is pure metaphysics, that is to say, it is not founded in logical nor empirical 'scientific fact', if there exists such 'facticity'.

This purely metaphysical presumption clearly begs the question of Metaphysical Scientific Physicalism. 'Begging the question' or 'assuming the conclusion' (*petitio principii*) is an informal logical fallacy wherein an argument's premises assume the prior truth of the conclusion instead of supporting it, a brand of hidden 'circular reasoning' (*circulus in probando*) that presumes a desired conclusion to be true when that conclusion does not at all follow from the premises. This classical fallacy of logic and rhetoric is often subtle and difficult to detect when complex logical syntax of language conceals that the conclusion is pretending to be a premise. It is often accidental, grounded in cognitive desire for a result.

Philosophical Mechanism is the view that all appearing phenomena may be explained in the physicalist deterministic terms of 'mechanical laws governing physical motion of matter-energy'. MWI is fundamentally an observer-independent 'reductive physicalist' ontology. Here, arising spacetime stuff objectively exists as a purely physical *observer-independent* 'real world out there' (RWOT), whether or not it is observed by a conscious observer. Thus is MWI immersed and encaged in the classical dogmas of Scientific Local Realism, and Scientific Metaphysical Materialism/Physicalism which is indeed its purpose to defend. Clearly, this 'global web of belief' (Quine 1969) is an exemplar of the old classical knowledge paradigm ontology that is Scientific Local Realism. We must now ask, if this be Realism, what remains of Reality?

Well then, just what is Metaphysical Scientific Local Realism?

'Scientific Local Realism' includes both 'Direct Realism' aka 'common sense Realism' or 'naïve Realism', and the subtler 'Indirect Realism' or 'Representative Realism' or 'Representationalism' where we perceive reality not directly but via perceptual representations called 'sense data' (*hyle*). Realist ontologies generally oppose the ontological views of Antirealism and of philosophical Idealism where the substantial physical world is not at all a physical process as 'scientific' reductionist physicalists believe, but ultimately a monism of the mental, or 'mind only', a 'very persuasive illusion' (*vidya maya*) created by our intrinsic mental process. Appearing reality is 'mind only'.

Please understand that all of these ontologies that arise in the noble history of Philosophy of Mind and its more recent cognizant issue known as Philosophy of Consciousness and Contemplative Science are entirely trans-empirical metaphysical speculative theoretical systems of human concept and belief. There exists not a whit of empirical 'proof' for any of them; not even our 'naïve common sense Realism' certainty in the belief of a physically substantial, permanent 'real world out there' (RWOT) experienced by a permanent, often all too real self-ego-I.

Professional philosophers of physics are quite aware of this tricky situation. Physicists and cosmologists are not. Perennial metaphysical realist philosopher and logician Bertrand Russell (co-author of *Principia Mathematica*) referred to this bit of common reality conjuring as "The metaphysics of the Stone Age". Or, more correctly, since the Realism and Physicalism metaphysic arose in the Near East with the 'proto-Semitic' tribes circa 15th century BC or earlier, we might choose to call it "The metaphysics of the Bronze Age".

That Scientific Local Realism admits of no compelling empirical grip beyond our habitual deep cultural background ideological belief in it, may explain how it is that realist/physicalist physicists fear philosophy, and the especially scary diaphanous 'philosophy of consciousness'. 'Consciousness' strikes terror at the heart of most physicists.

Yes. It is now becoming clear that Big Science, and especially quantum physics and quantum cosmology must at long last address as I have said the ontic elephant in the room, namely, the absence of a settled post-realist ontology of the universal quantum Ψ-wave function at the heart of Quantum Field Theory (QFT/QED).

The many 'foundational interpretations' of quantum mechanics are philosophical attempts to do just that. Strangely, MWI is a leading contender for that dubious honor. It seems that most quantum physicists have bought into this purely mechanistic metaphysic. The perfect subjectivity of the vast quantum *kosmos*, and hyper-objective mechanistic MWI: strange metaphysical bedfellows indeed. The cosmic irony is thick enough to drown in.

In any case, mechanist realist/physicalist MWI asserts that Schrödinger's inherently subjective 'superposed' universal quantum wave function is objectively real from the beginning and so there is no 'wave function collapse' into an objectively 'real world out there' (RWOT) at the instant of an observer's quantum measurement, or of a sentient macrocosmic perception. Indeed, there is no quantum wave function at all! The inherently subjective quantum Ψ-wave function of the entire universe is here only objective and purely physical. The vexing 'quantum measurement problem'—how, or if, wave function collapse occurs—is resolved! For MWI the orthodox quantum 'wave function postulate' is altogether denied. All possible 'superposed' observations/measurements are actualized in one of infinitely many possible parallel real worlds/universes. Well, let us suppose that I measure an electron as spin-up rather than spin-down.

> All other measurement outcomes still exist and are perfectly real, just as separate worlds ... Both parts of the final wave function are actually there. They simply describe separate, never-to-interact-again worlds ... The wave function of an electron can put it in a superposition of various possible locations, as well as in a superposition of spin-up and spin-down ... each part of the superposition [is] a separate world.
>
> —Sean Carroll 2019 p. 114

Thus, in one world Schrödinger's poor little cat is dead; in another world the lucky cat is alive. Two cats, two separate noninteracting parallel universes. No problem. Beyond belief? Yes. Logically consistent? Yes. Utterly inane? Yes. What's going on here?

A Schizoid Thought Experiment. At a special dinner I must make a pre-prandial choice from my modest wine cellar. My MWI guest assures me that there exists an entirely separate parallel objectively 'real world out there' (RWOT) in which I actually choose the 2004 Romanée Conti for our dinner; and an equally real world in which I choose the 1978 Chateau Mouton Rothschild for the same dinner. In the quantum MWI gloss, at the instant of my choice all other equally real universes are closed to me. The orthodox quantum wave function is denied. There is no 'wave function collapse' into a single real world. For MWI "Both parts of the final wave function are actually there." Whatever choice I might make is actualized in one of infinitely many parallel entirely real worlds. I exercise no spooky free will consciousness because all of my choices will actually happen in some alternate universe. At the moment of choice there are now "multiple copies" of me; one in each of all possible yet real parallel universes. For every quantum event moment a separate real world 'branches' or splits off. I am, by theoretical fiat, precluded from experiencing any of these other 'many worlds'. MWI does not reveal how.

By now I am feeling schizoid and dissociated even before drinking the wine! Moreover, sadly, I cannot enjoy both wines in these two worlds simultaneously. Just as well. Such a gluttonous ontology would surely qualify as wretched excess in any world.

This far fetched mega-pluralistic ontology of the MWI view of QFT with its many parallel 'alternate universes' constitutes a wholesale multiplication of reality-slice entities that desperately needs an *'Occam's Razor'* antidote, to wit, "Entities must not be multiplied without necessity." That is to say, when considering competing hypotheses about the same result the best solution is usually the one with the fewest prior assumptions. Occam's Razor is a pragmatic, extra-evidential heuristic guide rather than a final arbiter among competing theories. This old bromide, known to the wise as the 'Principle of Parsimony' has rescued many a theoretical explanation from scientific and philosophical oblivion. As Einstein told, "Everything should be made as simple as possible, but not simpler."

Confused and perplexed by MWI? Join the crowd. There remain many potentially disqualifying questions for the dubious Many Worlds Interpretation. Here are a few.

For MWI is there 1) a single spacetime or even a formless ontic universal metaphysical ground in which all of these separate branching proto-existing many worlds exist? If so, how? 2)

Do any of these many worlds possess any shred of locally real existence anywhere, other than a hypothetical posit of their nonexistence somewhere? 3) Why metaphysical mechanistic Physicalism? Is everything purely physical by mere ideological stipulation? 4) What if anything saves the theory from the nihilist ontic extreme of an ontic pessimism, or of an Absolute Idealism, that is to say, of utter nonexistence of any 'real' worlds at all? 5) What constitutes an 'observation' that splits one world from all others? 6) What is a choice? 7) MWI "is purely physical and mechanistic", so what is the physical mechanism that causes the absurd branching/splitting? Quantum decoherence we are told. How so that? Mereologically (part/whole relations), parts are perforce included and subsumed by a greater holonic more inclusive whole. 8) What is the relation of an infinity of many 'non-existent' existing worlds to the boundless primordial awareness-consciousness whole or ground in which all reality processes perforce arise? Is such a mereologically necessary all-subsuming perfectly subjective ground conceptually, theoretically reduced to the MWI purely objective non-collapsible, ostensibly non-existent quantum Ψ-wave function itself? What is that primordial ground in which or in whom these MWI many worlds arise? Does MWI ideologically preclude the monistic cosmopsychic all subsuming primordial awareness-consciousness ground of our noetic Primordial Wisdom Tradition, that vast infinite ultimate whole that grounds our relative human consciousness/experience?

Ontic and epistemic problems abound. Does MWI create more ontic, epistemic, and phenomenological problems than it solves? There's plenty of separability in MWI. But no unified ontology. I say mere logical consistency of an otherwise dubious and wholly *ad hoc*, unbelievable metaphysical belief system should not presume to be an adequate quantum foundational ontology, let alone a centrist integral Noetic Quantum Ontology.

Ostensibly, MWI salvages our metaphysic of physicalist Scientific Local Realism. But at what cognitive cost? The doctrine stipulates an adventitious infinity of duplicates of a real local quantum observer; many 'real copies' of me, in each moment of my awareness. Can such a schizoid belief system really qualify as a believable ontology, whether Scientific Local Realism, or anything else? If so what remains of the credible non-entangled relative Local Realism of Plato, James, Russell, Bohm, and Einstein? MWI does more epistemic harm than good. Conceptual clarity, even course common sense is utterly lost.

MWI conflates observer-dependence in the quantum measurement and observation of observer-dependent wave function collapse with observer-independent local 'quantum decoherence and branching' (of the wave function); that is, the loss of quantum coherence as observer and measuring apparatus interact and couple in the process of a quantum measurement in real time. Sean Carroll [2019 p.119]:

> That simple process—macroscopic objects become entangled with the [measurement] environment which we cannot keep track of—is decoherence ... Decoherence causes the wave function to split, or branch into multiple worlds. Any observer branches into multiple copies along with the rest of the universe ... In one of those worlds the experimenter will have seen spin-up and in the other they will have seen spin-down. But both worlds are indisputably there ... The price we pay for such powerful and simple unification of quantum dynamics is a large number of separate worlds.

Now that's a big price! So, quantum decoherence is crucial for MWI. It is the cause, so the theory goes, not of a wave function collapse, but of a 'branching' or proliferation of the quantum wave function into infinitely many separate, independent 'branching' parallel worlds. The coherent information in an entangled quantum system 'decouples' or 'leaks' into the instruments and measuring environment as quantum coherence decoheres. So, it is argued, there need be no 'problem of observer consciousness'. With no observed wave function collapse by a sentient consciousness there is no need to posit such a consciousness at all. There is then no 'quantum measurement problem'. The spooky subjectivity of consciousness is altogether 'explained away' by a highly speculative, question begging purely physicalist mechanism of 'non-interacting separate worlds', a denial of human consciousness altogether. Wow! If this be quantum ontology, God deliver us from quantum inanity; not to mention quantum humility.

MWI grounds its physicalist denial of Einstein's 'something deeply hidden' in human consciousness in the metaphysical purely mechanistic local realist and materialist/physicalist presumption that, as Professor Sean Carroll has told in his excellent best seller *Something Deeply Hidden* (2019):

> Consciousness arises from brains ... or 'nervous systems' or 'organisms'. These are assumed to be 'coherent physical systems' ... conscious observers branch [into parallel universes] along with the rest of the wave function.

So, for MWI the receding 'classical paradigm' metaphysical physicalist assumption of physics and neurobiology that "Consciousness arises from brains" is an essential pillar of the theory. But how does human consciousness/experience arising in its infinite primordial awareness-consciousness ground arise from merely physical human brains?

That human experience-consciousness arising from its mereologically necessary formless primordial awareness-consciousness ground is reducible to purely physical brain matter is a fallacy of the begged question of metaphysical Realism/Physicalism, as we have just seen.

Antirealist Philosophical Metaphysical Idealism in its sundry cognitive raiment—is perhaps the primary ontology of our great Primordial Wisdom Tradition—Eastern Hinduism, Buddhist

Dzogchen, Taoism, and the Abrahamic Western mystical voices of Judaism, Christianity, Islam. Metaphysical Idealism as an Antirealism asserts that the *ultimate* nature of arising appearing reality is not merely the conspicuous and indisputable appearance of *relative* physical spacetime stuff, including the structure and function of physical brains. Mind or consciousness is the ultimate nature of all appearing reality.

Instead, please consider here that the nondual ultimate Nature of Mind or Being Itself with its human experience/consciousness is fundamentally 'Big Mind' (Ultimate Truth reality dimension), the all-pervading ground, primordial awareness-consciousness being itself. It is That (*tathata*) in which, or in whom 'Small Mind' (Relative Truth reality dimension) human consciousness with its big physical brains, along with all the rest of physically real quantum stuff arises, participates and is instantiated.

There are antirealist, realist, and dualist ontologies in each of the above wisdom traditions. Mahayana Madhyamaka Buddhism has found a 'Middle Way' Two Truths view that affirms both a *relatively* real and existent spacetime that is *ultimately* absent and empty of any innate intrinsic solid existence. This view represents a centrist position between a false dichotomy of the ontic extremes of metaphysical absolute existence (Realism/Physicalism) and metaphysical absolute nonexistence (Antirealism/Idealism).

Has Sean Carroll failed to understand the ontological depth pointed to by quantum nominalist Antirealism and by Metaphysical Idealism in his consideration of ontology (p. 223-224)? Has he failed to fathom the depth of that very 'something deeply hidden', the nondual, trans-conceptual primordial awareness-consciousness ground itself—Tao, *dharmakaya, nondual Parabrahman, infinite Ein Sof, Abba* God the nondual Primordial Father—whence emerges these many "quantum worlds" of spacetime by limiting it to mere human conceptual purely physical brain consciousness (pp. 219-225)?

Unfortunately, such is the standard orthodox 'scientific' understanding, encaged as it is in a waning classical old Realism knowledge paradigm that is the foundational dogma of Scientific Local Realism, and mechanistic Metaphysical Scientific Materialism-Physicalism. MWI defends this fallacious bygone classical realist/physicalist view.

If present quantum mechanics—QFT/QED/QCD—is to evolve into its next more inclusive knowledge paradigm it must openly and assiduously engage these ontological questions of 'consciousness' in all of its depth. This means that interpretations of the universal quantum wave function—all of them inchoate ultimate ontologies—that ideologically adhere to a purely mechanistic Physicalism must, as David Hume told, "be committed to the flames for it is nothing but sophistry".

Fear of an arising physics post-empirical, post-Core Theory nondual ultimate ontology, with its reflexive dualist denial mechanisms, constitutes an intellectual failure to courageously engage such a providential Noetic Quantum Ontology by way of *both* objective conceptual/mathematical theory and praxis, *and* super-spooky subjective direct contemplative practice and experience (*yogi pratyaksa*) of our recent Science of Consciousness with its emerging Contemplative Science. Scary radical wisdom indeed.

Yes, we must engage both voices of our innate human *noetic cognitive doublet*—both faces of gnosis, both exoteric objective Science and esoteric subjective Spirit in which it arises. The ideological exclusion of either of these reality dimensions is a recipe for human ignorance (*avidya, ajnana, hamartia*/sin) and the suffering of living beings that is its inevitable result. Our knowledge, our innate love-wisdom mind with its human happiness is at stake. As Plato told so long ago, "For no small matter is at stake here; the question concerns the very way that human life is to be lived." [*The Republic*, Book I]

Perhaps the buddhas, *mahasiddhas*, saints, and sages of our great noetic wisdom traditions have wisdom to share that will illumine our quantum ontic grail quest for the *ultimate* nature of trans-conceptual 'post-empirical' nondual reality itself. Perhaps we might view the monumental universal quantum wave function as an inchoate conceptual ontology for that nondual ultimate nature of mind/consciousness—bright indwelling Christ nature Buddha mind Presence (*vidya, rigpa, christos*) of That—that is always already present within every human being, even ontic disoriented MWI acolytes.

Let us then expand our view of 'consciousness'—relative human and primordial ultimate—from the pessimistic epistemic ontic limit of a 'purely physicalist mechanistic' MWI understanding. That, so we may better understand this ultimate 'Nature of Mind'— 'supreme source' of the 'supreme identity' *Presence* that we actually are in its greater, all embracing, all subsuming ontic context. MWI fears such a post-realist/physicalist integral noetic knowledge ontology and epistemology process.

In the primordial wisdom (*jnana, yeshe, gnosis*) view it is the "something deeply hidden" *ultimate,* formless, timeless, selfless all embracing awareness-consciousness ground itself, by whatever grand name, in which or in whom human consciousness, and indeed all *relative* conventional spacetime physical and mental form arises and participates. Embodied physical brain is ultimately enfolded in that prior timeless formless ground while continuously unfolding and 'emerging' in relative space and time. This remarkable relational interdependent (*pratitya samutpada*) *process* is experienced but not caused by the human brain's prodigious physical neuronal processing and operations.

To be sure, human consciousness has its 'neural correlates' in physical brain. Yet clearly, from that physical fact we cannot conclude that "Consciousness arises from brains". Indeed, it is the other way round. In a larger more inclusive view the awareness of our physical brain, human consciousness, and everything else arises from its prior all subsuming primordial awareness-consciousness ground. Primordial awareness-consciousness ground itself is ontologically prior to and embraces and subsumes human consciousness with its all of its neurophysical qualities arising therein.

Hence, we shall conclude that human consciousness does not *ultimately* arise from, nor is it caused by mere relative physical brain nor an 'organic' central nervous system. Brain, Sean Carroll's "coherent physical nervous systems", human consciousness, and the whole cosmos of spacetime physical and mental form, including the universal quantum Ψ-wave function, arise from natural timeless boundless whole itself, all pervading primordial awareness-consciousness ground itself, by whatever grand name. As that *mereological* process is necessary, how could it be otherwise?

Mereology (part-whole relations) is the logically compelling holistic reasoning that points to an urgent ontological, epistemic, and phenomenal truth. Separate relative spacetime parts—the microcosmic in the macrocosmic—are necessarily included in an even greater more inclusive all subsuming whole—whether or not we choose to see it or name it. Just so, that ultimate "implicate order of the vast unbroken whole" (David Bohm) is itself the ground of the infinite multiplicity of its participating and instantiating parts. Wholes embrace and include their parts; parts are nested in their wholes—all the way down to microcosmic parts; and all the way up to the ultimate kosmic primordial ground itself. That whole or ground is known to those who know as the prior ultimate one truth unity that is nondual Reality Being Itself in which, or in whom all of our appearing realities, and all of us, arise and participate.

This concludes our exploration of the many worlds of the MWI attempt to 'save the appearances' via rescue of objective Metaphysical Scientific Realism/Physicalism from the subjective jaws of Irwin Schrödinger's global 'universal quantum Ψ-wave function'. The absurd Many Worlds Interpretation of Quantum Field Theory. May it rest in quiescent peace, never again to bewitch the minds of desperate if well meaning physicists.

Prelude to Grounding Quantum Field Theory in its Prior Ontic Base

Yes, Irwin Schrödinger's prodigious quantum wave function is a still inchoate conceptual, mathematical approach arising from its 'grounding relation' in basal noetic nondual ultimate ground, primordial metaphysical ontic ground of all appearing phenomena and our theories about it. Quantum mechanics can no longer dodge an ultimate ontology that engages both its dimensional voices—objective relative mathematical, and perfectly subjective ultimate. Thus do we require a middle way integral Noetic Quantum Ontology.

In Mayayana Buddhist philosophy this ultimate ontic and relative epistemic mereological (part-whole relations) 'grounding relation' is known by way of our perennial Two Truths trope—the relative truth of form (*dharmata*), physical and mental human consciousness/experience—'emerging' within, and never departing, its formless basal ultimate primordial emptiness/*shunyata* base (*gzhi rigpa*), its Basic Space (*chöying*) *dharmakaya* ground. Perhaps it is the always already present indwelling 'primordial Presence' of this nondual 'supreme source' that is that very "something deeply hidden" of Einstein, Bohm, Carroll, and Quantum Field Theory. I believe that this is so. [Boaz 2020 Ch. V] Let us then continue our approach to a centrist integral Noetic Quantum Ontology.

Toward A Noetic Quantum Ontology: The Wave Function and its Ground

Noetic Prelude. So the prodigious inherently subjective superposed quantum mechanical universal Ψ-wave function of our emerging spacetime reality is a holistic, compelling, dualistic, relative, conceptual/mathematical description of Einstein's "something deeply hidden", that infinite nondual Ultimate Truth dimension (*paramartha satya*) that is the formless, timeless, selfless primordial awareness-consciousness ground in whom all the worlds—parallel branching or otherwise—arise and participate.

A One Truth Unity. That vast all-pervading boundless luminous cognizant consciousness whole—indwelling noetic bright Presence of That—abides utterly beyond even the greatest intellectual virtuosity of a conceptual and mathematical grasp of our objective human consciousness arising and participating therein. All the buddhas, saints, *mahasiddhas,* and love-wisdom masters of the history of our species have told it. For our noetic Primordial Wisdom Tradition the non-conceptual nondual invariant *one truth unity* of our perennial Two Truths—relative and ultimate—is the very Nature of Mind, 'Big Mind'. It cannot be grasped by Relative Truth, the dualistic conceptual cognition of 'Small Mind'. Try as we may. That it can is an exemplar of a philosophical 'category mistake', as we have seen. That is to say, our perennial Two Truths cognitive trope represents unique qualitative *relative* dimensional differences, even as they participate together as a noetic nondual invariant *ultimate* prior one truth unity.

Yet, "wonder of wonders", this noetic nondual ultimate wisdom may be recognized, then realized by way of our indwelling contemplative love-wisdom Christ-Buddha mind (*christos, buddhajnana*), the aboriginal Tao that is beyond the biases of our concept-belief understanding. "The Tao that can be told is not the primordial Tao." Bright always already present nondual primordial Presence (*vidya, rigpa, christos*) of That (*tathata, satchitananda*).

Knowing-Feeling Awareness of That Unity. Yes. Primordial awareness wisdom (*jnana, yeshe,* gnosis) is naturally, always already innately present to trans-conceptual direct contemplative human cognition, the *yogi pratyaksa* of the well-trained mind. That that is 'deeply hidden' to human conceptual cognition is revealed, breath by mindful breath, to our direct contemplative cognition. [*Appendix A*] That love-wisdom mind "energy is eternal delight" [William Blake], always present cognizant love-wisdom knowing-feeling clarity and bliss. But don't believe it. It is entirely beyond belief. Should you desire to know the whole story, do check it out, if you are not already doing so. I have come to call such 'mind training' *awareness management.*

Fraught 'scientific' denial of this subtle 'spiritual' voice of our human *noetic cognitive doublet*—objective conceptual and subjective contemplative—is no longer possible for an open authentic inquiring mind. Now we know better. Indeed, all of the antinomies of both classical and quantum consciousness have pointed to this great noetic syncretic truth, yet we have too often chosen to remain caged in our classical 'scientific' "global web of belief".

Therefore, let scientists of all stripes who would know the truth of that that lives as/in the very ultimate 'Nature of Mind' venture here. Human cognitive evolutionary development is perforce woefully incomplete without this adventure of relative conceptual mind awakening to the ultimate trans-conceptual 'supreme identity' of its noetic 'supreme source'. As the great Neoplatonist Plotinus told, "Development is envelopment" at ever deeper levels of Spirit embrace. Let Big Science understand that.

Indeed, the mindful process of that awakening to the prior ontic and ever present phenomenal unity of the original noetic nondual unity of our perennial Two Truths—relative form and ultimate emptiness, Science and Spirit—is the purpose of this book.

Be all that as it may, the spooky 'quantum enigma', the abstruse quantum mathematical formalism brings to mind an infamous pith of Richard Feynman, polisher of Paul Dirac's QED: "I think I may safely say that *nobody* understands quantum mechanics."

And nobody does. How so? QFT/QED with its logic-defying subjective superposed 'universal quantum wave function' essentially exceeds the cognitive limit of our mere conceptual understanding. That is the prodigious 'quantum enigma' that is a conscious finite knowledge-wisdom portal into the infinite spooky 'cloud of unknowing', the 'uncertainty relations' that veil our conceptual understanding. QFT opens that door.

The 'quantum measurement chain' requires—in most 'foundational quantum interpretations'—the 'consciousness' of an observer to 'collapse' the subjective infinite 'superposed' quantum Ψ-wave function and reveal microcosmic 'matter waves' of a macrocosmic objectively 'real world out there' (RWOT). And consciousness—as *relative* human awareness arising and abiding always in its vast *ultimate* awareness-consciousness ground—transcends yet embraces the universal quantum Ψ-wave function, and for that matter all of the objective laws of physics of this present spacetime universe of ours.

The *relative* classical two-valued truth functional logic of our semiotic logical syntax of language is not adequate to an *ultimate* understanding of the trans-conceptual nature of consciousness—conscious human awareness—let alone its all subsuming primordial awareness-consciousness ground. That providential thuth opens into a liberating new post-empirical, post-quantum, even contemplative human Science/Spirit knowledge paradigm. And yes, we discover the knowing-feeling lucent Presence That (*tathata*) upon trans-conceptual quiescent 'mindfulness of breathing'.

MWI Many Worlds Revisited

We are not quite finished with MWI. However, if you are fed up with MWI, as I confess I am, please skip this section.

Our past ideologies are no longer adequate to present needs. We have seen that MWI's ostensible objectification of Schrödinger's subjective universal quantum wave function offers hyper-objective Big Science solace to the thorny subjective paradoxes of quantum consciousness. The present spooky 'lucid mysticism' (Pauli) of observer-dependent quantum interpretations—Schrödinger's Cat, Einstein's EPR Paradox, and von Neumann's Boundary Problem—are resolved in a classically comfortable—if *fantasque*—logically consistent objective, 'purely mechanistic', entirely physical, super-deterministic, observer-independent, local realist 'real world out there' (RWOT). Except, for MWI, it has quite strangely become 'many real worlds out there'. We then lamented the price paid, namely, a reasonable believable interpretation of our appearing quantum realities.

Physics' uncomfortable comfort zones that are the 'scientific' dogmas of Metaphysical Physicalism and Scientific Local Realism is saved. But at what cost? MWI epistemology exceeds in absurdity even the 'quantum weirdness' of the inherently random nonlocal universal quantum Ψ-wave function itself that it purports to clarify and objectify.

Hence, MWI now faces the disquieting conclusion that to make logical sense of Quantum Field Theory (QFT) with its causality busting entangled nonlocality, we need an even more contrived theoretical absurdity. What could be more contrived than a subjective quantum wave function that magically 'collapses' into an objective spacetime 'real' electron? How about infinitely many 'branching' *alternate universes* wherein an infinity of choices of a quantum micro-measurement outcome, or of a macro-observation outcome by the 'consciousness of an observer' is somewhere, somehow physically, objectively realized! Yes, this means an infinite number of such objectively physically real experiencing conscious selves experiencing an infinite number of quantum universes, *ad infinitum.* In short, a logically unsound 'infinite regress', an endless circle of reasoning. 'Many worlds' indeed. But the epistemological problem runs deeper than that.

We've seen that MWI purports to be a logically consistent but rather torturous way to 'save the appearances' of the much beloved Metaphysical Local Realism and Metaphysical Physicalism of Modern Science, bound as it is, and should be, by the core 'principle of causality'—cause and effect reasoning. And what is Science if not causality? We saw that MWI objectifies and reifies Schrödinger's inherently subjective quantum Ψ-wave function superpositions through a denial of the orthodox Ψ-wave collapse at the instant of a quantum measurement. Ψ-wave function is here an objective entity. Yes, the price paid is the logically unsound 'infinite regress' of the infinitely 'many worlds' inanity. While such a metaphysic is 'logically possible', it utterly exceeds the reach of empirical possibility, not to mention our naïve common sense Realism notions of what appearing realities can be.

Therefore, the MWI hypothesis, along with its Ψ theoretical rivals, is beyond the reach of empirical verification. And physics is an empirical science. You can see the problem. Are we here grasping at ontological straws? Recent Quantum Field Theory (QFT/QED) just gets "curiouser and curiouser". [Lewis Carroll] But maybe that's not so bad. After all, metaphysical ontology is, almost by definition, 'post-empirical'. QFT and its scary MWI interpretation open new and uncommon doors of perception. So why bother?

Because it gets worse. We saw above that the purely physicalist/mechanistic, super-deterministic MWI is fast becoming the prevailing quantum ontology among frantic quantum physicists and quantum cosmologists desperate for an objectivist realist/physicalist ontology—not because it makes good sense, or is believable, or is empirically sound, or verifiable—but because it is 'logically consistent' and cannot be proved logically contradictory. But neither can the absurd hypothesis that the notorious 'quantum measurement problem' can be resolved by application of quantum zero point vacuum energy (ZPE) fairy dust sprinkled upon a Geiger counter at precisely the instant of Ψ-wave collapse at the moment of a quantum measurement be proved logically contradictory. Empirically absurd yet logically sound. An adequate quantum ontology?

Nor should QFT in any of its other interpretations be construed as violating the axioms of formal Aristotelian logic. They do not. QFT is counterintuitive and bizarre but not logically contradictory. It works perfectly in making stochastic predictions that give us computer technology, our TV and smart phones, and intercontinental laser communications. The inherent subjectivity of QFT is connecting our objective conceptual and subjective contemplative worlds! MWI is little more than a 'spanner in the works'.

Formal logic is as Kant would say 'analytic'. It gives us no new practical 'synthetic' information, or understanding. Classical deductive logic is two-valued and truth functional. Either true or false. Either A or not-A but not both. It only tells us if a conclusion logically follows from its premises. It yields no further information as to the truth or the utility of a formal logical syllogism's basic premises. Human knowledge requires a broader, deeper understanding than formal logic can provide. Just so, MWI needs more than formal logical soundness. Logically unsound infinite regress or not, MWI requires a bit of veridical empirical possibility, not to mention empirical non-absurdity.

That the quantum wave function of QFT is monumentally counterintuitive is not the real problem. Copernican heliocentrism breeched common sense in 1543 when it replaced geocentrism. As did Einstein's revolutionary counterintuitive Special Relativity Theory in 1905. The *London Times* called relativity "An assault on common sense."

Perhaps we are overly troubled by the merely conceptual random non-causal aberrations of QFT. Perhaps there are more things in heaven and earth than are dreamt of in our dualistic conceptual philosophies. Going beyond our ideologically comfy observer-independent Scientific Local Realism to a nominalist antirealist, even idealist observer-dependent panpsychic 'monistic primary cosmopsychic' metaphysic requires, in the 21st century, an all too unfamiliar human intelligence, and real cognitive courage. [Not likely if we are to have any hope of tenure.]

Yes. The classical ideologies and cognitive biases of the past are no longer adequate to our emerging post-quantum 'post-empirical' cognosphere.

The extreme objectivism of MWI has pointed out that subjective quantum mechanics needs a centrist foundational quantum ontology that exceeds and grounds its mathematical formalisms. Both MWI and QFT have unwittingly opened a 'post-empirical' door to the nondual infinite—noetic primordial awareness-consciousness ground of our being here in space and time—that we may enter in a contemplative cognitive dimension where noetic primordial wisdom outshines our no longer adequate 'scientific' beliefs and formal logical strategies of the past. Let us now further explore that brave new world.

Post-Quantum Paraconsistent Logics and the Wisdom of Uncertainty

Expanding Our Logical Horizons. We are not stuck with empty classical logic. Clearly, we require a three-valued quantum deductive logic (3VL), or a multi-valued quantum logic (MVL) which replaces the 'law of excluded middle' of the Greek classical deductive two-valued (either true or false)'truth functional' logic of Aristotle.

For example, the Hindu Nyala system—A, not-A and a third indeterminate value—permits *both* A and not-A; e.g. both relative existence and ultimate nonexistence to manage the cognition of both quantum and Buddhist acausal reasoning. Lama Professor Anne C. Klein (2006) has referred to such postformal cognition as "the logic of the non-conceptual". Yes, human cognition must include both faces of our *noetic cognitive doublet*—both objective Science and its perfectly subjective Spirit ground.

Thus do we require a contemplative 'logic of the non-conceptual' as it has arisen in Hindu and Buddhist philosophy that altogether transcends classical deductive and inductive logic. Spooky indeed.

[We shall continue our exploration of paraconsistent logic below in our discussion of Kurt Gödel's Incompleteness, and Intuitionist logical systems.]

The Quantum and the Wisdom of Uncertainty. Therefore, until we move beyond the obsessive classical scientific bias for naïve extremist absolute objective certainty—*either* A *or* not-A, either absolute existence or absolute nonexistence—to interpret the non-objective quantum theory's view of a random observer-dependent cosmos, no propitious Quantum Gravity Theory (QGT) shall arise to unify the presently mathematically incommensurate two pillars of physics, namely, Einstein's classical General Relativity Theory (GRT), and Bohr's and then Dirac's post-classical Quantum Field Theory (QFT) with its Quantum Electrodynamics (QED). The very real 'quantum measurement problem' with its 'problem of an observer consciousness' requires a cognitive breakthrough that shall be facilitated via an integral Noetic Quantum Ontology that permits a QGT, if that is mathematically possible at all. The scary 'logic of the non-conceptual' that is the 'wisdom of uncertainty' shall here be required. Niels Bohr, revealer of the quantum Principle of Uncertainty/Indeterminacy and student of Taoism and of Buddhism understood this well.

Once again, we must somehow unify our *noetic cognitive doublet* that includes both the exoteric objective and the esoteric subjective dimensions of our human cognitive experience. Might the centrist Mahayana Buddhist Two Truths ontic dominant trope—the invariant *one truth unity* of Relative Truth and Ultimate Truth—offer a providential Middle Way? We have already seen that indeed it does.

Let us then approach this intimidating 'wisdom of uncertainty' and explore a centrist 'middle path' philosophy between the ontological extremes of absolute existence and absolute nonexistence; to wit, objective substantialist Scientific Local Realism/ Physicalism, and the antirealist, often subjective nihilism that is both Eastern and Western Absolute Idealism in our grail quest for the 'post-empirical' prior unity of relative objective Science and the perfectly subjective ultimate Spirit ground in whom it arises. Our quest for an inclusive centrist integral Noetic Quantum Ontology requires it.

Uncertainty, Complementarity, and Quantum Field Theory. We've seen that Bell's Proof (1964) or 'Bell's Inequalities', and eighteen more recent analogous 'Bell Tests' through 2022 have all shown that 'spooky' acausal quantum nonlocal entanglement is the truth of the matter. The classical causal Newtonian view of Scientific Local Realism—an *observer-independent* purely objective local absolutely existing separate, substantial, permanent spacetime real world out there (RWOT)—is in point of fact, *observer-dependent.* That is to say, our RWOT is not *ultimately* existent in the absence of the presence of a *relative* observing sentient consciousness; sometimes with quantum measuring instruments with decoherent pointers in hand.

We have also seen that in the quantum 'collapse' orthodoxy a conscious observer is required to 'collapse' Schrödinger's universal quantum Ψ-wave function revealing an objectively real electron entity eigenstate that somehow—QFT can't explain how—pops or fluctuates into 'zero point energy' spacetime stuff from a hitherto infinite 'superposed state' of ZPE quantum emptiness in which all possible states exist simultaneously. [Recall, matter is 'borrowed' gravitational energy.] 'Spooky' scientific metaphysic indeed. And yes, quantum pioneer Albert Einstein hated it to the end of his life.

For relativistic QFT/QED, the proto-physical light energy ($E=mc^2$) that fills the worlds with motion-matter-form clearly exists relatively, conventionally, yet it is entirely 'objectively random', non-causal, nonlocal and non-objective. It requires an observer's consciousness to reify and objectify it—make it really real. Thus does this quantum view parallel the causal Buddhist Two Truths centrist Middle Way Prasangika Madhyamaka ontology that is foundation of acausal *Ati Dzogchen*. [Boaz 2020 *Ch. VIII*]

Niels Bohr's philosophically antirealist quantum 'entangled nonlocal behavior' of light—Einstein's "spooky action at a distance"—violated Einstein's inner local realist sensibilities as expressed in the 1st postulate of his 1905 Special Relativity Theory (SRT), namely that the speed of light is relative, local and finite, not nonlocal entangled and infinite. Recall that for relativistic physics it is the relative finite velocity of an electromagnetic signal that bestows our really real *local* spacetime reality.

Yet, quantum nonlocality/entanglement seems to theoretically permit superluminal, faster than light transmission of an electromagnetic signal between two 'space-like separated' particles in a single quantum system separated by many light years. Yes, Einstein hated it—although he was, along with his pal Max Planck one of the founding fathers of the original 1900 quantum theory. In 1928 Bohr and Heisenberg, with a little help from Pauli and Dirac formulated what we now know as the Copenhagen Interpretation of the 'old quantum theory' that was to become Quantum Field Theory.

The seismic knowledge paradigm shift from Newton's and Einstein's classical mechanics to quantum mechanics occurred with the advent of the two complementary foundational theories of the Copenhagen Interpretation. These were Warner Heisenberg's Principle of Quantum Uncertainty, and Niels Bohr's Principle of Complementarity with its classical logic defying wave-particle duality. Our microcosmic reality is *both* a wave spread out in space, *and* a point-like particle; both A and not-A. These two great scientific minds forever changed 400 years of our Modernist 'classical view' of a local realist, perfectly objective, purely physical spacetime reality.

Indeed, Paul Dirac's Relativistic Quantum Electrodynamics (QED) enhancement of Quantum Field Theory (QFT) is the theoretical basis of the prodigious physics Standard Model of Particles and Forces (Λ-CDM); arguably the greatest intellectual achievement of humankind, although it is now generally considered by those in the trade to be essentially incomplete. QFT/QED remains to be unified with Einstein's classical GRT. And yes, we need a post-empirical centrist integral Noetic Quantum Ontology for that, and through that, in due course and by grace, a unifying Quantum Gravity Theory (QGT).

Be that as it may, 2021 discoveries in the physics Muon Sector seem to have demonstrated that our beloved Standard Model is not only incomplete, but it cannot be essentially correct. No surprise here. It is, as most theoretical physicists have suspected, a provisional theory. Indeed, are not all scientific theories provisional and incomplete, apprehensively awaiting that next more inclusive theory? The cognitive 'gestalt shifts' that resulted in the scientific revolutions of Copernicus, Kepler, Galileo, Descartes, Newton, Einstein, Bohr, Heisenberg, Gödel, Freud, and Jung are all cases in point.

Many theoretical physicists and most philosophers of physics understand that our providential Standard Model of Particles and Forces now abides upon the cusp of a new paradigmatic physics revolution. That QFT has failed to quantize the gravity of Einstein's GRT is a root cause of this new evolutionary 'Scientific Revolution' (Thomas Kuhn 1962, 1970, *The Structure of Scientific Revolutions*) in science and culture that is now upon us.

Middle Way Madhyamaka Buddhist philosophy seems destined to play an ontological role in this arising *Noetic Revolution in Matter, Mind and Spirit.* [Boaz 2023] It should come as no surprise that both Bohr and Heisenberg, as well as Schrödinger, were inspired students of our Eastern noetic wisdom traditions—Hindu Vedanta, Buddhism, and Taoism.

Niels Bohr went so far as to include the Taoist black and white *tai chi/yin yang* symbol in his coat of arms placing it prominently upon the front gate of his Copenhagen estate. This ancient icon represents the interdependence, complementarity, balance, and prior unity of dualistic opposites arising within the vast boundless whole itself (Tao, *Wu, Mu, dharmakaya*). To wit: light and dark, positive and negative charges/forces, true/false, active (*yang*), receptive (*yin*), unity/duality (the Two Truths), existence/nonexistence, nonlocal wave and local particle, A and not-A. The dualities perforce appearing in relative conditional space and time are inherently complementary, like light and dark. You can't have one without the other. That ancient wisdom was revealed to recent science as a result of Niels Bohr's prodigious Principle of Complementarity.

We have seen that Herr Professor Einstein engaged his intellectual equal and philosophical nemesis Niels Bohr in the justly famous two decades long 1936 Einstein-Bohr (EPR) debate over the fundamental nature of reality—Einstein's observer-independent objective causal substantialist Local Realism against Bohr's observer-dependent subjective acausal nonlocal nominal Antirealism.

Does the random-acausal, nonlocal, quantum uncertainty and complementarity of Heisenberg and Bohr trump Einstein's and Newton's classical GRT causal, objectively certain local RWOT? It seems *prima facie* that these two rough hewn ontologies—Realism/existence and Antirealism/nominal or even nonexistence—are mutually exclusive metaphysical views. But are they? Might they be complementary views of a greater, holistic more inclusive post-quantum ontological ultimate reality awareness-consciousness ground of the formless, all subsuming vast unbounded whole itself?

Mereological Clarity. Clearly, there necessarily 'exists' such a nondual ground. Mereologically (part-whole relations) relative space-time parts are perforce embraced in a more inclusive primordial ultimate whole, or nondual 'Basic Space' (*dharmadhatu, chöying*) by whatever grand name, in which or in whom this all arises. Parts are subsumed in a greater more inclusive whole; wholes necessarily embrace their participating parts. Simple basic part-whole logic.

Spacetime form arises in some ontologically prior yet phenomenally present formless, timeless, selfless ground—unbounded all subsuming whole itself, name it as we will. Indeed, that trans-conceptual nondual primordial awareness-consciousness ground is the basis of human awareness-consciousness arising and instantiated therein. The perennial noetic wisdom tradition of humankind is an expression of our species wisdom quest to realize our always already present connection to That (*tathata*). Let physicists consider this great truth in their inchoate ontological 'foundational interpretations' of the universal quantum wave function, arising in that aboriginal ontological ground.

As we deepen our understanding of Buddhist Two Truths philosophy/practice we begin to see how it is that the primordial boundless whole of nondual reality itself as it arises in space and time both relatively exists, yet ultimately does not exist. That is to say, all this arising spacetime located local and nonlocal stuff exists relatively, conventionally, scientifically, yet not ultimately or absolutely, as we have so often seen in these pages.

To accomplish such an understanding we shall need to enhance Aristotle's Greek dualistic two-valued logic—*either* true *or* false, either A or not-A—with an Eastern or Western three-valued logic (3VL) that logically permits *both* of these truth values: both A and not-A, both existence and nonexistence, both wave and particle. Why? Because this is the way that the things of our dualistically experienced reality appear to exist.

The semiotic logical syntax of language is necessarily 'truth functional' or two valued—either true or false, either existence or nonexistence, but not both. We require a cognitive currency that permits the subtle nuances of our actual human experience. Our perennial Two Truths—relative form and its ultimate emptiness ground—are a prior ontic, changeless, nondual, noetic, invariant through all cognitive references frames *one truth unity* (*dzog*)—an ontological constant as it were. Such a view is a nondual 'post-empirical' wisdom view. It may suffice as an antidote to the false dichotomies of our all too common sense perspectival dualistic concepts, beliefs, and cognitive biases—particularly those beliefs that we are certain are 'true'. As American truth expert Mark Twain told:

> It's not what you don't know
> that gets you in trouble.
> It's what you know that just ain't so.

Einstein's Local Realism versus Bohr's Antirealism. Please consider that Metaphysical Local Realism/Physicalism—that physical stuff is a relation of one to one 'correspondence' to an external, objective, *observer-independent* reality—is a view as to the way that the dimension of spacetime Relative Truth exists. For Bohr's Copenhagen nominalist Antirealism and for Metaphysical Idealism this relative external objective 'real world out there' (RWOT) is hypothetical, nominal and not assumed necessarily to exist absolutely/ultimately 'from its own side'.

Thus is Antirealism, Philosophical Idealism, and primary 'priority monistic panpsychic *Dzogchen Kosmopsychism* a conceptual condition or cognitive bridge that is a path to the perfectly subjective dimension of all-embracing one truth unity of our perennial Two Truths (relative and ultimate). That is nondual primordial *ultimate* Spirit ground in whom *relative* Science arises and participates. Is it a cognitive bridge too far? For scientists and philosophers conceptually steeped in the cognitive bias that is Metaphysical Scientific Local Realism/Physicalism it is strong medicine—a bitter ontic pill indeed.

The great Einstein never overcame his cognitive bias for an objective Local Realism metaphysic, a non-random, inherently cause and effect proto-theistic locally real spacetime cosmos. Quantum 'objective randomness' meant for Herr Professor Einstein that his theistic Creator God had no choice. He famously told Bohr in one of their heated exchanges, "God does not play dice with the world!" Bohr is reported to have retorted, "Oh Einstein, stop telling God what to do with his dice!" The completeness of the new 'always correct' quantum theory, and therefore the future of 20th century physics was at stake.

The 1936 EPR (Einstein, Podolsky, Rosen) debate between Bohr and Einstein persists in the quantum physics of today. The core issue—quantum entanglement, nonlocality, 'space-like' faster than light speed universal interconnectedness—remains still in our century long grail quest for a viable objective interpretation or foundation of an inherently subjective quantum theory description of spacetime reality. Recall that in 1964 Irish physicist John Stewart Bell 'proved' the existence of nonlocal quantum entanglement.

What has become abundantly clear is that we require a centrist, even noetic view that mediates or bridges between undoubtedly relative conditionally 'real' causal observer-independent Scientific Local Realism (cause and effect phenomenal reality), and acausal observer-dependent quantum entangled nonlocal reality. A viable, multidimensional centrist integral Noetic Quantum Ontology must provide that.

Let us then further explore the ontological unity of the Buddhist Two Truths, relative and ultimate—a *relative* objectively really real world out there (RWOT), arising in a nondual perfectly subjective primordial awareness-consciousness *ultimate* Spirit ground.

The Unity of Science and Spirit: Toward a Centrist Noetic Quantum Ontology

A Middle Way Buddhist Foundation. Middle Way Prasangika Madhyamaka Buddhist ontology views all appearing spacetime form as relatively, objectively, causally, scientifically real; yet subjectively, ultimately it is nondual Spirit itself—formless, timeless, selfless 'Basic Space' (*dharmadhatu, chöying*) emptiness/ *shunyata* ground state—*Perfect Sphere of Dzogchen,* essence and nature of nondual primordial *dharmakaya* ground. Just so, spacetime 'form' is relatively, conditionally, conventionally objectively really real—just not ultimately real. That is to say, appearing stuff is "absent and empty of any iota of absolute intrinsic existence ... Yet we must respect its relative conventional existence." [Nagarjuna]

Now the Middle Way Buddhist absence of form's absolute ultimate existence is still an absence of *something*. And that absence exists. The absence of an elephant in the room logically implies the existence of at least one elephant somewhere. So, boundless emptiness exists! It's not a mere negative 'void'. Empty spacetime form has a vivid lucent quality—an inherent luminosity over and above its inherent emptiness. Buddhist emptiness is itself 'empty of *ultimate* existence'.

How then does emptiness (*shunyata*) exist? We are told by the wise that emptiness exists not 'intrinsically', ultimately or absolutely as a vast cosmic container of spacetime stuff, but by way of human conceptual imputation and reification—in short, boundless Buddhist "emptiness is established by human conceptual minds". [H.H. Dalai Lama] That is known to Buddhist scholars as "the emptiness of emptiness", as we have seen.

Therefore, this Buddhist view of the Two Truths that are ultimate emptiness and its relatively arising form is not a philosophically idealist nihilistic denial of objective spacetime form altogether. If human beings in embodied form did not objectively exist who is it that enters in and practices the Buddha's Eightfold Path to liberation from suffering? Who is it that practices compassionate *bodhicittta* for the benefit of living beings? Who is it that ponders Buddhist emptiness and quantum ontology in the first place?

Antirealist emptiness/*shunyata* represents a profound aspirational centrist mean between the relative existence (Scientific Local Realism) and ultimate nonexistence (Absolute Idealism) of emerging spacetime form. As we have so often seen, objective *relative* quantum form arises, appears, and participates within its perfectly subjective all-pervading *ultimate* quantum emptiness ground. An inchoate Noetic Quantum Ontology!

These Two Truths, relative and ultimate, of the Buddhist centrist 'middle path' must be viewed as an ontic prior yet phenomenally present unity. Our perennial Two Truths—Ultimate Truth, prior perfectly subjective, nondual, enfolded all-inclusive primordial awareness emptiness Spirit ground of everything (*paramartha satya*)—and Relative Truth, the Science of objective physical and mental emerging quantum spacetime matter/energy form (*samvriti satya*) continuously unfolds in and as this ultimate 'groundless ground', vast boundless all subsuming *kosmic* whole itself (*dharmadhatu*) in which or in whom all these cosmic participating parts mereologically arise. These two *kosmic* reality dimensions are always already a nondual primordial unified *one truth unity* (*dzog*), utterly indivisible, interdependent, and invariable throughout all relative-conventional human reference frames. A settled fundamental quantum ontology requires such a holistic unified view in order to move beyond its habitual realist/physicalist ideological bias; does it not?

The present state of quantum ontology—the foundational 'quantum mechanics interpretations' that we have briefly surveyed—have failed even to consider such an ultimate ontology, choosing instead to dwell in the comfy cognitive abode that our disconnected superficial physicalist concepts *about* the deeper reality have for 400 years of Modernity proto-religiously conceptually fabricated.

The Buddhist Madhyamaka centrist Middle Way ontology path is not at all dissimilar to some of the objectivist 'quantum interpretation' views. But a viable quantum ontology must balance the inherent ontological subjectivity of the nonlocal nondual acausal interconnectedness of the primordial superposed 'universal quantum Ψ-wave function'—quantum emptiness—with the objective reality of the seemingly separate infinitely abundant causal 'matter wave' functions of local spacetime form.

A robust quantum epistemology must build upon such a nondual primordial ontological basal ground. It must then transcend yet include its own inherently dualistic mathematical quantum formalisms in a centrist theory that describes the interdependent relationship of the ontological identity of the nonlocal nondual ultimate subjectivity of the universal quantum Ψ-wave function with its many objective physical and mental wave functional entities as they unfold in real relative spacetime existence from their prior timeless nondual primordial Spirit ground.

Hence, the spacetime *relative* global 'universal quantum wave function' arises and functions within, and has never departed its formless, nondual *kosmic ultimate* primordial ground, boundless whole of all cosmic appearing physical and mental spacetime form.

Such an ultimate quantum ontology is prior to, yet not separable from its relative unfolding epistemology which is its natural cognitive extension in a real relative space and time. We can no longer split, ignore, or deny the natural interdependence of these two utterly interdependent dimensional cognitive functions. To do so is a pernicious example of the infernal 'taboo of subjectivity'; and a profligate logical 'category mistake'.

We've often seen in these pages that until a settled nonlocal integral Noetic Quantum Ontology emerges, there shall be no Quantum Gravity Theory (QGT)—the great mathematical consummation that quantizes gravity, finally unifying the hitherto incommensurable two great theoretical pillars of Modern physics, namely, Albert Einstein's General Relativity Theory (GRT) and Dirac's and Feynman's Quantum Electrodynamics (QFT≈QED)—if such a purely formal mathematical intention is indeed logically possible at all. And if it is not, so much the worse for mathematical logic.

My Reader may have noticed that there is at work here in the process of the arcane discipline of physics a rather humorous *kosmic* irony. The imperious laws of physics work well in practice. Spacetime reality always spontaneously shows up, rough hew it as we will. No problem whatsoever. Now if only we could make these laws work in theory!

Well, is the *ultimate* nature of appearing reality local, observer-independent, objective and physical; or is it nonlocal, observer-dependent, subjective and immaterial? Such a false dichotomy has now become cringe-worthy. So how about a nice centrist middle way? Mahayana Madhyamaka Buddhists have done a good job with it.

The Nondual Primordial Ground of the Quantum Ψ-Wave Function. The real unity of objective Science and its perfectly subjective Spirit ground—the assiduous practice of that unity—will engage a subjective, integral Noetic Quantum Ontology that transcends yet embraces its objective, relativistic quantum mathematical formalisms. Irwin Schrödinger's 'universal quantum Ψ-wave function' perforce emerges from a more inclusive, mereological all-pervading, fundamental ontic ground; name it as you will.

The infinitely superposed quantum Ψ-wave is finally a proto-subjective *probability wave*! Indeed, spacetime itself with its atavistic foundational primal nonlocal entangled ZPE zero point vacuum energy field quantum fluctuations—formless aboriginal physical ground of spacetime form—arises from that basal nondual primordial awareness-consciousness Spirit ground itself, the 'Basic Space' (*chöying*) of all that appearing spacetime phenomena. Legendary quantum physicist and author of nonlocal 'hidden variables' quantum ontology David Bohm bespeaks it beautifully:

> The vast implicate order of the unbroken whole [is] the ground for the existence of everything . . . .In this flow mind and matter are not separate substances. Rather, they are different aspects of one whole and unbroken movement [the 'holomovement'] . . . .Wholeness is what is real, and fragmentation is the illusory response to [it] . . . . The notion that all these fragments are separately existent is an illusion . . . .Relativity and quantum theory agree, in that they both imply the need to look at the world as an undivided

> whole, in which all the parts of the universe ... merge and unite in one totality.
>
> —*Wholeness and the Implicate Order*

That holistic view is indeed the foundation for our post-classical integral centrist Middle Way Noetic Quantum Ontology. Thank your lucky stars for this great truth of nondual Basic Space (*chöying*) ground of reality Being Itself. Everything arises and depends upon That (*tathata*). All of it, and all of us are interdependently interconnected in this vast *kosmic* sea, unbounded whole itself, *ultimate* primordial dharmakaya ground of this great phenomenal gift of *relative* really real 'ontologically relative' spacetime stuff. We are not separate and alone but embraced in that natural *kosmic* womb of form. Great joy!

But please don't *believe* this! It is entirely beyond belief—our habitual objective cognitive concept/belief systems. Yet it is readily present via contemplative mindfulness meditation as the luminous Presence of that formless timeless original 'groundless ground'. The aboriginal ground of reality itself is 'groundless' because it is empty of a substantial conceptual base, e.g. an anthropomorphic theistic Creator God of concept and belief; and it transcends yet includes our realist/physicalist scientific biases. The paraphysical 'quantum zero point energy field' (ZPE) is as close as mathematical physics can come to this naturally arising formless timeless trans-conceptual fundament that is primordial "ground for the existence of everything". [Bohm] Noetic recognition, then nondual 'spiritual' realization of that basal matrix ground is accomplished through the practice of the compassionate love-wisdom mind of the contemplative path. [*Appendix A*]

Such a holistic view may remind us that this noble aspiration to noetic unity is always already accomplished deep within us! Objective Science and its perfectly subjective Spirit ground are always an ontological prior and phenomenally present one truth unity. It has always been thus.

From an ultimate view the primordial Two Truth reality dimensions—relative spacetime form and its nondual ultimate spacious emptiness ground—are utterly indivisible. And now we can see it (*samadhi*)! It is That (*tathata*) *buddhic* Presence to which we awaken via Buddha's 'mindfulness of breathing'. It is in this liminal cognitive boundry space that the real work begins.

Is Spooky Quantum Nonlocal Entanglement Really Real? We have seen that beginning with John Stewart Bell's 1964 Bell's Theorem, and 50 years of numerous (18) nonlocality physics 'Bell Test' experimental loophole closing confirmations, quantum nonlocal entanglement is now considered by the protagonists of this 100 year scientific drama to be 'scientifically' proven. The Einstein-Bohm last gasp conjecture for a nonlocal 'hidden variable parameter' to 'save the appearances' of Scientific Local Realism now stands refuted for theoretical physicists who bother to study it, and for nearly all philosophers of physics. [Boaz 2022 *Ch.* 7]

Or does it? What is the relative seed of common sense empirical truth in Scientific Local Realism that cannot be credibly denied? Whether spacetime stuff is *ultimately* real—Middle Way Buddhists deny that it is—appearing objective reality is at least *relatively* really real! After all, here we are, along with real trees and stars. Middle Way Buddhists agree. Whether or not we impute and reify quantum abstractions into existence, spacetime stuff and beings are everywhere! What could possibly be the existential reality status of an embodied mind that denies its own existence? The yoke of the burden of rejoinder for antirealist and idealist skeptics is heavy indeed. Still, there exist 18 null Bell Tests.

We must avoid the false epistemic dichotomy that insists that the nature of appearing reality be *either* ultimately real (Metaphysical Scientific Realism/Physicalism), *or* ultimately illusory (antirealist Metaphysical Idealism)? There is a centrist 'middle path' between these metaphysical extremes of absolute existence or absolute nonexistence. That path was expressed by Gautama the Buddha of this age. And now we understand it!

Recall that from the metaphysical ontology you choose arises the cause and effect karmic phenomenal reality you deserve. Perhaps it is better to err on the side of ontic and epistemic holism; and the altruistic, compassionate happiness-inducing *bodhicitta* conduct that arises here in relative time.

The mundane academic history of religion and philosophy, both West and East, may be seen as a dispirited history of such a false absolutist distinction between *either* objective, local realist, monistic Metaphysical Local Realism/Physicalism; *or* subjective, antirealist monistic Metaphysical Absolute Idealism. But this bedeviled history of our species' quest for some absolute, even theistic objective certainty that will bestow human happiness has been necessary in order to arrive, individually and collectively, at our present liminal developmental juncture—the turning point toward scientific and spiritual wholeness—and the happiness—both human flourishing and harmless ultimate Happiness Itself that abides herein.

A Centrist Middle Way. Clearly, our Noetic Quantum Ontology requires a centrist middle way ontology between these ontic metaphysical extremes that present to dualistic thinking mind as *either* objective existence *or* subjective nonexistence. We have seen that Mahayana Prasangika Madhyamaka philosophy and practice—conceptual causal foundation of acausal nondual highest *Ati Dzogchen*—has profoundly accomplished such a pragmatic yet integral (all the parts necessary to complete a whole) centrist understanding.

Quantum Field Theory has unwittingly pointed to such a middle way through its distinction between the spooky *subjective* nonlocality of the superposed universal quantum Ψ-wave function prior to its collapse—'consciousness causes collapse' (Wigner)—into a safe and sane local *objective* micro and macro reality, our 'real world out there' (RWOT) that we have all come to know and love. Here our objective *observer-independent* realities are neatly objectified/reified via an inherently subjective process of an 'ontologically relative' *observer-dependent* consciousness. Two Truths one truth unity indeed.

Our East-West human wisdom project has—for at least 10,000 years—been to both conceptually and contemplatively understand the prior, already present unity of these two all too human cognitive modalities, our *noetic cognitive doublet* that is objective conceptual, and nondual perfectly subjective contemplative. Without conscious beings to consciously ponder such questions there is no 'problem of consciousness', no 'quantum measurement problem', no form, no emptiness, and indeed no sentient experience/consciousness at all.

Thus is our primordial awareness-consciousness emptiness ground a "groundless ground". Recall that for Mahayana Buddhist philosophy the ultimate nature of emptiness/*shunyata* is that, while it is *relatively* conventionally real, it is in itself absent and empty of any whit of intrinsic *ultimate* existence. Emptiness is 'groundless' not as an important cognitive source of both conceptual and contemplative understanding about the inherent basal nature of appearing reality, but as a substantial kosmic entity 'out there' that creates spacetime stuff. Buddhist emptiness is therefore 'established' not by metaphysical necessity, but by pragmatic human conceptual designation.

Basic Space Beyond Emptiness. For great 14th century *Dzogchen* master Longchenpa emptiness is a metaphor for the prior all subsuming "infinitely extensive, naturally occurring timeless awareness, primordial basic space of phenomena" that transcends yet embraces it. Longchenpa speaks in his monumental *Chöying Dzod* (2001):

> Basic space is buddha nature—buddhahood that is spontaneously present by nature ... beyond supreme emptiness,

> sublime knowing, ancestor of all buddhas... unborn awareness... All arising form and wisdom is the adornment of unborn basic space... single mandala of naturally occurring timeless awareness... spontaneous equality... self-arising original wakefulness... Awakened mind, marvelous and superb...

Basic Space (*chöying, dharmadhatu*) may be understood as the noetic nondual primordial awareness-consciousness emptiness ground of all arising spacetime phenomena, beyond human concept and belief about it, but not beyond the direct contemplative feeling/knowing experience of our "original wakefulness"—"naturally occurring timeless awareness"—that is our already present "awakened mind".

To be sure, objective Science—physics, cosmology, neurobiology—and its perfectly subjective Spirit ground require a propitious Two Truths ontology, objective relative and nondual ultimate that is the 'grounding relation' of such an objectively fluent epistemology. We shall further engage this relation below. As David Bohm explained, "The unbroken whole is the ground for the existence of everything... an undivided whole in which all the parts of the universe merge and unite in one totality."

Thus is Heisenberg's transitional *schnitt* or 'explanatory gap' between objective local phenomenal experience and subjective, noetic, nonlocal, nondual contemplative experience bridged, at least conceptually. And it is unified non-conceptually through the mindful contemplative, often nondual direct experience (*yogi pratyaksa*) that is the subject of the emerging discipline known as Contemplative Science. [Boaz 2022, *Ch. IV*]

Perhaps in this monumental grail quest for human absolute objective certainty we shall discover a unifying, if ever incomplete theory that unites the objective classical relativity of Einstein's local GRT with the inherent subjectivity of quantum nonlocal QFT/QED. I have here and elsewhere suggested the cognitive architecture for such a Noetic Quantum Ontology. No doubt that the formalist mathematics of such a consummation—if logically possible at all—shall add greatly to the relative human happiness of theoretical physicists; and as well add confidence to the quest of contemplatives for ultimately subjective nondual certainty, which on the accord of the Buddhas and *mahasiddhas* of our wisdom traditions is "already accomplished from the very beginning". [*Dzogchen* founder Garab Dorje]

Our perennial Two Truths—relative and ultimate—beget two corresponding and interdependent quests for certainty; until that is, the seeking subject ends the fruitless goal (*Wu-Wei*) for a desired object of its quest. Now, in the ultimate realization and peace of *buddic* 'no more learning' dawns the wisdom of non-seeking. Told Gautama the Buddha of this present age, "Wonder of wonders all beings are already Buddha." Told Jesus the Christ, "That which you seek ... the Kingdom of God is already present within you ... and it is spread upon the face of the world, but you do not see it." [Luke 17]

As to Einstein's "something deeply hidden", the interdependent centrist relationship between relative Science and ultimate Spirit—this prior and present unity of our perennial Two Truths—John Stewart Bell, the most profound quantum physicist since Irwin Schrödinger has pointed to the truth of the global universal quantum wave function that abides at the nondual heart of the quantum enigma. He shall have the last word:

> Suppose that quantum mechanics were found to resist precise formulation. Suppose that when formulation beyond FAPP [for all practical purposes] is attempted, we find an unmovable finger obstinately pointing outside the subject, to the mind of the observer, to the Hindu scriptures, to God, or even only Gravitation? Would that not be very, very interesting?

The Standard Model of Particles and Forces

In order to understand our foundational integral Noetic Quantum Ontology it will be useful to know the relation of the present Standard Model of Physics to its Quantum Field Theory.

We have seen above that the Standard Model of Physics (*lambda* Λ-CDM or LCDM) is an elegant yet incomplete mathematical program based in the fundamental symmetries of nature that describes the interactions of microscopic elementary particles in a context of the fundamental forces or interactions of nature, namely the electromagnetic, strong, and weak nuclear forces/interactions. Sadly, the Standard Model of postmodern physics is unable to include the fourth force/interaction, namely Einstein's entropic gravity.

The Standard Model is the foundational conceptual bridge into a quantum mechanical understanding of physical spacetime reality, namely Relativistic Quantum Field Theory (QFT, QED). It presumes to describe three of the four known fundamental forces or interactions of physical reality, and the discovery and classification of all present and future elementary particles that instantiate, connect, and animate these physical forces/interactions.

However, the gravity force/interaction with its force carrier graviton particle is conspicuously absent in our Standard Model of Physics. *Thus does the noble Standard Model fail to explain roughly 95 percent of the physical stuff—dark matter and dark energy—in our known universe!* It seems we need a Super Post-Standard Model that includes gravity; to wit a Quantum Gravity Theory (QGT) that unifies the mathematics of our 'two perfect theories' of physical form, namely, Standard Model Quantum Field Theory (QFT) with the geometric gravity interaction of Einstein's General Relativity Theory (GRT).

Well, what are these wondrous 'four forces of nature'? The electromagnetic force (symmetry U(1), infinite range); the strong nuclear force (symmetry SU(3), range of a proton length); the weak nuclear force (symmetry SU(2), range 1/1000th of a proton distance); and the gravity force (infinite range) with its subatomic force carrier the graviton particle, discomfitingly absent in the Standard Model of Physics.

Gravity is the weakest yet most pervasive of the four forces. It acts upon all forms of mass-energy, and thus upon all subatomic particles, including the gauge bosons that carry the forces. Great gravity is the geometric interaction that binds together the worlds—"creator and destroyer of worlds".

Matter particles (fermion quarks) are effected by all three subatomic forces while matter particles (fermion leptons) are effected only by electromagnetism and the weak nuclear interaction, but not the strong nuclear force. Photons are bosons with zero charge.

The physics Standard Model is buttressed by experimental confirmation of the existence of fundamental quarks, including the 'top quark' (1995), the tau neutrino (2000), the Higgs boson (2013), and the hopeful unification of the electromagnetic and weak nuclear forces via the 'electroweak theory' (Glashow, Feinberg 1967) to give us the now 'renormalizable' electroweak force/interaction (1971 Gerard 't Hooft).

However, while theoretically self-consistent, the Standard Model does not explain Einstein's GRT and so does not presume to be a complete theory. It has failed to include a description of the gravity interaction of Einstein's General Relativity Theory (GRT) and thus of the accelerating expansion of this universe due a most mysterious 'dark energy'—probably Einstein's 'cosmological constant lambda Λ'—virtual ZPE 'zero point vacuum energy of empty space', which constitutes fully 68.3 percent of the universe. [Is dark energy constant, or does it increase in cosmological time?]

Nor does the Standard Model provide a viable 'dark matter' particle that may well constitute 26.8 percent of the matter of this universe. Ordinary baryonic matter—atoms, trees, people, stars, galaxy superclusters and vast galactic filaments—comprises only 4.9 percent of the physical whole cosmos shebang! Has the present Standard Model of physics relegated quantum cosmology to the outer darkness?

So, the wondrous Standard Model of 21st century physics, while including quantum mechanics, has failed to describe Modern physics' 'dark sector'—dark energy and dark matter—which comprises fully 95 percent of our 'observable universe'! Perhaps then we should refer to such Standard Model hubris as the Substandard Model.

The finite velocity of light and the finite age of this present universe of ours create an absolute cosmological 'observation horizon' beyond which we cannot see. Cosmologists can never observe the physics regimes beyond this horizon. Although the James Webb Space Telescope (JWST) is quickly changing this game.

Our visible universe is 'homogeneous'; it appears about the same everywhere. Is it the same old cosmic stuff beyond our observation horizon? Here we enter in the highly speculative universe of discourse known as the 'cyclic multiverse', as we shall soon see.

Moreover, the Standard Model of Physics fails to predict the masses of quarks and leptons; and the strength of its three forces relative to distance. Still, it remains the speculative foundation of the physical stuff of appearing reality upon which the 20th century quantum paradigm, namely, QFT/QED/QCD, constructs its formalist mathematical description of that reality.

The Standard Model works quite well for our microcosmic subatomic world. And Einstein's GRT equations work quite well with the vast intergalactic macrocosmic world of 'Big G' gravity. Bad news again. These two wondrous foundational theories—the Standard Model and Gravity—produce 19 physical parameters which we cannot derive from the basic principles we are given by nature. Of these 19, 18 arise from the woefully incomplete Standard Model. The 19th is the parameter of the strength of gravity as a function of distance. The 18 as yet to be derived Standard Model parameters include the unknown masses of the six quarks and three charged leptons; the strength of the Higgs field (its mass is now known); four parameters regulating quark decay; the strength of the three subatomic forces; and finally the QCD empty field when matter is not present.

What happens if the actual but unknown values of these parameters changes a tiny bit? With most of them, not much. But if the masses of the 'up' and 'down' quarks were different than they are, heavy elements like carbon and oxygen could not form in stars because stars could not form. And thus the supernova heavy element ejecta (oxygen, carbon, iron) stardust dimension of matter, and smart embodied people who speculate about it, could not arise in the first place. [The Anthropic Principle below]

Dark Energy: The Cosmological Constant?

Dark energy—theorized to be the resurrection of Einstein's cosmological constant (*lambda* Λ) of his pre-1932 GRT field equations—is the vast basal energy field that pervades this whole universe of ours. It has its primal origin in the ZPE 'quantum zero point vacuum energy field' (aka 'vacuum energy' or 'quantum foam'). This vacuum energy is not the result of a dynamical field. It is an intrinsic geometric property/quality of 4-D spacetime itself. Thus it is constant and does not increase or decrease as the space of the physical universe and its stuff expands, which we now know that it does (Hubble's Law and Einstein's GRT). Indeed, we now know that this expansion of spacetime is accelerating exponentially! [1998 Riess; and Schmidt-Perlmutter; 2011 Nobel Prize in Physics].

On the accord of Quantum Field Theory (QFT/QED), within the proto-physical foundational universal ZPE zero point vacuum energy field arise minute uncaused, utterly 'objective random' quantum fluctuations in this primordial universal field that continuously gives rise to 'virtual particles' that "pop in and out of spacetime existence". Thus do quantum vacuum fluctuations contribute to the energy density value of the cosmological constant. (Carroll 2003)

Einstein's cosmological constant Λ *lambda* is here seen as a constant energy density for this entire universe that is constant or changeless in time and in space. Vacuum energy is the absolutely constant volume of energy present in every cubic centimeter of space—the timeless energy of space that emerges and persists through all cosmic time. Its value is very, very small, but not zero. So the vacuum of 'empty space' is not entirely empty. Heisenberg's Principle of Uncertainty precludes the existence of a 'perfect vacuum' in nature. How? Quantum uncertainty permits nearly eternal, uncaused (beyond the law of cause and effect) utterly 'objective random' vacuum fluctuations giving rise to matter-energy. They are not only predicted by QFT, their effects have been observed. So, the ZPE quantum vacuum energy is smoothly distributed and constant in its density.

Although we cannot calculate the precise contribution to the vacuum energy density by the quantum vacuum fluctuations, we can estimate its value. Sadly, our best theory estimates are 120 orders of magnitude greater than the *observed* dark energy vacuum density! This is the prodigious *cosmological constant problem*, as we shall soon see.

Well, what if it turns out that our constant vacuum energy density is not so constant after all? What if the quantum vacuum energy were zero instead of slightly positive, and not cosmologically constant, but dynamical and subject to change in time. Then dark energy must be something else entirely.

Quintessence. This idea is grounded in the theory known as *Quintessence* (1998 Steinhardt and Caldwell)—an alternative to the ZPE vacuum energy—posited as a 'fifth fundamental force' as it couples with ordinary matter. Dark energy is here generated by a vast, slowly evolving dynamical bosonic fundamental field of nature wherein the energy density is nearly constant as the universe evolves and expands. It differs from the slightly positive 'cosmological constant' explanation of dark energy in that it is not a constant of nature but a natural dynamic force changing in time.

Quintessence may be gravitationally attractive or repulsive. It is said to have become repulsive—accelerating the recession of the universe—3.5 billion years following the empirically impossible 'Hot Big Bang' singularity that ostensibly began this ever-present universe of cosmic spacetime stuff in which we all arise.

If the Quintessence dark energy density of the cosmos is very slowly decreasing the universe may still be accelerating but the Hubble Constant H (Hubble Parameter) will gradually diminish.

[*H* is the ratio of galactic recession velocities to distance; the greater the distance away, the faster the recession.] In this 'fate of the universe' scenario our spacetime universe gradually expands until its thermodynamic energy properties become zero— 'heat death'— the proverbial *Big Chill.*

If dark vacuum energy density is increasing (w = -1 'phantom energy') the Hubble Constant increases and the accelerated expansion of this universe will increase slowly, finally overcoming the gravitational attraction of local galaxies. The dismal result is that all mass is torn asunder, right down to its very atomic structure. This terminal state of cosmos we have come to know and love as the *Big Rip.*

If the dark vacuum energy density is decreasing (w = +1) the universe will slowly contract under the force of gravitational attraction ending in a *Big Crunch,* and perhaps conditions that facilitate a reflexive new Big Bang and a bright new universe within an infinite cyclic multiverse (conformal cyclic cosmology). Indeed, our present universe may be a timely exemplar of this natural gravitational process. So, in dependence upon the prodigious vacuum energy density universes may expand or contract.

Let us hope that our cosmological constant remains more or less constant. In any scenario self-conscious beings, embodied or not, have three trillion years (give or take a trillion) to ponder it.

The Cosmological Constant Problem. Einstein's non-zero cosmological constant *lambda* Λ is the simplest and the prevailing explanation for the mysterious dark energy vacuum energy density that drives the accelerating expansion of this universe. It acts to counter the attractive force of gravity—"to hold back gravity" as Einstein told it—driving large scale galactic structure apart rather than binding it together. So here its theoretical value should be consistent with astrophysical observations. Unfortunately, as we have seen, the quantum theoretical value is greater than the actual observations by as much as 120 orders of magnitude! No one has a clue how. "The worst theoretical prediction in the history of physics." [Steven Weinberg]

If the vacuum dark energy *lambda* Λ were large and positive the space of the universe would accelerate such that atomic structure could not exist. If Λ were large and negative cosmos would instantly collapse. So, if Λ were not just as it is, we would not be here to ponder its value. Once more we are led to the notion of a cyclic *multiverse* in which 'environmental selection' determines a universe with physical constants that permit a hospitable solar system 'goldilocks zone' that furthers the emergence and evolution of self-conscious beings who may inquire about such things. The Anthropic Principle again. Yes. We exist as a function of a universe that must be 'fine tuned' to support life forms.

Although we cannot precisely calculate the predicted vacuum energy contributed by ZPE quantum energy fluctuations, we can estimate its probable value. The bad news. The *cosmological constant problem* is the huge discrepancy between the small actual *observed value* of the vacuum energy density—the tiny value of the cosmological constant—and the very large predicted *theoretical value* of the zero-point energy field calculated by Quantum Field Theory (QFT). This fundamental dark energy quantum ZPE field, the Universal Quantum Vacuum in which matter arises—it constitutes nearly 70 percent of all the matter in the universe—is known to be extremely small. Yet we have just seen that it's energy density is as high as 10^{120} orders of magnitude greater than that predicted by QFT/QED calculations! Big problem! What's going on here?

From cosmological observations of type 1a supernovae, and from the cosmic microwave background radiation (CMB) physicists can directly measure the total energy density of the universe, including the ZPE vacuum energy. And yes, this observed, measured value is tiny in relation to QFT calculated predictions of the vacuum energy. Thus is revealed the 'cognitive gap' in physics between observation and theory. Too often we lack a cognitive bridge between scientific observation and scientific theory. Reality presents itself for our observation and measurement. If only we could get our conceptual theories about it to agree with what we observe! Discomfiting to physics, to say the least.

In short, QFT has demonstrated the existence of the vacuum energy—nonlocal entangled quantum emptiness—and that the vacuum must experience reality constituting quantum fluctuations. GRT tells us that those fluctuations are energetic and so add positive vacuum energy to the total energy density of the cosmological constant Λ. This tiny energy drives the acceleration of the universe. Yet when we calculate it, it's enormous. Even when Lorentz invariance is accounted for, the calculated discrepancy is still 60 or so times greater than the observed value. This anomalous "worst theoretical prediction in the history of physics" has become known to physicists as the quantum "vacuum catastrophe". Not so good for our confidence in recent Standard Model physics.

Paraconsistent mathematical 'renormalization'—beyond the scope of our present inquiry—is usually seen as an *ad hoc* dodge of the cosmological constant problem, not as a solution. What to do?

Proposed solutions to the *cosmological constant problem* have thus far failed. Modifying Einstein's sacrosanct gravity (GRT) equations is required, but mostly taboo. Gravity is 'too perfect', even though it does not work at vast intergalactic distance scales, such as the one we are now considering. Theoretical efforts to 'modify gravity' are therefore an important sector of theoretical physics and will perhaps one day succeed as we move toward a paraconsistent logic required by a Quantum Gravity Theory (QGT).

Mordehai Milgrom's 1983 Modified Newtonian Dynamics (MOND) is still a viable, if *ad hoc* explanation of *dark matter* for explaining the 'missing mass problem'; why large scale galactic structure does not obey Newton's and Einstein's Law of 'Big G' Gravity. Either there exists in galactic structure huge quantities of unseen matter (dark matter), or Newton's, and thus Einstein's gravity must be modified. The *ad hoc* 'Milgrom's Law' must now be integrated into a complete theoretical system. In 2004 Jacob Bekenstein took a big step through his TeVeS non-Einsteinian relativistic generalization of MOND. Yet conceptual problems still abound. Indeed, all useful theories await that next more complete theory. This must include our most beloved theories, even GRT and QFT/QED.

We have seen that perhaps the most fundamental constraint in this regard is our mostly hidden unconscious metaphysical cognitive bias, namely locality and reductive Physicalism as embodied in Metaphysical Scientific Local Realism and Metaphysical Scientific Materialism/Physicalism.

Modified gravity? Einstein himself modified his own 'perfect' GRT field equations—twice. In 1907 he inserted the non-zero cosmological constant (*lambda* Λ) into his field equations to "hold back gravity" and ensure the static, non-expanding 'steady state universe' of the 'global web of belief' of early 20th century physics. Then in 1931 he deleted Λ upon Edwin Hubble's 1927 discovery that our universe is indeed expanding. Perhaps Einstein should have trusted his original 'perfect' gravity equations. Einstein is said to have called his tweaking of what is often referred to as 'a perfect theory', "the greatest blunder of my life". Natural near universal cognitive 'confirmation bias' regularly infects even the greatest minds. Present company excluded, of course.

But the old tired cosmological constant lambda Λ was not yet finished. In 1998, upon the discovery of the acceleration of Hubble's expanding universe, non-zero *lambda* Λ arose again, phoenix-like from the ashes of the prodigious GRT field equations to outshine as the dark energy cause of our rapidly accelerating receding universe.

One hundred years of unwavering ideological fealty to Lord Einstein's remarkable GRT field equations still obstructs our progress.

In any case it is becoming clear that closing the cognitive gap between our astrophysical observations and our quantum theory about them—an impossible gap of 120 orders of magnitude—requires a modification of Newton's 'Big G' gravity *constant* as it appears in Einstein's GRT field equations such that it functions as a *variable*; thereby negating the quantum vacuum density contribution to the cosmological constant. The Cyclic Universe cosmology and cosmogony, and Roger Penrose' conformal cyclic cosmology (CCC) claim to nullify the pernicious 'cosmological constant problem', as we shall see. [Steinhardt and Turok 2007] This and other recent work in quantum gravity will further this never ending evolutionary *process.*

The Anthropic Cosmological Principle

Our Spaceship Earth, so this theory goes, lives in a galaxy who abides in a vast and infinite multiverse with countless different sectors, some of which have different vacuum energies. Sectors of the multiverse with a small vacuum energy, like ours, possess the age and the 'fine-tuned' fundamental physical constants that support life forms that have evolved into embodied beings that are capable of pondering such things as 'quantum vacuum energy density'. In any other alternative universe with different physical constants such beings could not have arisen in the first place.

Post-Standard Model Superstring Supersymmetry are friendly toward the now respectable *Anthropic Principle.* [Carroll 2003; Boaz 2022 *Ch.* 7]

We have seen that if the cosmological constant Λ dark energy density in our region of the multiverse was greater than what we observe, then the accelerating expansion of the universe would be too great to allow large scale galactic structure in the first place, including pretty blue planets like ours that arise from the heavy element ejecta (carbon, oxygen, iron) of dying stars. 'Cosmological constant problem' indeed.

Kuhn's Scientific Revolution. It is this conceptual 'problem', perhaps more than any other physics anomaly, that points to the paradigm busting incompleteness of our current Standard Model, General Relativity, and even the prodigious Quantum Theory itself, resulting in an inevitable Kuhnian 'scientific crisis' followed by a profound cognitive 'gestalt shift' toward a scientific 'paradigm shift' and a new inchoate scientific and cultural noetic knowledge paradigm. Such a radical consciousness knowledge paradigm shift is indeed now upon us. It represents nothing less than a Kuhnian Scientific Revolution.

Philosopher of Science Thomas Kuhn in his monumental 1962/2012 *The Structure of Scientific Revolutions* has pointed out the obvious: Science cannot live wholly upon the bread of scientific objectivity alone. We must utilize both voices of our inherent human *noetic cognitive doublet*—objective conceptual scientific, and subjective non-conceptual contemplative. We shall see below that a paraconsistent 'intuitionist' logic and mathematics is the conceptual bridge between these two human cognitive modalities.

There remains an indwelling still small voice of deep subjective primordial wisdom (*jnana, yeshe,* gnosis) that continuously informs our rational, objective conceptual cognition. Let us recognize, and amplify it. The great minds of our kind have, in myriad ways, practiced connecting to this 'innermost' wisdom voice. The cognitive contributions of this intuitive spontaneous love-wisdom mind—Buddha mind, Christ mind—may be analyzed in the subjective scientific laboratory of the mind using both conceptual and trans-conceptual contemplative skillful method. This is the domain of our emerging academic discipline known as the Science of Consciousness, which includes Contemplative Science with its profound 'mindfulness of breathing'. [*Appendix A*]

For example, when we examine with the 'skillful means' of 'analytic penetrating insight' (*vipashyana*) a conceptual or an emotional 'problem' or 'contradiction' or 'question', we bring to bear the two primary capacities of human cognition: 1) *objective cognition* (conceptual Science, intellectual, theoretical, logical, mathematical) which we then proceed to approach from a deeper cognitive strata of formation; 2) *subjective cognition* (Spirit, intuitive, feeling/felt sense experience, concentration, non-conceptual, contemplative, *shamatha*/mindfulness, even nondual meditation of our human noetic wisdom traditions.

First, we dwell upon the problem one-pointedly, conceptually. Then we drop the question entirely and relax into our breath as it rises and falls in the *jnanaprana* lifeforce in the belly, and rest in the peace and non-conceptual 'clear light' clarity of the very noetic nondual Nature of Mind. [*Appendix A*] This is the practice of *shamatha,* the foundation of the 'analytic penetrating insight' of *vipashyana.* After a few minutes, we reenter our conceptual cognition and unpack what news we have received.

In any case, Thomas Kuhn reminds us that our objective scientific paradigms, not to mention our religious paradigms, are relatively grounded in the subjectivity of our collective sociocultural worldview—our 'perspectival' inter-subjective 'global web of belief' (Quine 1969). Just so, the culture of modern Science is steeped in 400 years of the Western Enlightenment Age of Reason—our Greek Apollonian (and Dionysion) hyper-rational legacy that has become habitual Metaphysical Scientific Local Realism, with its ontic consort Metaphysical Scientific Materialism/Physicalism.

Thus is our Western mind beset with a classical, observer-independent, objectivist cognitive bias that requires a substantial, material, local 'real world out there' (RWOT). The inherent subjectivity of the 'superposed' universal quantum *Ψ*-wave function has shaken us from our objectivist "dogmatic slumber" as Kant might have put it if he were here today. We have now been forced by Relativistic Quantum Field Theory to engage the suppressed 'unscientific' subjective, intuitional, spiritual side of our human nature. Science and Spirit unified at last! And now we can see it! Splendid *kosmic* irony indeed.

Just so, most of our 'scientific' ideological totems—Francis Bacon's "idols of the tribe"—shall be shaken as we approach a Quantum Gravity Theory and a Grand Unified Theory. Soto School founder 11th century Zen master Dōgen Zenji told it well: upon the wisdom path, "All that can be shaken shall be shaken."

The physical parameters of quantum cosmology are fine-tuned such that self-conscious beings may arise to ponder such questions. From this fortunate state of cosmic affairs arises the profound, if tautological and anthropomorphic biased Anthropic Principle in both its strong and weak versions (SAP, WAP), and Barrow and Tipler's Final Anthropic Principal (FAP). We must consider well this all too obvious anthropic tautological yet revealing necessary truth.

That said, critics of the Anthropic Principle have referred to it as the Completely Ridiculous Anthropic Principle (CRAP). [Barrow and Tipler, *The Anthropic Cosmological Principle*, 1986]

Be all that as it may, our very notion of 'universe' necessarily implies a self-reflexive conscious participation in it; an embodied conscious experiencing mind. No universe, no experience, no inquiring minds. The conspicuous truth of the fine-tuning of the fundamental constants of nature, of our appearing physical and mental realities, must sooner or later be addressed from the view of the spooky Anthropic Principle, and its logically and empirically necessary 'cyclic multiverse'.

The Cyclic Multiverse

For one who would engage the subjective theistic undertones of the Anthropic Principle in order to enter in its objective philosophy and physics, and its subjective 'spiritual' truths, one is perforce led to the 'cyclic multiverse'—the now nearly mainstream notion of multiple universes arising beyond the limit of the objective visible horizon of our own universe; beyond which we can objectively know nothing. The cyclic multiverse is a necessity presented by our prevailing 'inflationary Big Bang cosmology'.

In 1930 Albert Einstein hypothesized an "oscillating universe" wherein this cosmos is an eternal cycle of repeating oscillations, each cycle beginning with a 'Hot Big Bang' and its expansion into space and time, then in the fullness of time ending in a gravitational collapse into a 'Big Crunch', and then a reflexive 'Big Bounce' into a new scintillating Big Bang universe. Expansion followed by reflexive gravitational contraction. Big Bounce cyclic models were endorsed by Einstein and his collaborator Willem de Sitter, George Gamow, Roger Penrose's "conformal cyclic cosmology", and in Loop Quantum Cosmology (the cosmology of the popular post-Standard Model Loop Quantum Gravity theory).

In Loop Quantum Cosmology a prior existing universe collapses, not to a singularity, but to a timeless point just prior to it wherein the quantum effects of gravity have become so strongly repulsive that the collapsing universe reflexively rebounds again outward as a bright new beginning cosmos. Thus does Relativistic Quantum Electrodynamics (QED) rescue us from a mathematically impossible Big Bang singularity, and salvages what remains of General Relativity, and of a busted Second Law of Thermodynamics, the inexorable physical law that states that entropy can never decrease.

Well, what happened *before* the Big Bang/Big Bounce singularity at the beginning of time? Such a singularity of zero volume and infinite energy transcends the laws of any physics known to human beings. Here Einstein's General Relativity utterly 'breaks down'. Quantum mechanical effects appear to dodge this untimely end of classical physics. What was present Pre-Big Bang? Did the proverbial mathematically impossible 'Hot Big Bang' singularity at the beginning of our universe arise as a Big Bounce following a 'Big Crunch' of an earlier universe as it contracted under the binding force of gravity? The now quite respectable recent 'ekpyrotic Pre-Big Bang' *Cyclic Model* of the Universe argues—against the prevailing Inflationary Multiverse Model—that it does. (Steinhardt and Turok 2007)

This cyclic model of the physical cosmos describes a universe that cycles timelessly and forever. It has evolved from the main elements of Superstring M-Theory, the notion of vast cosmic 'branes' and extra dimensions beyond the four dimensional spacetime continuum of Einstein and his math teacher Minkowski which predict a Big Bang beginning of time wherein the density of matter and radiation may be finite.

Equally important in this cyclic model's development is the idea of constant unchanging dark energy, which is utilized by the theory to explain the geometrically flat and smooth universe that we observe today. Finally, the possible decay of dark energy may provide enough energy at the end of a cosmic cycle to power a new Big Bang beginning and furnish the new expanding universe with large galactic structure, exploding supernovae whose heavy element ejecta (carbon, oxygen, iron) beget planets, life, and in due course and by grace, embodied minded human beings.

In this 'pre-bangian' model two vast perfectly symmetrical empty, flat, parallel cosmic 'branes' are assumed to exist in an infinite space and time. These branes gradually drift toward one another, in timeless infinite space, and finally collide creating a primordial Big Bang. The question immediately arises, "Is it possible to find a 'smoothing' mechanism for the ekpyrotic model that does not entail a period of [high energy cosmic] inflation after the Bang? [Steinhardt and Turok 2007] Dark energy should do the trick.

A universe dominated by dark energy would stretch the two branes causing them to be flat, smooth, and parallel. This is the same dark energy that is the cosmic cause of the stretching branes to accelerate our receding Hubble universe, just as we observe today. Such a cosmogony would be beginningless, eternal and never ending—a variation on the theme of our premodern primordial wisdom tradition's venerable cyclic Pythagorean *kosmos.* Infinite primordial awareness *kosmos* enfolds embraces and subsumes the merely physical cosmos arising within it. Good wisdom company indeed.

Moreover, we have here a reasonable cause as to how and why dark energy must exist; not as an *ad hoc* theory inserted in order to make the Standard Model fit the astronomical observations, but:

> as an essential and fully integrated element of the cosmological picture ... and a way to turn the single collision between braneworlds into a remarkable kind of cyclic event, allowing a transformative view of cosmic history ... With one key assumption—that branes attract [one another] with a springlike force ... colliding at regular intervals ... that becomes weaker as the branes separate—everything worked in principle ... It is now possible to translate that description into a dynamic theory of extra dimensions and braneworlds.
>
> —Steinhardt and Turok 2007

However, the accelerating dark energy does not dominate forever. As the two branes slowly come to meet, the springlike force causes an accelerating universe to begin to contract until, after a trillion years or two, the potential energy reaches zero and the accelerated expansion of the universe enters its contracting phase; the attractive force between the two cosmic branes grows, finally ending in the Big Crunch and its reflexive Big Bounce, the next cyclic Big Bang 'creation' event. And because the maximum temperature reached at the brane collision of the new Bang is much lower than that required to form magnetic monopoles, the cyclic universe has emerged without the problems of super high energy and temperature that the Hot Big Bang cosmic inflation model is fabricated to provide. So who needs inflation anyway? Following the ekpyrotic cyclic Bang the evolution of the universe proceeds as per the standard inflation model—if indeed Alan Guth's cosmic inflation really exists.

Here, at long last is a cogent resolution of the three primary problems of the present inflationary Big Bang theory of the origin of this universe of ours, namely, the 'magnetic monopole problem', the 'flatness problem', and the 'smoothness problem'.

Big Bang inflationary theory operates on two cosmic assumptions: 1) the space of the universe is *isotropic*, it is the same in all directions, and 2) space is *homogeneous*, cosmos is the same everywhere. These two assumptions are known as the 'cosmological principle'.

The first Big Bang problem is the superfine-tuned 'flatness problem'. A geometrically flat universe is one in which matter/mass density is precisely fine-tuned such that it is sufficient to halt its expansion, yet not great enough to gravitationally collapse it. This is known to cosmologists as the *critical density* of the universe. We seem to live in a flat universe (critical density $\Omega = 1$). Cosmologists have been unable to discover a reason or a cause for the present density of our universe to almost exactly equal its critical density. Strange stochastically remote fine-tuning flatness 'coincidence' indeed!

The second Big Bang problem is the 'horizon problem' aka the 'homogeneity problem' or the 'large scale smoothness problem' which is another Big Bang Theory fine-tuning problem; namely, the absence of an explanation for the observed homogeneity or sameness of *causally disconnected* regions of space in the absence of a mechanism that establishes the same initial conditions everywhere in the cosmos. How can vastly different regions of space have almost identical temperatures when they have never been in 'local' physical contact; that is to say, they have never been causally connected?

Our Standard Model *lambda* Λ-CDM (cold dark matter) cosmology sees the solution to all three of these primary cosmological problems in the now prevailing Cosmic Inflation Theory of Alan Guth. In Cosmic Inflation during the first second after the Hot Big Bang the tiny volume universe is in thermal equilibrium, is homogeneous, and is physically and causally connected. In the wink of an eye this tiny volume universe is timelessly, mysteriously instantly inflated to an enormous volume stretching the originally homogeneous, smooth, flat, causally connected volume of the universe over vast distance scales. A happy end of thorny Big Bang Theory cosmological problems. Unfortunately, the solution is mathematically and empirically impossible! It violates the laws of physics as we now understand them. Not to mention, it is not falsifiable. Worse still, many cosmologists working in this physics sector don't buy into such a contrived 'inflationary theory'.

Enter stage left, the Cyclic Multiverse Model which offers a dark energy resolution to these urgent questions of the origins of our improbably fine-tuned home in the vastness of intersteller space.

The Multiverse and It's Noetic Primordial Ground

If our present beloved universe with its law-like physical parameters somehow came to be, why not other universes, a multiverse, or sectors of this universe beyond our visual horizon, with similar or even different physical laws and 'creation events'?

The physics of some of these possible universes may well preclude carbon or silicon based intelligence, but not necessarily other immaterial modes of awareness-consciousness. Indeed, our Primordial Wisdom Tradition, both East and West, views all physical and mental spacetime form arising and emerging in, without ever separating from, the interdependent vast boundless whole itself. That formless, timeless, selfless whole is the ontologically prior yet phenomenally present trans-conceptual, noetic nondual primordial awareness-consciousness 'groundless ground' itself. That basal indivisible *ultimate* all-ground mereologically subsumes, pervades, is always present to, and never separate from all of this *relative* physical and/or mental form that is its spacetime instantiations. That is the ultimate, invariant across all cognitive reference frames, one truth unity (*dzog*) of the Buddhist Two Truths, relative and ultimate.

Broadly construed, that is our Primordial Wisdom Tradition view. All of the love-wisdom masters and *mahasiddhas* of the 'three times'—past, present, future—have told it. Now don't *believe* this. It's utterly beyond belief and conceptual scientific cognition. Please consider it well.

Physicist Max Tegmark (2014) has suggested 'four classes' or possible universes of the cyclic multiverse. All of them utterly *fantasque*, but perhaps no more so than what we have already encountered in our cosmological adventure thus far.

The first class is our own unlikely universe which apparently supports the appearance of what is often construed as self-conscious 'intelligent' life forms. Not so speculative since we live in that one.

The second class is small cosmic inflationary 'bubble universes' (Alan Guth) that due to random 'quantum fluctuations' inflate and slough off from our universe to form other entire universes, with physical constants not always like ours.

The third class is the mind-numbing 'many worlds interpretation' (MWI) quantum ontology which we have explored in considerable detail above, and found wanting.

The fourth universe class is the ideal Platonic universe of discourse that is pure abstract mathematics, presumably engaged from time to time by either an embodied or disembodied intelligence. Or else just hanging out here in a universe of pure potential Platonic Forms awaiting such a sentient engagement. [Max Tegmark, *Our Mathematical Universe*, 2014]

Why bother to consider such scarcely 'possible' fantasy universes? Cyclic multiverse theory may provide empirical, testable theories that illumine the understanding of our current Standard Model of Physics and Cosmology that seeks to know this present universe of ours that is home to a smallish galaxy with an average Population I GV yellow dwarf star in its main sequence out here in one of its galactic arms, that warms our little blue spaceship earth. We wish to know this vast 'unbroken whole' in all of its objective, subjective, and even perfectly subjective spiritual raiment. Nothing must be taboo.

Multiverse theory is often said to require a fabulous, dubious Theory of Everything (TOE) to explain 'everything' in our own ordered universe yes, but in 'all possible universes' as well. A tall order indeed for mere human objective conceptual cognition.

Standard Model Particles and Fields: the Higgs 'God Particle'

As we proceed in our exploration of unification in physics, and toward a viable centrist Middle Way integral Noetic Quantum Ontology, it is cognitively refreshing to note that our current physics 'Standard Model' mathematically reduces all particles in this endless subatomic 'particle zoo' to just two fundamental (non-composite, empty inside) particle types: 1) *fermions*, the matter particles (half-integer spin 1/2), which include baryons (protons and neutrons), and leptons (electrons and quarks); and 2) massless gauge *bosons*, the force-carrying particles (integer spin 1 or 0), which include massless photons and strong nuclear force gluons, massive weak nuclear force W and Z bosons, and the recently discovered (confirmed March 13, 2013) massive *Higgs Boson*. The massless gauge bosons are a formless symmetrical situation. This formless primordial perfect symmetry must be 'broken' for form/matter to arise in space and time. How is this 'symmetry breaking' process naturally accomplished?

The Higgs particle is a highly unstable scalar particle (spin 0), with even parity and 0 charge; it's surprisingly low yet substantial mass is 125 GeV, with non-zero 'vacuum expectation value'. It breaks stable electroweak symmetry which then generates the masses of quarks, leptons, and W and Z gauge bosons of our realities via the mysterious mass/matter bestowing '*Higgs Mechanism*'. The Higgs Boson presumably 'completes' the 'incomplete' physics' Standard Model; at least as to its presumed subatomic particles.

Hence, from the electroweak symmetry busting 'Higgs Mechanism' arises the masses of leptons and quarks—the mass/matter of the whole universe. The *Higgs field* bestows upon subatomic particles their mass. And the Higgs field gives mass to the Higgs boson. The 'Higgs mystery' is that the Standard Model still cannot predict the masses of its fundamental particles. The Higgs mechanism should allow this; but has failed to do so. Something is still missing. Another challenge for a TOE 'hope for a miracle'.

This providential 'Higgs field' is said to pervade every particle-field of spacetime physical cosmos, including all of us, whether we choose to believe it or not. Indeed, the Higgs boson particle has been dubbed by Nobel laureate Leon Lederman, a bit facetiously, the '*God Particle*'. It is after all the creator, sustainer, and physical ground of all the world systems in our physical cosmos. Such cognitive nondual intuition as to the prior and present unity of Science and its Spirit ground is quite natural. Let us not deny the wisdom of such spontaneous cognitive activity because it is "post-empirical". After all, the "lucid mysticism" (Pauli) that is the QFT/QED Ψ-wave function is thoroughly post-empirical.

What then is the cause, the source and ultimate ground of this wondrous 'God particle' with its all-pervading reach and presence? Whence the Higgs particle, this 'force carrier' that bestows the gift of spacetime existence and embodied human consciousness/experience?

The Higgs boson perforce emerges and arises from the continuous *relative* broken symmetry of the *ultimate* perfectly subjective 'unbroken symmetry' of the boundless whole'—formless primordial awareness-consciousness ground of everything. [Again the Two Truths, ultimate and relative.] The 'Higgs Mechanism' process breaks that formless perfect symmetry and arises and is instantiated in spacetime as matter-energy form; all, it should be pointed out, without actually separating from the boundless whole itself in which, or in whom that all arises.

As Gautama the Buddha of this present age told, "Form is empty; emptiness is form." Spacetime form and its superposed quantum emptiness ground are already an indivisible unity. This objective scientific and subjective 'spiritual' understanding begets the harmless human compassionate happiness that is the always present unity of objective Science and its perfectly subjective all subsuming Spirit ground.

In other words, from an ultimate view the Higgs Mechanism may be seen as an indivisible relative causal, yet ultimately acausal cognitive *kosmic process*. A timeless, infinite tantric continuum. From this boundless timeless, formless *ultimate* whole or ground continuously arises *relative* conditional form in time. May I say it again? This ultimate emptiness ground and its continuously arising relative spacetime form are an ontologically prior and phenomenally present unity—the invariant one truth unity (*dzog*) of our perennial Two Truths, relative quantum form and its ultimate emptiness ground.

In short, that perfect *process* is ultimate unity of dualistic objective Science and the noetic perfect subjectivity of its nondual Spirit ground. The recognition (*samadhi*), and realization (liberation/enlightenment) of this wondrous process results in present relative and ultimate happiness of the human being abiding here in this great gift of relative phenomenal space and time.

Be all that as it may, the Higgs boson—all bosons—are the 'force carriers' of the physical subatomic process. An 'interaction' is an exchange of bosons in, for example, the photons of the electromagnetic force/interaction, or the gluons in the strong nuclear force/interaction.

Brief Review of the Standard Model of Physics. We've just seen that the four fundamental forces/interactions in nature described (except for gravity) by the Standard Model are 1) gravity (mediating particle is the energetic but massless graviton); 2) electromagnetism (mediating particle is the massless photon); 3) Weak Nuclear Force (mediating particles are massive W and Z bosons); and 4) the Strong Nuclear Force (mediating particle is the massless gluon) that binds together the foundational quarks of the physical stuff of all the worlds. We've also seen that gravity is conspicuously absent in the Standard Model. It is the prodigious Higgs Mechanism with its Higgs boson that breaks stable electroweak symmetry thus generating the masses of quarks, leptons and bosons that we are.

Gödel's Incompleteness Theorems: Toward Unification in Physics

Historical Prelude. Early in the last century the early 'old' quantum mechanical theory of Planck, Einstein, Bohr and Heisenberg unified classical Newtonian mechanics with classical atomic theory, and offered us post-classical Quantum Field Theory (QFT). Einstein's 1905 Special Relativity Theory (SRT) unified mass and energy, and space and time. Space and time as the 4-D spacetime continuum were unified by Einstein and his old math professor Hermann Minkowski. Einstein's geometric generalization of SRT, namely, General Relativity Theory (GRT) quickly followed late in 1915.

Next, QFT and SRT were unified by Dirac and Heisenberg with the Electromagnetic Force (itself Maxwell's union of electricity and magnetism) giving us Quantum Electrodynamics (QED). The Electromagnetic Force has now been tentatively unified with the Weak Nuclear Force via the 'electroweak theory' which governs electroweak interactions yielding a newly unified Electroweak Force. The discovery of the missing final particle in this glorious process of unification is the 2013 discovery of the Higgs boson which ostensibly confirms electroweak unification. But wait; there's more!

The next step in this astonishing process shall be the mathematical unification of the Electroweak Force with Quantum Chromodynamics (QCD) producing a Grand Unified Theory (GUT). We are close.

Work to be done: we must now unify the quantum world of QFT/QED/QCD with Einstein's General Relativity (GRT) resulting in the much desired quantization of gravity—a unified Quantum Gravity Theory (QGT). What then? We simply unify a consistent QGT with a consistent GUT to produce the great physics consummation to be wished: an impudent, logically and empirically impossible, necessarily incomplete Theory of Everything (TOE). Necessarily incomplete?

Is a Theory of Everything mathematically, logically possible? Can a *fantasque* TOE even begin to account for the subjective phenomena of primordial awareness-consciousness, nondual spirit, human and divine love, free will, and the rest? It cannot. Can a TOE even account for all *objective* phenomena? It cannot.

There is still much that is known to physics but missing from the TOE unification effort. Dark energy, dark matter, cosmic inflation, Big Bang and black hole singularities, the Higgs field mystery, the redundancy of quarks and leptons in the Standard Model, and many others. There are six quarks and six leptons. A proper TOE 'should' be founded upon a single explanatory principle; a single wave-particle. There is much left unexplained by our grail hunger for absolute objective, conceptual, mathematical certainty; much that is absent in the TOE 'idols of the tribe' unification strategies.

What shall we name a Theory of Everything that fails to include everything? A TOE pipe dream? Please permit me to here employ an *argumentum ad verecundiam*—a blatant appeal to authority as a foil against the impossible notion of a Theory of Everything. [Without *verecundiam* we shall indeed be poor in both Science and Spirit.]

Stephen Hawking was disabused of his own idealized TOE 'quest for certainty' upon reading Kurt Gödel's two landmark 1931 Incompleteness Theorems. A non-trivial theory of everything must necessarily be a consistent mathematical theory. Gödel's Incompleteness Theorems have proven that all mathematical theories are necessarily incomplete. Therefore, any TOE must be incomplete.

Moreover, in physics no physical theory is believed by theoretical physicists to be perfectly accurate, and entirely complete. Physics is by its nature provisional and fallible. Physics proceeds by a process of ever more inclusive cumulative approximations of ever increasing, possibly infinite mathematical and physics complexity which evolve into new ever more inclusive theories for the duration of the embodied conscious cognition of thinking beings. This fundamental process of learning and knowledge is perforce never ending, never complete. Thus is an ultimate Theory of Everything logically and empirically, necessarily incomplete.

Well and good. Let this suffice as a definition of a useful but limited relative 'theory of everything'. And because such a limited theory of everything is not truly a theory of everything, perhaps we may at last commit it to the proverbial trash bin of the noble history of an ever incomplete science of physics. May it rest in peace.

Let us then once again explore Kurt Gödel's wondrous Incompleteness for its enormous impact upon our all too human objectivist cognitive biases in science, religion, and culture.

Gödel's Incompleteness Negates Our Quest for Absolute Objective Certainty. Incompleteness in mathematical logic is concerned with the limits of provability in formal axiomatic theories. The '*Completeness Theorem of First Order Logic*' states that a proposition is universally valid if and only if it can be deduced from the axioms of the system using the 'rules of logic', namely, the 'rule of inference' or the 'inference rule'. Examples of the inference rule: *modus ponens*—if the premises of an argument are true then the conclusion is necessarily true—and *modus tollens* (denying the consequent); if a propositional statement is true, then its contrapositive is true. Gödel challenged this foundational Completeness Theorem and thereby the entire logical foundation of mathematics. His two Incompleteness Theorems of 1931 revolutionized our understanding of formal logic, mathematics, and philosophy of mathematics, and therefore of mathematical physics.

Kurt Gödel professed to be a mathematical Platonist. He defended the view that mathematical truth is real and objective, and that mathematics is a descriptive science. The objects of mathematics (numbers, sets, etc.) objectively exist 'out there' as ideal Platonic Forms abiding in an ideal Platonic dimension. A bit ironic, since he undid the ideal objectivist 'Hilbert's Program'. Gödel, in the beginning, supported what he called David Hilbert's "original rationalistic conception" of mathematics as a pure objective science.

In his 1929, at the tender age of 23, Gödel's doctoral dissertation proved the incompleteness of the sacrosanct First Order Logic Completeness Theorem—an astounding paradigmatic breakthrough. He proved that this foundational Completeness Theorem fails for Zermelo-Fraenkel Set Theory with its infamous Axiom of Choice (ZFC) now generally accepted as the more or less paradox free (e.g. Russell's paradox) foundation of all mathematical logic.

Zermelo-Fraenkel Set Theory (ZFC), later known as First Order Logic, prohibits the existence of a universal set—a set that includes all sets—thereby dodging the 1901 devastating Russell's Paradox (any set theory that contains an 'unrestricted comprehension principle' results in a logical contradiction). Georg Cantor, founder of modern set theory, was aware of this most problematic paradox as early as 1898, as was the great David Hilbert. Russell also proved that Gottlob Frege's monumental effort to reduce mathematics to formal logic was not logically possible.

Later Kurt Gödel was to challenge the quest for absolute logical certainty of the great David Hilbert; the logicism (all mathematics is reducible to formal logic) of Russell, Whitehead, and Frege; and thereby cast the noble disciplines of mathematical logic and philosophy of mathematics into intellectual chaos from which they have yet to recover. Objective scientific logical mathematical objective certainty is now kaput! A shock to the system of foundational Metaphysical Scientific Local Realism/Physicalism, to say the least.

Paraconsistent intuitional logical theory has of late come to the rescue, as we shall revisit below. Indeed, Gödel published on 'intuitionist logic', and proved that it is pragmatically valuable. He is himself often considered a logical intuitionist. No easy fete for an avowed mathematical Platonist.

Gödel became a professor at the Princeton Institute for Advanced Study where he befriended Einstein. They talked relativity, ontology, and philosophy of mathematics on long walks on campus. In the late 1940s Gödel published peerless papers on Einstein's GRT and cosmology, and on relativity and philosophical Idealism. He then concentrated for three years on the Philosophical Idealism of Leibniz. Kurt Gödel was a polymath who bridged the disciplines of mathematics, physics, and philosophy. As to his religious and spiritual life, "My belief is theistic, not pantheistic, following Leibniz, not Spinoza... Religions are, for the most part, bad—but religion is not." He considered himself religious and read the Christian Old and New Testaments on Sunday mornings. Theism and Logical Incompleteness. For a logician, strange metaphysical bed fellows indeed!

Yes. Kurt Gödel's work is essential in both current mathematical logic, in philosophy of mathematics, and in philosophy of mind. His Incompleteness Theorems are generally though not universally considered to be a refutation, or else a demonstration of the "infeasibility" of 'Hilbert's Program' to discover a complete and logically consistent set of axioms for all of mathematics—a mathematical TOE as it were.

Indeed, the great quantum mathematician John von Neumann, who was aware of the incompleteness of the orthodox Completeness Theorem before Gödel, took the position that Gödel's Incompleteness demonstrated the "infeasibility" of classical mathematics altogether. In a letter to Gödel, "Your result has solved negatively the foundational question: there is no rigorous justification for classical mathematics." In a 1931 letter to his early teacher Rudolf Carnap Gödel opines: "I am of the opinion 1. [I] have shown the unrealizability of Hilbert's program. 2. There is no reason to reject [logical] intuitionism." Indeed, we shall see that Gödel accepted the alternative logical intuitionism of Brouwer.

Gödel's Proofs were followed by 1) Tarski's 1933 *Undefinability Theorem* on the formal 'undefinability' of logical/mathematical truth; 2) by Alonzo Church's proof that first-order predicate logic is undecidable and so David Hilbert's 'decision problem' (the third of Hilbert's 'three questions') is not solvable; and 3) by Alan Turing's theorem that there can be no algorithm to solve the computational 'halting problem' (whether a program will finish, or run on forever). These three were shown to be equivalent and are usually referred to as the *Church-Turing Thesis* which is the recursive backbone of a proposition's undecidability, that is, its incompleteness. It is beginning to appear that 'mathematical truth' is indefinable. What then may be said of *that* truth?

Clearly, classical mathematical logic was, at mid-century, undergoing a dramatic Kuhnian 'scientific crisis' that portends a radical 'scientific paradigm shift', and a 'scientific revolution. Gödel's Incompleteness was the liminal cognitive fulcrum of this revolution in logic and mathematics. Let us then identify these two theorems of Gödel that have forever changed the world of mathematics, and profoundly impacted the 'global web of belief' that defines the systems of concept and belief of Western science and culture.

Gödel's First Incompleteness Theorem ('Gödel 1931') proves that in any axiomatic system beyond basic arithmetic (e.g. Peano's arithmetic) there are true propositions about the 'natural numbers' in the system that can be neither proved nor disproved from the axioms within that system. For any such consistent formal system there will always be statements about natural numbers that are true, but that are not provable within that system. Incompleteness applies to any first order logic theory. Gödel explained in a letter to Hao Wang:

> By enumeration of symbols, sentences and proofs within the given system, I quickly discovered that the concept of arithmetic truth cannot be defined in arithmetic. If it were possible to define truth in the system itself we would have

> something like the liar paradox, showing the system to be inconsistent…

Gödel discovered that such logical paradoxes do not arise if the notion of 'truth' is replaced by the notion of 'provability'. Truth and provability are not 'co-extensive'. From this simple logic arose Gödel's monumental First Theorem of Incompleteness ('Gödel 1931'). His stronger Second Theorem appeared as 'Theorem XI' in his 1931 paper "On Formally Undecidable Propositions in *Principia Mathematica*".

This occult 'undefinability' of logical truth in arithmetic was also present to the brilliant mind of Alfred Tarski well before he published his *Undefinability Theorem* in 1933. This great discovery of the incompleteness and indefinability of the axioms of mathematical logic ran counter to the prevailing cognitive bias for David Hilbert's much valorized logical absolutes of his noble 'Hilbert's Program'. So the heterodox views of both Gödel and Tarski received much illogical abuse from their perplexed peers steeped as they were in the prevailing orthodoxy and intellectual cognitive biases that insist upon absolute objective mathematical certainty.

Hilbert soon grasped the significance of Gödel's proofs. Wittgenstein did not. If the cognitive purity of mathematical logic cannot provide such conceptual certainty, what hope do we have for it at all? Devastating to the very mathematical foundations of Big Science itself.

Gödel's Second Incompleteness Theorem follows from the first; no such axiomatic system can *prove* its own logical consistency (if it *is* consistent). Such a system cannot self-reflexively prove its own mathematical consistency. This theorem is 'stronger' than the first theorem because the first theorem does not address the consistency, or inconsistency, of the of the *whole system* itself. Gödel, the math prodigy and student of both Western and Eastern philosophy was 24 years old when he published his two Theorems in 1931.

[For the cognitive comfort of all but the most logically hardened of my readers, and for my own psychiatric stability, I have chosen to omit the quite abstruse formal logical notation of the simple elegant proofs of Gödel's two incompleteness theorems.]

Gödel's Incompleteness Theorems are among the most important advances in the history of mathematical logic. They demonstrate the limit of provability in formal axiomatic logical theories. They have greatly impacted philosophy of mathematics and logic, epistemology, philosophy of mind, and philosophy of consciousness. Yes. There are inherent limits as to what can be proven within any first order mathematical theory.

Gödel's Incompleteness had a huge deflationary impact upon mathematical logic by showing that logic alone cannot provide a foundation for certainty in mathematics, or in anything else. This was devastating to 'Hilbert's Program' to prove the consistency of the foundational theories of mathematics; and to the work of Russell, Whitehead, and Frege, as we have seen. Logic and mathematics have proven to be riddled with inconsistency.

Paradoxically—for a mathematical Platonist—Gödel supported the paraconsistent constructivist (math is a cognitive construct, not an objective truth) Intuitionist Logic of Brouwer, Heyting, and Weyl; not to mention the Eastern Hindu and Buddhist paraconsistent deductive logical systems. [Boaz 2020]

And if mathematical logic fails to provide us with absolute objective certainty, can there be any *conceptual* certainty? After all, our objective conceptual certainty is nested in the semiotic logical syntax of language. Global knowledge paradigm shift indeed.

"What hath God wrought?!" Does the truth of the existence of the timeless primordial awareness ground, boundless whole itself, the very Nature of Mind lie in spooky scary subjectivity? This great perennial truth of the inherent limits of human objective, logical, purely conceptual understanding has become known to those who know as 'the wisdom of uncertainty'. For Alan Watts it is the fortuitous 'wisdom of insecurity'. It represents an aperture, or cognitive 'gap/*schnitt*', indeed a conscious cognitive portal into our trans-conceptual, 'innermost secret', contemplative, perfectly subjective wakefulness of our love-wisdom Christ-Buddha mind—timeless, formless primordial awareness-consciousness ground in whom this all arises, participates, and is instantiated.

That is the mathematical and logical foundation of the prior unity of objective Science and its perfectly subjective Spirit ground.

'Rosser's Trick'. In 1936 J. Barkley Rosser greatly enhanced Gödel's two original 1931 Incompleteness Theorems by providing a most elegant simplified proof of the theorems. 'Rosser's trick' clarifies Gödel's original incompleteness by altering his famous sentence, "This sentence is not provable", to "If this sentence is provable, there exists a simpler proof of its negation". This provides a 'stronger' version of Gödel's First Theorem by replacing Gödel's rather cumbersome ω-consistency to consistency by way of a tricky 'provability predicate' [*Prov* (x)*].

Gödel's Impact on the Western Intellectual Tradition. It was the monumental 1925-1927 2nd edition of *Principia Mathematica* of Whitehead and Russell, despite its objectivist biases, that laid the meta-logico-mathematical foundation for Gödel's work. The splendid irony is that Gödel used *PM* to disprove *PM's* primary aim to explain all mathematical statements in the terms and notations of symbolic logic; that is, to reduce all of mathematics to formal logic. This program is known as 'logicism'.

Gödel's work in the logic of mathematics is generally considered by those who care about such things, the end of our 400 year Enlightenment grail quest for absolute objective certainty—a quest that has fortuitously begat our engagement with the spooky trans-conceptual '*wisdom of uncertainty*' in both Science and Spirit. That conceptual uncertainty is reflected in the inherent subjectivity of Heisenberg's quantum Uncertainty Relations, Schrödinger's superposed quantum 'universal Ψ-wave function', and in the perfect subjectivity of Tibetan Buddhist *Perfect Sphere of Dzogchen.* That cognitive balancing act continues into the 21st century *Noetic Revolution in Matter, Mind and Spirit* that is now upon us. [Boaz 2023]

Just so, the ambitious TOE unification quest for absolute objective certainty fails to address, or even attempt to explain away, our human inherently non-reducible subjective experience—human love, emotional, compassionate, altruistic ethical, intuitive, creative, mythopoetic, artistic, and religious and spiritual. This subtle subjective experience dimension is *ipso facto* beyond the conceptual dimensional reach of even our most spirited intellectual philosophical, scientific, and mathematical virtuosity. And yes. In recent Buddhist philosophy this is known as the pragmatic 'logic of the non-conceptual'.

Can a Theory of Everything explain this inherent limit of logic and mathematics?

Gödel proved that David 'Hilbert's Program' or 'grand plan' for a formally consistent, logically complete foundational basis for all of mathematics was not logically possible! We've seen that Tarski's *Undefinability Theorem* further proved that 'truth' in logic and mathematics is formally undefinable. Church and Turing demonstrated that formal mathematical truth is ultimately unfindable by any possible computer program (the 'halting problem'). Thus was Hilbert's lifelong quest for an axiomatized mathematics which would in principle free us from cognitive uncertainty thereby formally demolished. May it rest in peace, never to arise again. It has been adeptly replaced by post-empirical 'paraconsistent and intuitionist logic and mathematics'.

The astonishingly smart and resourceful David Hilbert, the pure mathematician, upon realizing that his 'Hilbert's Program' was dead in the water of Gödel's Uncertainty, then took up physics and developed his own geometrical theory of the general relativity field equations which was the equal of his friend Albert Einstein's new 1915 GRT theory. Indeed, his elliptical geometry was decisively subtler and simpler. The two great physicists published their work almost simultaneously. It was the noble Hilbert who publically deferred to Einstein as the originator of GRT, avoiding any priority disputes about this great physics revolution.

The amazing mind of David Hilbert then took up quantum mechanics with its monumental mathematical formalisms, and thereby contributed to von Neumann's work on the mathematical equivalence of Heisenberg's matrix mechanics and Schrödinger's quantum wave equation which had been the subject of considerable dispute. Hilbert saw that Schrödinger's version was more elegant and easier to apply. His resulting elegant and infinite 'Hilbert space' continues to play an important role in the quantum mathematical formalisms that have arisen in a century of quantum mechanics.

Through this process both Hilbert, and Einstein's benefactor, friend, and non-Euclidian Riemannian geometry tutor Marcel Grossman realized the colloquial truth, which he kindly shared with Einstein, "mathematics is far too difficult for physicists".

But wait! What about geometry? Geometry has for 3000 years been our solid foundation for elusive absolute conceptual certainty. Immanuel Kant was sure of it.

Euclid's geometry, our Greek paradigm of absolute objective certainty was, in the middle of the 19th century, replaced by the non-Euclidian geometry of Gauss and Lobatchevsky working independently. Bernard Riemann, a student of Gauss, in 1854 used the infinitetesimal calculus of Newton and Leibnitz to develop a new non-Euclidian hyperbolic geometry that bears his name. This is the geometry that Einstein needed (the Einstein tensor) to solve the problems of differential topology that arise in the quite complex four dimensional spacetime of General Relativity. It was this self-consistently certain geometry that Marcel Grossman taught to his needy friend Einstein that now appears in the GRT gravity field equations as the Riemann curvature tensor **R.**

Well, if geometry is the paradigm case of absolute objective certainty, and these geometries are all equally certain, then which one is *really* absolutely certain? Clearly we require a new set of super-certain axioms. These axioms were provided by the great David

Hilbert, as we have just seen. And that grand program was a fortuitous failure. It was not until 1931 that Kurt Gödel demonstrated that mathematics and formal logic cannot be our guide for absolute objective certainty, try as we may. However, while the *ultimate* 'truthfulness' of mathematics remains ontologically elusive, fortunately its pragmatic *relative* truths still rule our objective world of theoretical physics and electronic technology.

Meanwhile, the Quantum Revolution with Schrödinger's inherently subjective 'objective random' universal quantum Ψ-wave function has added a further bitter pill of subjective uncertainty (Heisenberg's Principle of Uncertainty) to our eternal grail quest for absolute conceptual certainty. Our habitual, "all too human" (Nietzsche) demand for perfectly objective cognitive certainty has indeed fallen on hard times. If one requires absolute certainty it will not be in objective Science, but in the contemplative, even nondual view and practice of nondual Spirit.

Matter, Measure, Utility, and Unity

Matter and Measure: The Quantum Standard Model. Well, what about non-problematic, course physical atomic spacetime stuff? How might we further complicate that? We have just seen that hadrons (means robust) are subatomic composite particles made of two or more 'empty' foundational quarks bound together by the strong nuclear force/interaction, much as molecules are bound together by the electromagnetic force/interaction. Hadrons include baryons (protons and neutrons), and bosonic mesons (pions and kaons), and bosonic photons of light. The mass of ordinary matter is composed of two hadrons—baryonic protons and neutrons—and is caused by the attractive, 'contained' short range binding energy of these quarks by the strong nuclear force. Thus are hadrons comprised of baryons (usually three valence quarks), and mesons (one quark and one anti-quark not bound in the nucleus).

Got all that? If not, don't worry too much. You will never need it; unless you decide to take up quantum physics as a meditation to distract your otherwise anxious mind.

Recall that a physical micro system (e.g. leptons/electrons) evolves deterministically in classical time (*t*) following the Schrödinger Ψ-wave equation when it is not measured/observed. And such a system evolves non-deterministically, following the 'collapse postulate', when it *is* measured/observed to 'create' an objectively 'real' physical electron. Such a measuring "consciousness causes collapse" (von Neumann, Wigner). This non-objective postulate is the quantum creator of our objective realities.

Well, what precisely is the definition of 'measured/observed' and how in heaven and earth does that cause 'collapse' of the superposed quantum Ψ-wave function into real stuff? Or does the wave function collapse at all? That is the *'quantum measurement problem'*.

Objective quantum measurements are executed by the subjective 'consciousness' of an observer-experimenter reading a 'decoherent' measuring apparatus (computer, camera, Geiger counter), as we have seen. Here the conceptually untidy subjectivity of 'consciousness' creeps into our hitherto classical, ostensibly purely objective, mathematical 'scientific' quantum measurement process. Due to its inherent subjectivity, 'consciousness' is, in the obsessively objective gloss of modern physics, in a word, anathema! 'Consciousness', whatever it is, is still taboo in physics and cosmology. The antirealist nominalist Taoist Niels Bohr loved it. The local realist, Judaic Christian Einstein hated it. They battled over it for many years in the famous EPR debates.

Thus arises that eternally vexing quantum conundrum we have come to know and love as the quantum measurement problem. Once again, just how does this "lucid mysticism" (Pauli) universal quantum wave function collapse occur? Alas, our question remains unanswered. How does a conscious measurement observation of a quantum physical micro-system radically change that state from a subjective 'probability cloud' of all possible 'superposed wave-like states' of a physical quantum system such that it will 'collapse' into an objectively really real particle-like *thing* with a definite 'eigenstate of position'? Curious indeed. We have come to know this paradox as the 'quantum enigma'.

Our inability to observe and explain objectively such a subjective quantum wave function collapse has delivered a plethora of 21st century philosophical 'foundational interpretations' or 'quantum ontologies' of the wondrous 20th century quantum mechanics of Bohr, Dirac, Heisenberg, and Schrödinger. We have explored the seven most viable above. All of them fail as a fundamental integral Noetic Quantum Ontology that rises above the classical mechanics of Newtonian physics.

What is the prior ontic *ultimate* ground of the present epistemic *relative* Ψ-wave function conceptually arising within that perfectly subjective ground, vast all subsuming nondual boundless whole itself? We are herein perhaps too gradually constructing such a 'post-empirical' Noetic Quantum Ontology.

Unifying progress in physics and cosmology has stalled. The great physics desideratum that is a Quantum Gravity Theory (QGT) to unify hitherto mathematically incommensurable Quantum Field Theory (QFT) and Einstein's General Relativity Theory (GRT) of gravity—the prodigious quantization of 'Big G' gravity—is as remote as ever.

We must now venture beyond our classical, observer-independent biases for a logically impossible pure objective certainty and both contemplatively and analytically engage the timeless formless, nondual noetic primordial awareness-consciousness emptiness ground of all objectively appearing spacetime quantum form. It is here that objective Science and subjective Spirit hang together.

We have just seen that until a settled quantum ontology arises from the theoretical ashes of quantum measurement uncertainty there shall be no quantized gravity, the holy grail of a Quantum Gravity Theory (QGT) that finally unifies hitherto incommensurable Quantum Field Theory (QFT) with Albert Einstein's classical General Relativity Theory (GRT). If this is mathematically possible at all.

Well, is the quantum consciousness sentient observer/experiencer/experimenter objective and local, or subjective and nonlocal non-temporal? Middle Way Buddhists might argue that it is both, depending on the view, relative or ultimate. Is both local relative and nonlocal ultimate existence at once—the prodigious Mahayana Buddhist Two Truths—a logical contradiction? I have argued here and elsewhere that it is not. It is two relative views of one trans-conceptual nondual ultimate reality whole. How is this so?

Yes, 21st century quantum theory has some unfinished ontological business to attend to before it may contribute to our 100 year old grail quest for a mathematically consistent unification of currently incompatible GRT with QFT/QED; in short, that prodigious physics quantification of Isaac Newton's and Albert Einstein's classical gravity 'G' in a Quantum Gravity Theory (QGT). We now think we know that everything is ultimately quantum in its nature. So we must somehow quantize non-quantum (classical) great gravity—"the creator and destroyer of worlds" (*Bhagavad Gita*).

Has this illusive 'epistemological problem' of a subjective nonlocal quantum reality that is dependent upon an objective observer consciousness to explain its measurement 'collapse' into an objective real world out there (RWOT) presented a logical paradox that contributes to the inherently vexed theoretical incommensurability of the subjective QFT/QED and objective 'classical' GRT mathematical formalisms?

Indeed it has. The inherent subjectivity of 21st century 'spooky' antirealist quantum nonlocality/entanglement has perforce demonstrated that the quantum nature of spacetime reality is, in the *relative* view, robustly objective, yet *ultimately* non-objective, even perfectly subjective. Sounds like the 'Two Truths' of Mahayana Middle Way Prasangika Madhyamaka Buddhism, conceptual foundation of nondual *Ati Dzogchen*.

Utility and Unity: The New Paradigm. There is no *ultimate* contradiction, no conundrum, no dilemma here at all. As Niels Bohr's quantum Principle of Complementarity has correctly told, the stuff of spacetime reality appears to human perception dualistically, as private mental properties or subjective consciousness, and public objective physical properties—in the old platitude, mind-body dualism, or the 'mind-body problem'. These two faces of our human experience/consciousness abide in the diaphanous context of an experiencing subject, a self-ego-I. And yes, all of this Science process arises and plays in the perfectly subjective nondual noetic primordial awareness-consciousness Spirit ground of all spacetime being.

Thus does our explanatory understanding require two utilitarian complementary views, two interdependent cognitive modalities—quantitative, relative, objective, and qualitative ultimate subjective—to both conceptually then contemplatively understand the deepest nondual nature of being here in time and space. And yes, I have come to call these two qualities of human knowing our *noetic cognitive doublet.*

We must skillfully utilize our human interdependent noetic cognitive doublet: relative objective conceptual, and subjective trans-conceptual, even ultimate nondual. Let us utilize both lest we become cognitively caged in one or the other of these two complementary indivisible and unified dimensions. Let us then assume the prior noetic *one truth unity* of Buddha's prodigious Two Truths—*relative* Scientific, and its inherently spooky (to the course mind) *ultimate* primordial Spirit ground.

We have seen many times in these pages that this perfectly subjective ultimate primordial 'Nature of Mind' and its human consciousness/experience of arising appearances cannot be fully understood, let alone 'spiritually' realized, through the objectively biased methodologies, doctrine, logic and mathematics of the physical sciences.

That the ultimate primordial Nature of Mind and its gross and subtle contemplative experience can be entirely grasped by mere conceptual discursive mind is the cognitively biased impudent presumption of our Greek philosophical legacy—incarnate in modern Metaphysical Scientific Local Realism and its epistemic consort Metaphysical Materialism/Physicalism—all appearing spacetime reality is always only locally real and physical. Profoundly pragmatic doctrinaire monistic Realism/Physicalism dogma indeed.

Still, it must be admitted that Metaphysical Local Realism/ Physicalism is relative-conventionally, if not ultimately correct. There's a bunch of physical spacetime stuff out there, and in here. It's just that the physical dimension of reality cannot account for everything arising to the senses and the heart. Is that not abundantly clear? Both quantum science and contemplative science have demonstrated this for us.

The wise have told it well: ultimate reality itself—nondual primordial awareness-consciousness ground of all that arises in relative space and time—is beyond the scope of mere human conceptual cognitive and mathematical processing. The wisdom masters of our noetic Primordial Wisdom Tradition have taught this great truth for millennia. Yet the prior one truth unity of that reality ground is contemplatively knowable.

Einstein and Bohr, in different ways, both understood and wrote about this profound wisdom truth. We have seen that Einstein followed the objective wisdom of the Abrahamic monistic Idealism of Spinoza. Bohr followed the subjective wisdom of Lao Tzu and Taoism. Both read Hindu Veda-Vedanta philosophy, and both were familiar with at least some Buddhist philosophy. Would that lesser physicists follow their sterling ecumenical example.

The dogmatic, obsessively objectivist, realist, materialist European Enlightenment ideology that for 400 years has colonized the Western heart and mind is no longer adequate to our deeper physics and cosmology growth; nor to awakening to our innermost subjective primordial Christ/Buddha love-wisdom mind. Bright indwelling, happy Presence of That. And most of us know it. We require a new, more inclusive integral noetic theory of mind and nature. Indeed, we need the emerging scientific, cultural, and spiritual unifying noetic knowledge paradigm that is now abroad in the worlds of both Science and Spirit as they come to meet in our 21st century Noetic Revolution in matter, mind and spirit.

The inherently subjective wisdom of Max Planck's 1900 'quantum of action' has at long last objectively revealed this unifying truth to physicists and mathematicians. None but the best have seen it. As Thomas Kuhn has pointed out (*The Structure of Scientific Revolutions* 1962, 2012), a couple of generations is required for the acolytes of a waning 'scientific paradigm' to expire so that a new more inclusive but ever incomplete knowledge paradigm may arise. How may we human beings expedite this tedious evolutionary wisdom process before we self-destruct?

How indeed? Once again, let us utilize fully our innate human *noetic cognitive doublet*—objective, conceptual, *relative* theoretical Science; and the subjective, intuitive, Contemplative Science of ultimate all subsuming Spirit—in this 21st century Noetic Revolution that is now upon us. [Boaz 2022; 2023]

Paraconsistent Logic: Gödel, Incompleteness, and Intuitionist Logic

High on our list of wisdom priorities—just below training the fearful, angry 'wild horse of the mind' in contemplative mindful peace and selfless *bodhichitta* conduct—must be a post-quantum theory quest for a post-classical, post-Aristotelian alternative deductive logic that liberates an integral Noetic Quantum Ontology.

As we saw above, we must replace classical limited binary two-valued, true-false deductive logic with a formal 'paraconsistent' three-valued (3VL), or multi-valued (MVL) deductive logical system. Such a logic, whether West or East, will modify 'truth functional', two-valued, true-false either-or bivalence, thereby permitting truth values beyond the limiting binary cognitive extremes of *either* true *or* false—for example, either absolute existence or absolute nonexistence; either particle or wave. Such a logic shall allow for a logical proposition to have a pragmatic 'paraconsistent' truth value that may be *both* true *and* false; both relative existence and ultimate nonexistence. Aristotle and Descartes would think it schizoid sophistry. Pythagoras and Plato would understand it immediately. We must in this our postformal, 'post-empirical' Noetic Revolution move beyond such pernicious habitual false dichotomies that indulge this destructive sophistry.

Recent quantum logic, along with the 20th century 'logical intuitionalism' of Brouwer and Gödel—who drop the 'law of excluded middle' of classical logic (either 'A or not A' with no third possibility)—have accomplished such a result. Let Western logicians and Buddhist and Hindu logicians together enter in jocular symposia on paraconsistant logic over pizza and ale. That such symposia are nonexistent demonstrates the cognitive distance that still obtains between the wisdom of the West and the wisdom of the East.

The limits of classical Aristotelian formal logic to accommodate the recent discoveries in quantum physics, 'paraconsistent/inconsistent mathematics', as well as the problems in 'natural language', have motivated new adventures in formal deductive logic; to wit, paraconsistent and intuitionist deductive logical systems. Upon the failure of an objectively certain foundation for logic and mathematics based in the work of Gödel, Tarski, and Church, 'paraconsistant logic and mathematics' represents a successful effort to construct a 'non-explosive' enhancement of classical mathematics upon a foundation of post-classical paraconsistent intuitionist logic. How shall we understand this?

Paraconsistent formal axiomatic logical systems accommodate classical logical inconsistency in a formal controlled way that treats inconsistent propositions as potentially informative and valuable. Any deductive logical system is paraconsistent if it is not 'deductively explosive' (*ex falso quodlibet,* from falsehood anything follows); that is, if once a logical contradiction has been discovered, *any* proposition/statement can be inferred from it. From contradictory premises *anything* can follow. So 'deductive explosion' buries any residual truth value and renders the argument utterly useless.

We require a middle way paraconsistent deductive formal logic that does not throw out all deductively inconsistent propositions as a matter of course. Pragmatic truth is valuable, and much more subtle. Moreover, a centrist integral Noetic Quantum Ontology requires such a logic. Thus do we seek a deductive logical system that provides and reveals knowledge beyond the nostrums of noble classical truth functional deductive logic.

Let us then not throw out the 'baby' of inherently random logically inconsistent quantum mechanics (the Ψ-wave function), and logically impossible Big Bang and black hole singularities, and the rest, with the 'bathwater' of formal 'explosive' logical contradiction. Indeed, most of the theories and the discoveries of 'post-empirical', post-Standard Model physics arises above and transcends the realm of classical truth-functional (either true or false) formal mathematical logic.

Yes. Classical formal logic is 'deductively explosive'. It is 'old paradigm' radical logical extremism. It very effectively nails contradictory statements, and as David Hume told, "commits them to the flames" of what is presumed to be not but "useless sophistry". The Western heart and mind have long believed that logical inconsistency is always catastrophic because there is nothing further that can be coherently objectively reasoned about. Yet there remains much pragmatic truth in a cognitive paraconsistent grey regime between the idealized logical truth functional extremes of either true or false, black or white. Not only the extreme physical environments of quantum cosmology, but our every day logical reasoning is a good example of such pragmatic logic.

In any case, it seems to me that we need both the cognitive 'software' of paraconsistent logical understanding to complement the 'hardware' of formal mathematical logic. These two modalities viewed as a pragmatic unity advance the noetic wisdom that altogether transcends logical conceptual thinking and belief. We must not fear to tread where 'there be demons' for here abides the noetic nondual 'mind of God'.

Paraconsistent ("para" here means "beyond") logic challenges the waning classical logical orthodoxy. Here a paraconsistent 'consequence relation' does not 'explode' into absolute falsehood, nor into 'logical trivialism' (the belief that everything and every contradiction is somehow true) in the face of inconsistent information in a formal argument's premises. The classical notion of logical consistency—absolute logical certainty—is relaxed toward the idea of pragmatic and intuitive 'logical coherence'.

That some contradictory statements may both be paraconsistently true (e.g. both relative existence and ultimate nonexistence) does not entail the view that there exist 'true contradictions' (dialetheism). Asian logical systems are often paraconsistent logics wherein some seemingly contradictory statements may both be true. Hindu Nyaya Yoga and the Buddhist logic of Dignaga and Dharmakirti may be seen as paraconsistent systems where statements/propositions of the form "*S* is *both P* and not-*P*" are permitted. For example: "Appearing spacetime reality is both relatively, conventionally existent, yet ultimately nonexistent." Just so, every quantum entity or particle may be truthfully described as both a point like particle, and a wave spread out in space. The classical concepts of 'particle' and 'wave' cannot fully describe physical reality at the quantum scale. Yet together they do. Niels Bohr told that this "duality paradox" is a fundamental metaphysical "fact of nature". Wave and particle, existence and nonexistence, are complementary aspects of a unified reality—Bohr's foundational Principle of Complementarity. Such statements express the meaningful subtlety of the paraconsistent nature of objectively 'real' spacetime existence arising from its perfectly subjective formless, timeless ground.

I should note here that Nagarjuna's *reductio ad absurdum* (*prasanga*) argument is not included in the set of paraconsistent logical systems. It partakes of classical Greek reasoning as seen in Aristotle's *Prior Analytics* in his "demonstration to the impossible" argument; and is grounded in the Law of Non-Contradiction of Aristotle's Three Laws of Thought which form the foundation of Western Greek classical deductive logic. In paraconsistent logics East and West cognitively come to meet.

In the West, paraconsistent logic is now mainstream. Why? The motivation for research in paraconsistent logic is that it offers nontrivial, valuable knowledge that transcends the wasteful inherent limits of classical, two-valued logic: for example, Bohr's contradictory model of the atom, Maxwell's elegant electromagnetic equations; the fundamental quantum *Ψ*-wave function; the 'zero-point quantum vacuum energy field' (ZPE); dark energy; the Anthropic Principle; the cyclic multiverse; not to mention logically and empirically impossible black hole and Big Bang singularities. All of these logically contradictory theories are fortunately pragmatically managed by logical paraconsistency. Praise be!

Moreover, paraconsistent dialetheia (a contradiction that is true) manages diabolical logical paradox—e.g. the Liar's Paradox ("This sentence is false"), and the infamous Russell-Zermelo Paradox which devastated the logical foundations of mathematics, and our atavistic grail quest for absolute objective certainty in the mode of David Hilbert and Bertrand Russell. The noble discipline of classical deductive logic bears the seeds of its own destruction. This ancient program for absolute objective certainty is self-reflexively self-destructive. Russell's Paradox logically proves it! Cosmic irony indeed.

Further, paraconsistent logic manages the natural inconsistencies, and the 'problem of vagueness' inherent in 'natural language'; and in 'belief revision' in our mostly inconsistent cultural 'global web of belief'; in hidden contradictions lurking in the programs of Artificial Intelligence; in the intrinsic logical paradoxes in the semiotic logical syntax of language and semantics, in 'paraconsistent mathematics'; in set theory; and in the shocking paradox of Gödel's revolutionary two Incompleteness Theorems ('Gödel 1931').

How is this all logically possible? A paraconsistent logical system, for example 'Adaptive Paraconsistent Logics', may be used to neutralize the inconsistent arithmetic pointed out in Gödel's Incompleteness Theorems permitting the 'logical provability' of his hitherto contradictory incompleteness paradox. A paraconsistent application neutralizes the inherent limits of arithmetic certainty that naturally arise from Gödel's Theorems. Without abandoning classical logic, paraconsistent logical application isolates classical logical inconsistency and contradiction without rejecting or negating the hitherto 'explosive' but still informative truth values that may still be present therein.

This process models the dynamics of our actual normal human semiotic pragmatic, intuitive thinking-reasoning cognitive processes. Human reason naturally, reflexively adapts itself in the presence of a contradiction such that any previous faulty inference is cognitively revised; notwithstanding our powerfully delusional universal 'conformation bias'. Thus do we hopefully avoid destructive "deductive explosion" with its valuable lost truth values. That is the unspoken goal of paraconsistent logical practice, and the saving grace of alternative paraconsistent and intuitive logical systems.

Many Valued Logical Systems (MVL). In classical 'truth functional' logic there are only two truth values—true or false. The logical syntax of human language is inherently truth functional. But there is a lot of reality between true and false. MVL drops this classical presumptive pretense and permits more than two truth values. The simplest strategy is three truth values (3TV logic): only true, only false, and *both* true and false where this third truth value is an indeterminate value—*neither* true nor false, or *both* true and false. Here we may view an assignment of a truth value not as a *truth function,* but as a *truth relation* among several possible logical functions.

Therefore, as to the relation of our prior and present unity of objective Science and subjective Spirit there exists a relationship of Niels Bohr's Complementarity—in the very midst of Heisenberg's Uncertainty—a convergence between dualistic objective Science and its nondual even mystical perfectly subjective ontic Spirit ground. And it is, and has always been thus. It is that nondual primordial wisdom knowing feeling Presence (*vidya, rigpa, christos,* gnosis)—far beyond or above the dualistic logical syntax of language—to which we awaken through prodigious unifying view and practice of our innate noetic cognitive doublet that is 1) objective, logical, conceptual Science, *and* 2) subjective contemplative, even nondual noetic direct yogic experience (*yogi pratyaksa*), or Spirit.

About as close as modern logical theory gets to perfectly subjective nondual Spirit is 'Cantor's Absolute': the set-theoretical class/set of all possible infinite sets—the all-inclusive unity of all possible infinite sets—a noetic, all-embracing superduper set! Problem. This supreme empyrean entity cannot itself be a set! If it were a set it would *ipso facto* necessarily include itself. But, as we have seen, self-referential sets crash and burn upon Russell's Paradox—the post-logical horror of the supremely annoying Russell-Zermelo Paradox. Yet we muddle through somehow. Paraconsistent logic helps.

Thus do we encounter the disconcerting Gödelian limit of our inherently dualistic human reason. We finally come to understand that ultimate absolutes, let alone all-pervading Ultimate Truth itself—perfectly subjective nondual Spirit—cannot be known by mere relative conceptual discursive logic and reason alone, no matter how subtle. This is indeed a logical sticky wicket for our European Enlightenment's (the Age of Reason) much valorized human reason, and for our futile discomfiting rational grail quest for absolute conceptual objective certainty.

The good news? The primordial wisdom door of our innate human trans-objective, 'post-empirical' *contemplative cognition* perforce opens wide to the nondual infinite 'supreme source' that is the always already indwelling luminous Presence of our 'supreme identity' revealed by analytic penetrating insight (*vipashyana*), the paraconsistent and contemplative view and praxis of our great noetic wisdom traditions. Trans-conceptual mindfulness meditation is the finite skillful means and conscious awareness portal that opens into such infinite nondual primordial wisdom (*jnana, yeshe,* gnosis, *epinoia*).

Minimal Logic. Norwegian Ingebrigt Johansson (1904-1987) created a post-classical subsystem of 'intuitionist logic' that precludes negative 'deductive explosion'. Minimal Logic is then a paraconsistent logic that suppresses the classical Greek logical Law of Excluded Middle, and the destructive Principle of Deductive Explosion (*ex falso quodibet*). Minimal Logic thus retains the positive aspect of intuitionist logic, and of classical logic, without the dualistic burden of the Law of Excluded Middle: "for every proposition either that proposition or its negation is true". Here there is no 'middle' or third possibility, no logical middle way. And we require a 'middle path' between the naïve extremes of *either* true *or* false, of either existence or nonexistence, of either particle or wave.

The Law of Excluded Middle is one of Aristotle's Three Laws of Thought that form the foundation of 24 hundred years of classical Greek deductive logic. The other two laws are the Law of Identity, each object is identical only with itself; and the Law of Contradiction, contradictory propositions/statements cannot both be true at the same time and in the same sense.

Thus Aristotle's wasteful Law of Excluded Middle has of late fallen on hard times. For Buddhist Mahayana Middle Way Prasangika Madhyamaka logic of Dignaga and Chandrakirti, and for many modern deductive logical systems, both East and West, including paraconsistent logical systems, the either/or but not both of the Law of Excluded Middle is replaced with some version of the 'principle of negation as failure'. That is to say, instead of a proposition being strictly truth functional—only true or only false—a proposition is either true, or it is unable to be proved true. Thus was David Hilbert's limited classical formal 'consistency' augmented by practical intuitional/experiential factors in determining new pragmatic axiom systems in the hitherto overwrought formalism of logic and mathematics.

Logical Intuitionism. Logical intuitionists L.E.J. Brouwer, his student Arend Heyting, and Hilbert's former pal Hermann Weyl over against the logical formalism of the logicists (mathematics may be reduced to formal logic) Hilbert, Russell, and Frege, had also questioned the pragmatic efficacy of the Excluded Middle Principle. Indeed, it was Brouwer who founded the prodigious mathematical meta-philosophy called Logical Intuitionism, approved by Kurt Gödel, as a foil to the prevailing absolutist logical formalism of David Hilbert, John von Neumann, Gottlob Frege, and Bertrand Russell.

The primary epistemology of Logical Intuitionism is its view of the nature of mathematical truth. What do we mean when we assert a mathematical truth? Quite astonishingly, for Brouwer mathematical truth is inherently *subjective*—a mental 'construction' whose validity is confirmed by way of human *intuition*—to wit, via a deductive but 'intuitionist logic'. Any mathematical object is essentially an 'ontologically relative' conceptual *construction* of an embodied thinking mind. Buddhist 'Universal Ontological Relativity' writ large.

In contradistinction, classical two valued logic asserts that the existence of a mathematical object can be proven only by formally, syllogistically refuting its nonexistence. Intuitionism holds that this does not constitute a proper 'construction' of the object, which is required for its admission to the *fantasque* dimension of Platonic mathematical existence. Thus is Logical Intuitionism a brand of 'mathematical constructivism'.

Simply but not simplistically stated, the contentious battle between the hard logical formalists and the soft intuitionists was about David Hilbert's use of the formal logical Law of Excluded Middle as applied to Cantor's infinite sets. Intuitionists allow application of the Law of Excluded Middle when confined to finite sets, but not infinite sets. Hilbert at first emphatically protested the Logical Intuitionist's subversive denial of what he termed the "mathematician's primary tool"—namely, Aristotle's classical Law of Excluded Middle. He opined, "It is the task of science to liberate us from arbitrariness... and to protect us from the specter of logical intuition." Hilbert, prior to his Gödel Incompleteness epiphany, demanded the extremist absolute logical certainty of his absolutist 'Hilbert's Program'. His rude awakening to 'logical mathematical incompleteness' is indeed the mark of a great mind. Hilbert went on to contribute to quantum logic his crucial notion of 'Hilbert space'.

Brouwer, in his seminal paper "The Untrustworthiness of the Principles of Logic", questioned the orthodox 'global web of belief' (Quine 1969) of the time, that the axioms of classical logic possess absolute objective validity, entirely independent of the subject matter to which they are applied.

While the hopes of the ambitious 'Hilbert's Program' and of the *Principia Mathematica* of Russell and Whitehead for an absolute logically complete, certain, and decidable logical foundation for *all* of mathematics have been forever dashed by the logical intuitionism of the Incompleteness Theorems of Gödel and Rosser, the debate over the legitimacy of paraconsistent logical systems continues.

Still and all, while it is not possible to formalize *all* mathematical systems, the math systems currently in use, namely, Zermelo-Fraenkel set theory in First Order Logic provides a practical completeness formalism for all the mathematics that matters, namely, the mathematics we all use.

We can now begin to see that even the desideratum of objective certainty that is formal logic, geometry, and mathematics has been tamed and "relativized" by the truth of 'ontological relativity', to wit the prodigious *Principal of Universal Relativity*. Objective spacetime stuff and our cognitive efforts to comprehend it has been happily reduced to the Antirealism and the quantum 'uncertainty relations' of our own human subjectivity. We conceptually impute, reify and construct our observer-dependent perspectival realities—our hitherto local 'real world out there' (RWOT)—via our deep cultural background (preconscious) inter-subjective 'theory laden' global web of belief. [Quine 1969] Bitter medicine indeed for the acolytes of our waning but still prevailing global scientific and sociocultural ontology that is Metaphysical Scientific Local Realism and its natural epistemic accomplice Scientific Materialism/Physicalism.

Prelude to Progress in 21st Century Physics. It is now clear that a profound epistemic logical/mathematical and ontic metaphysical healing is required to accomplish the great physics desideratum of a unifying Quantum Gravity Theory (QGT), the quantizing of Newton's and Einstein's 'Big G' gravity interaction. The continued absence of a consistent QGT has utterly stalled the progress of 21st century physics and cosmology; and cast an epistemic pall upon our hallowed Standard Model of Particles and Forces with its *lambda* Λ-CDM quantum cosmology. [Carroll 2003] On this, many theoretical physicists and most professional philosophers of physics now agree.

What to do? Super-Symmetric Super String M Theory? Its epistemic nemesis Loop Quantum Gravity? Alas, purely objective theories grounded only in classical logic are now *ipso facto* precluded! The inherent spooky subjectivity and acausal 'objective randomness' of the quantum theory cannot be so easily tamed. Until the quantum folks get their metaphysical house in order and construct a settled radical 'post-empirical' post-classical centrist or Middle Way integral Noetic Quantum Ontology, grounded in a 'paraconsistent' post-quantum logic, a QGT quantized gravity theory is but a pipe dream. Thus do we proceed with our quest for such a 'foundational interpretation' of Quantum Field Theory.

Unlikely Symposia. Yes. Let quantum physicists, philosophers of physics, and Middle Way Buddhist scholar-practitioners dialog over pizza and ale and together engage the ontic metaphysic of a viable, noetic (matter, mind, spirit subject-object unity) settled integral Noetic Quantum Ontology—the *ultimate* fundamental, foundational primordial awareness ground in which arises the *relative* 'universal quantum wave function' of QFT—along with all the rest of microcosmic and macrocosmic physical and mental reality. Such a process shall, one may hope, illumine the utter mystery of gravity—thermodynamic entropic "creator and destroyer of worlds" (*Bhagavad Gita*), and perhaps facilitate the quantization of Einstein's great gravity in a new paradigm paraconsistent Quantum Gravity Theory (QGT).

I am here suggesting that the metaphysical key that unlocks the occult secret of a unifying QGT—if it is mathematically possible at all—is to be discovered in a paraconsistent, post-classical, post-empirical Noetic Quantum Ontology that dares to tame the terrifying math, and extend its cognitive reach beyond our conceptual mathematical understanding into the trans-conceptual, nondual primordial awareness-consciousness ground that is the perfectly subjective metaphysical ontological basis of all appearing spacetime reality. Never mind that such heresy may well preclude academic tenure.

The Idols of the Tribe

The Metaphysics of Modern Science. We have just seen that the belief systems of the physical and social sciences are firmly grounded in the ontology of classical Metaphysical Scientific Local Realism and the epistemology and phenomenology of Metaphysical Scientific Materialism/Physicalism grounded in our prodigious *lambda-Λ* CDM Standard Model of Physics which includes Quantum Field Theory (QFT). What is conspicuously absent in the Standard Model is Einstein's gravity theory, namely his geometric General Relativity Theory (GRT), as we have seen.

That 'global web of belief' (Quine 1969) that constitutes the metaphysical ontologies or systems of belief of Big Science we now know admits of no logical or empirical certainty; no absolute truth. Scientific, even mathematical theories are fallible, provisional, relative, and inherently uncertain (Heisenberg's Uncertainty Relations), always evolving toward that next more inclusive but always incomplete theory.

So we need not fear 'spooky' trans-empirical metaphysical knowledge. Science is already steeped in its own brand of *a priori* metaphysical presuppositions, value assumptions, and conformation biases that underlie scientific ideology, methodology, and metaphysical ontology. These metaphysical "idols of the tribe" of Francis Bacon have become the 'false absolutes' of Big Science that belie the truth of the interdependent unity of mind and body, matter and spirit, subject and object, relative and ultimate reality dimensions, and of course, objective Science and its perfectly subjective Spirit ground.

The inherent acausal randomness and subjectivity of the quantum theory has compelled classical Science to plumb the depths of its metaphysical ontology. The strange ontic bedfellows that are Science and Spirit are always already a prior and present unity!

These scientific biases are clearly revealed in/as the Fundamental Laws of Physics that provide an astounding array practical uses. These laws include ontological, epistemological, and phenomenological principles. Ontology may be understood as the *ultimate* nature of appearing reality. Epistemology is the *relative* knowing of it, how we may know it; and phenomenology is our human perceptual, physical, and mental *experience* of it.

Well, exactly what are these highly practical and productive cognitive biases of the miracle of Modern Science that are pointed out by philosophers of science? These "idols of the tribe" are:

1) *The Ontological Principle of Monistic Physicalism.* Appearing spacetime reality is a 'pre-given', quantitative, observer-independently existing purely physical, 'real world out there' (RWOT). All appearing physical and mental spacetime reality is in its ultimate nature only purely physical mass/matter. [See 3 below]

2) *The Epistemological Principle of Objectivism.* That purely physical appearing reality is in principle completely knowable to a human observer via objective, quantitative, empirical scientific observation, experiment, and classical mathematical analysis. Subjective personal and transpersonal experience are largely taboo in proper science, thus the destructive 'taboo of subjectivity'.

3) *The Ontological Principle of Material Substance Monism.* Material, physical substance is all that exists, ultimate ontological ground of all appearing spacetime existence. There can be nothing other than, greater than, or inclusive of this one substantial, permanent, eternal, all-pervading physical substance. There can be no truth, being, ground, or qualitative dimension beyond material substance. [See 1 above]

4) *The Epistemological Principle of Scientific Reductionism.* All subjective experience—private, first person, mental, emotional, spiritual—is inherently reducible to objective, physical electrochemical neural correlates in purely objective electrophysical brain matter and brain function. Mind, love, beauty, mythopoetic wisdom, God (whether theistic or nondual) are no more nor less than an 'emergent property', an 'epiphenomenon', an 'artifact' of purely physical brain and its

electro-chemical processes and functions. Causality is always 'upward', from microcosmic physical to mental brain states. Holistic 'downward causality' from prior formless timeless primordial awareness-consciousness ground, to embodied mental, then physical form is ideologically precluded. Human consciousness is properly reduced to material central nervous system structure and function (Metaphysical Functionalism). The human being is here little more than Lewis Carroll's (Alice's) "bag of neurons".

5) *The Epistemological Principle of Local Universal Causal Determinism.* All spacetime physical and mental events are entirely determined by local, objective, quantitative, purely physical cause and effect. Holistic, 'top-down' nonlocal, acausal, indeterminate, qualitative explanation of events is ideologically precluded. This in spite of quantum indeterminacy. Here the sacrosanct causal Principle of Locality—Scientific Local Realism—is at odds with the acausal, indeterminist quantum entangled nonlocality of QFT, representing a knowledge gap between our two empyrean scientific pillars: Quantum Field Theory (QFT, QED), and the gravity of Einstein's General Relativity Theory (GRT).

6) *The Epistemological Closure Principle.* The above causal explanation of a purely physical, ultimately objective, quantitative, universal material substance reality ground of being is "causally closed" to any non-physical, non-objective causal or acausal explanation. Ironically, this bottom-up (physical to mental) epistemic closure precludes the nonlocal entangled acausal Quantum Field Theory, the most successful, most predictive scientific theory in 400 years of Modern Science.

7) *The Epistemological Principle of Universalism.* The preceding 'scientific' ideology is the only correct explanation of the relative and ultimate nature of all appearing spacetime reality.

> No differing view can possess truth. All differing views are in error. ["The Idols of the Tribe" is excerpted from Boaz 2021, *Appendix D*]

A Quantum Leap Into the Nature of Mind

What is now required is a 'quantum leap' in fear and trembling into a post-empirical, paraconsistent, antirealist, even noetic quantum mechanical metaphysic that Warner Heisenberg's Principle of Uncertainty has forced upon our human conceptual cognition. Should we become mired in a bygone grail quest for classical absolute objective certainty we shall miss the profound quantum centrist point of both the relative uncertainty (Heisenberg), and the ultimate complementarity (Bohr) of this lovely gift of the ground of reality that is the very Nature of Mind in which we all arise and participate. [More on such a Science-Spirit brave new world below.]

Whatever the metaphysical view-belief as to the ontological reality status of an observing self-ego-I—that it is absolutely existent (Metaphysical Materialism/Physicalism, absolutely non-existent (Metaphysical Idealism); or relatively existent but ultimately non-existent (the Buddhist centrist Middle Way)—what is clear is that our being here in time perforce requires the presence of an objectively present sentient observing consciousness who *experiences* something. No experiencing observing sentient self-presence means no sentient experience of stuff, means no existent real stuff at all. Both Buddhists and quantum mechanics require the awareness-consciousness of an embodied spacetime present experiencer of *something* that appears. Without an experiencer who is it that perceives and knows? We intuitively understand this.

As to such an objective experiencing self-consciousness, a self-ego-I, being here in a now scientifically dubious spacetime—what saith our noetic Primordial Wisdom Tradition—Hindu, Buddhist, Taoist, Abrahamic?

We are perennially told by the wise that this conscious physically embodied self, illusory or otherwise, that we all experience so vividly, interdependently emerges and arises from—and is indivisible from—our transpersonal, trans-conceptual 'supreme identity', our innermost indwelling nondual 'noself' (*anatman*) love-wisdom Buddha mind Christ nature, formless, selfless, timeless, already present luminous clear light Presence (*vidya, rigpa, christos*) of That (*tathata*) great boundless whole/ground.

And that indwelling 'clear light' love-wisdom mind 'instant pure Presence' is a manifest instantiation of the great enfolded ultimate primordial awareness-consciousness ground, unbounded whole itself, name it as you will: *dharmadhatu, dharmakaya, Nirguna Brahman, Tao, Wu/Mu, Yahweh, Ein Sof, Abba* nondual God the primordial Father—always unfolding now as relative spacetime particulars. It is 'That I Am Presence' (*tathata, satchitananda*) formless, selfless subject-object unity with noetic nondual ultimate awareness-consciousness ground in whom arises all this *relative* conventional embodied 'real' beingness participating in an *ultimate* illusory relative space and time; whether or not we, as egos, understand it conceptually, or believe it. We abide somewhere in the middle, awkwardly bestriding absolute existence and absolute nonexistence. Indeed a precarious, radical, 'ontologically relative' (we create our realities via our beliefs), rather off-putting, directly experiential centrist Middle Way metaphysic.

Thus, from the metaphysical ontology you choose, arises the phenomenal, karmic cause and effect reality you deserve. "What you are is what you have been; what you will be is what you do now." [Gautama Shakyamuni, the *nirmanakaya* Buddha of this Age.]

Once again, here arises the Mahayana Buddhist Two Truths dominant trope—the arising and manifesting *relative* reality of form from its ontologically prior emptiness, formless, timeless, selfless *ultimate* 'supreme source' ground, by whatever glorious name, concept, or belief. Just so, the prodigious quantum Ψ-wave function with its impenetrable mathematical formalisms, not to mention the mathematicians, arises from that same primordial ground. We require an integral Noetic Ontology to understand it.

Recall, the Buddha's inherently indivisible Two Truths—relative, objective, conceptual, scientific; and ultimate, subjective, contemplative, spiritual—are always already a prior and present invariant *one truth unity.* It is That (*tat, sat*) to which we awaken upon each mindful breath. These are the two unified reality dimensions in which we live. Wisely, compassionately and creatively balancing these two is our difficult lot, our joyous and confusing human condition. Is it not?

As we conceptually recognize and begin to contemplative directly realize (*yogi pratyaksa*) this utterly interdependent great *one truth unity (dzog)*—invariant through all physical, emotional, and mental human reference frames—we see our world not 'through a glass darkly', as dualistic judgments of attraction and aversion, approach and avoidance, but holistically, as the interconnected relatively imperfect but ultimately perfect unbounded whole that it actually is, beyond belief and atavistic dualistic thinking about it.

I have come to call this unifying cognitive pattern of reality recognition "holistic wisdom view". It is the 'pure view' from nondual primordial wisdom (gnosis, *jnana, yeshe*). It is grounded in relative 'discriminating wisdom' (*prajna, sherab*). It changes our lives for the better. Holistic wisdom view is the primary cause of surrendering and freeing our negative judgments of 'self and other' that we may skillfully express our innate *bodhichitta*—altruistic thought, intention, and engaged human action for the benefit of living beings. And that we are told by the wise is the very cause of cognitive peace, love and human happiness. Our human predicament and our endless quest for happiness made abundantly simple.

It is upon this holistic fundamental nondual Two Truths view that we establish our centrist one truth unity (*dzog*) that grounds a centrist Middle Way integral Noetic Quantum Ontology. We have seen that this vast nondual ultimate boundless awareness whole necessarily, mereologically (part-whole relations) embraces its relative instantiated parts—electrons, atoms, trees, stars, and all of us. And these interconnected parts necessarily, interdependently arise and participate (Buddha's 'Interdependent Arising') in that formless, timeless primordial ground. There is an apparent *relative* separation, but no *ultimate* separation. We are perforce always intimately connected to indwelling Presence That (*tathata*). We awaken to that great process via 'mindfulness of breathing'. This relation is one of primordial identity. We should feel better already! So let's unpack that process.

Concepts and Beyond. Well, these are all nice concepts. But how do we actually *connect* in our busy lives to indwelling Presence of that 'groundless ground', vast noetic boundless whole in whom this all arises? But we are always already connected! The bright Presence of That abides already within the spiritual Heart (*hridyam*) of the human being. No need to seek it elsewhere. We realize that miracle through assiduous practice—Ground, Path, Fruition always already accomplished View, Meditation, and skillful compassionate Conduct—under the guidance of a qualified meditation master. Instantly awaken now to that 'innermost secret' connection via a brief mantra prayer, for example *OM AH HUM*, or the Jesus Prayer. Once again, don't believe it! This urgent consideration is well beyond belief. As Buddha told so long ago, "Come and see (*ehi passika*)." That is a matter of trans-conceptual yogic *direct experience (yogi pratyaksa)*. [*Appendix A*]

Mahayana Prasangika Madhyamaka Middle Way *Dzogchen* Buddhist practitioners refer to the continuous arising of stuff in time within its all embracing primordial wisdom base (*gzhi rigpa*) or ground as 'Interdependent Arising', or 'Interbeing' (*pratitya samutpada*). Once again, that cause and effect emergence of *relative* spacetime form for embodied self conscious beings arises from its *ultimate* selfless, formless, timeless, acausal, nondual emptiness awareness-consciousness ground. That vast unbounded whole arises and descends continuously as the foundational causal matrix of prior causes and conditions.

On the account of this intuitive, immeasurable 'logic of the non-conceptual', wonder of wonders, we are never separate from That! Intimate *buddic* Presence of That. It is that primordial love-wisdom mind Presence, that mind essence, the direct experience (*yogi pratyaksa*) of the very Nature of Mind that knows, and feels this great truth of reality being itself. We arise, participate and have never departed that vast trans-conceptual nondual whole, by whatever august name. What is your mind? *That* is your mind. That is the one who knows. Who am I? *Tat Tvam Asi.* That I Am! We do have this choice.

As we begin to *contemplatively connect* via 'mindfulness of breathing' (*shamatha, vipashyana, jnanaprana*) to that pristine fundamental selfless 'noself' essential intelligence that we actually are, the nondual ultimate ontic selfless noself ground of our dualistic relative conceptual 'self' outshines like the brilliance of the sun on a snow mountain. Buddha told so long ago, "Noself is the true refuge of self." So we 'take refuge' in That. Gradual surrender of self-ego-I is how we connect. That's how we know and feel it—the luminous feeling-knowing love-wisdom Nature of Mind, bright noetic awareness Presence of it. So simple. H.H. Dalai Lama once revealed, "That's it! Just open the door." Let it be that this precious noetic nondual teaching of Gautama the Buddha (*shunyata,* emptiness of self), and of Jesus the Christ (*kenosis,* self-emptying) shall flourish on Western ground, ever outshining our biased dualistic concepts and beliefs about it.

A Noetic Quantum Ontology: An Integral Science of Matter, Mind, and Spirit

Physics and cosmology are quantitative. 'The qualitative' (value, volition) is active yet largely suppressed and denied in the common orthodoxy of the physical and social sciences. At long last, physics is now beginning to recognize and strategically develop this inherent qualitative dimension in science. The prodigious result shall be our growing recognition of the prior unity of objective Science and its perfectly subjective Spirit ground.

Prologue to a Foundational Centrist Integral Noetic Quantum Ontology. We've seen that what is urgently required for recognition of the prior and present interdependent unity of Science and Spirit/Spirituality is a settled integral Noetic (matter/mind/spirit subject-object unity) Quantum Ontology with a centrist epistemology and methodology that accounts for our human *noetic cognitive doublet,* both faces of our human experience—objective conceptual, and subjective contemplative/spiritual. We require a foundational quantum ontology that includes both an objective quantum mathematical description of quantum phenomena, and of the subjective trans-conceptual primordial emptiness of its ground.

In other words, we must utilize the methods and noetic technologies of Contemplative Science to engage and explore qualitative subtle subjective phenomena that are inherently 'hidden' from, and beyond the reach of our mainly quantitative objective conceptual sciences.

Such established contemplative praxis reveals a conscious, finite portal into superconscious infinite ground of the noetic, perfectly subjective, formless, selfless ultimate reality matrix emptiness base (*gzhi rigpa*) of all arising, evolving spacetime form, including our theories and beliefs about this whole noetic *process.* Quantum Field Theory with its 'universal quantum wave function'—the Ψ (*psi*)-wave—is such a relative epistemic theory in search of an ultimate ontological ground.

In due course all of this emerging spacetime matter/energy form ($E=mc^2$) evolves self-conscious human beings who desire to know and realize their relationship with that ultimate primordial awareness ground or 'supreme source' in whom their relative spacetime being arises. Human beings have evolved a cognitive life that bestows both objective conceptual, and subjective contemplative modes of understanding their experience of this ground. Clearly, we must engage and refine both.

However, in the West it is the 'scientific', objective conceptual 'global web of belief' that has almost entirely colonized the Western heart and mind. The mostly missing subjective contemplative technology and practice of the East restores a balance. Without such a balance inherently subjective ontology—the pursuit of untrammeled recognition then realization of the ultimate ground of being—remains encaged in mere concepts and beliefs *about* it; if it is considered at all. Sadly, that is the present state/stage of our collective human cognitive evolution—mirrored in our present 'global web of belief'.

As to the cognitive processional that is the four state/stages of our human cognitive life—1) direct attention-perception; 2) objective, conceptual, quantitative; 3) subjective, qualitative, contemplative; and 4) always already present perfectly subjective nondual unity—we remain substantially fixed in the first two. Entering in state/stage three marks the beginning of our grail quest for a clear, complete personal and collective ontological understanding that is ultimately realized in state/stage four, knowing/feeling direct yogic experience of this wondrous whole *process*, great gift of our being here in space and time.

And yet—"wonder of wonders" (Buddha)—"It is already accomplished from the very beginning" (Nagarjuna), deep within us. It is That to which we awaken upon each mindful breath.

Be that as it may, the ontic 'groundless ground' of everything—the perfectly subjective, 'implicate', enfolded, ultimate boundless awareness-consciousness whole and 'supreme source' of our wisdom traditions—may be seen as that all-embracing, all subsuming 'basic space' unity, *dharmadhatu,* or *chöying.* Within this 'implicate' vast Buddhist emptiness (*shunyata*) ground arises 'quantum emptiness', all unfolding objective, 'explicate' physical relative spacetime particulars—energy, mass, force, charge, particle-waves, the universal quantum wave function, and the continuous phenomenal experience of embodied beings. Within that timeless formless emptiness ground all of this arising stuff of being here in spacetime participates, interacts, and is instantiated as form. Buddha bespeaks it thus: "Form is empty; emptiness is form."

Our ordinary human cognizance is not other than that basal 'clear light' primordial awareness ground. We have just seen that embodied minds naturally and spontaneously arise unbidden and continuously from that vast expanse, that 'supreme source' that is our 'supreme identity'. It is the inherent nature of the timeless formless primordial ground to manifest or express as evolutionary spacetime form. In the fullness of time involutionary form may evolve sentient beings with self-consciousness, which in due course and by grace may be realized as Buddha consciousness (*buddha-jnana*). Let us now recognize, then realize the interconnected unity of this intrinsic ontic relationship of boundless emptiness ground, and spacetime form arising therein. "No small matter is at stake here. The question concerns the very way that human life is to be lived." [Plato, *The Republic Book* I]

We saw in Chapter I that when viewed mereologically (part-whole relations), the panpsychic cosmopsychic prior ontological unity that is this great awareness whole ground subsumes and embraces its *relative* parts, while the parts perforce participate in and instantiate the *ultimate* vast nondual primordial whole itself. In the Buddhist view the prior and present invariant *one truth unity* of the Buddha's Two Truths—relative form and its ultimate ground—describe this wondrous ontic process. A mereological 'proof' of the existence of post-theistic, non-creator, nondual, indivisible Godhead?

The prodigious 'universal quantum Ψ-wave function' may be seen as the 'objective randomness', quantitative, mathematical conceptual voice of qualitative Big Science, as it continuously arises in its trans-conceptual, 'post-empirical' perfectly subjective nondual whole. That unbounded whole is the vast formless, timeless primordial awareness-consciousness Spirit ground of Being Itself in whom our quantum Ψ-wave function and its many quantum wave function forms arise. And that indeed is the epistemic foundation of our present inchoate centrist integral Noetic Quantum Ontology—as we shall soon see.

Recall, our human *noetic cognitive doublet*—dualistic, quantitative, objective, conceptual; and qualitative, subjective, contemplative, even nondual. We shall come to know and directly experience this cognitive doublet as always an ontologically prior yet phenomenally present complementary interdependent invariant one truth reality Science-Spirit unity. We must know that in order to understand our Noetic Quantum Ontology.

Clearly, such a Noetic Science of Matter, Mind and Spirit requires a methodological, 'post-empirical' relaxing of the adventitious limits of obsessively objective positivist view and praxis of objective Science with its prosaic 'taboo of subjectivity' regarding *a priori* contemplative Spirit knowledge. This habitual classical 'old paradigm' Scientific Local Realism and Scientific Materialism/ Physicalism bias/dogma, while quite useful in doing the math and the theory, still obstructs our emerging nascent 21st century Noetic Revolution in Science and Spirit that now arises phoenix-like from the ashes of a failed Greek Metaphysical Scientific Realism/ Materialism ontology.

New Paradigm indeed. Such a Kuhnian 'scientific revolution' is the quantitative aspect of the qualitative Noetic Revolution in matter, mind and spirit that is now upon us.

We have seen that the basal nonlocal entangled quantum emptiness of the proto-physical Universal Quantum Vacuum, the 'quantum zero point vacuum energy field' (ZPE) is said to be constant density dark energy, Einstein's cosmological constant *lambda* Λ of recent Quantum Cosmology's *lambda* Λ-CDM (cold dark matter) Standard Model of Physics. And that relative truth of Modern physics' quantum emptiness is an analog of the parallel pre-modern wisdom of Buddhist ultimate truth boundless emptiness (*shunyata/dharmakaya/kadag*). A centrist Middle Way Noetic Quantum Ontology shall unify and embrace these two faces of boundless emptiness—relative quantum and the all subsuming ultimate emptiness ground in whom it arises.

Ontic Conclusion. Hence, it is the nondual unbounded 'implicate order of the unbroken whole' (Bohm)—formless, selfless, perfectly subjective ultimate noetic primordial awareness, quantum/ Buddhist emptiness ground of all arising relative physical, mental, and spiritual reality—that is the conceptual foundation of our unified objective/subjective centrist Middle Way Noetic Quantum Ontology. That metaphysic conceptually bestrides quantum dual nature of Max Planck's 1900 'quantum of action', namely, the relative objective monumental quantum mathematical formalism, and its noetic nondual perfectly subjective primordial awareness ground. That is the 'grounding relation' of Quantum Field Theory and its Ψ-wave function to its prior all embracing aboriginal source condition.

In other words, the quantitative, quantum, objective, interdependent, relative truth dimension is embraced and subsumed within the qualitative, subjective, ultimate truth dimension of nonlocal, nondual noetic original ground or base (*gzhi rigpa*), 'supreme source' that is all-pervading complementary unity of Tibetan Vajrayana Buddhist *Perfect Sphere of Dzogchen.* And that clear light bright Presence is always already embodied here and now in our human form—even in the intellectual virtuosity of quantum physicists and philosophers of physics.

Let us then more deeply engage the relative objective Science of the ultimate perfectly subjective Spirit 'groundless ground' in whom it arises.

Quantum Emptiness and Spacetime Form. The para-self-consistent mathematical formalism of the Quantum Field Theory is inscrutable to most of us. Mathematical theorems presume to describe slices of physical reality. These pieces of the whole reality pie may be described in holistic and quite specific non-mathematical terms. So we need not be overly troubled by arcane quantum mechanical annotations, which I have here avoided. In this spirit let us then continue our exploration of that wondrous gift of Max Planck's 'quantum of action' in a more comfortable, if still a bit technical idiom.

The recent physics Λ-CDM (*lambda* cold dark matter) Standard Model is a prelude to Quantum Field Theory (QFT/QED/QCD) that subsumes all of the known subatomic elementary particles and their interactions—with the notable exception of Einstein's gravity (GRT) with its recently discovered force carrying graviton particle-waves. Quantum Chromodynamics (QCD) is the part of QFT/QED that pertains to the Strong Nuclear Force interactions, and the ZPE quantum vacuum in which physical stuff arises.

According to Quantum Field Theory the fabric of ZPE nearly 'empty space' consists of *fields*, a proto-physical quantity represented by a number (tensor) that has a value for every point in space and time, occupies space and contains energy, thus precluding a classically empty 'true vacuum'. The entire physical universe is composed of such matter fields whose quanta are *fermions* (electrons and quarks), and force fields whose quanta are *bosons* (photons and gluons); and finally, a Higgs field (ϕ) whose quanta is the mildly massive (125 GeV), zero charge, scalar (spin zero), highly unstable Higgs boson which generates, by way of the 'Higgs Mechanism', the mass of all appearing physical reality—sparrows, trees, stars, people and Buddhas—that are given through the robust fermions (quarks and leptons) that we have come to know and love.

Matter fields and force fields are presumed to have 'zero point energy'. They 'exist' nominaly as the all-pervading quantum 'zero point energy field' fluctuations (ZPE), the para-physical Unified Quantum Vacuum, the 'quantum emptiness' which itself arises and plays in its ontic prior nondual primordial timeless, formless awareness-consciousness nondual Spirit ground that subsumes even the "implicate order of the enfolded vast unbroken whole" [David Bhom]

ZPE, aka the 'quantum vacuum state', or 'quantum vacuum' is the lowest energy state of any particular quantum field. This universal 'vacuum state' is then not mere 'empty space' but the sum of all quasi-physical zero point energy fields. This is known as the 'quantum vacuum energy'. I have come to refer to it as *relative* 'quantum emptiness' for it closely parallels natural formless *ultimate* Buddhist emptiness in which, or in whom our quantum realities arise and are instantiated. The quantum vacuum state has measurable effects (e.g. the Casimir effect). It is suspected to manifest as Einstein's cosmological constant (Λ), the very vacuum energy density of space. That is probably the diaphanous *dark energy* that impulses the speedy recession of this expanding universe to what is now presumed by most cosmologists to be its ultimate entropic thermodynamic 'heat death'—the 'Big Chill' fate of this present universe of ours—a few trillion years hence, give or take a trillion years or two.

The quantum ZPE vacuum energy is the ground state energy that exists throughout the space of the entire physical universe, or even multiverse. The 'average expectation value' is the 'vacuum expectation value' (VEV). For example, the all-pervasive Higgs boson particle has a non-zero vacuum expectation throughout the whole of spacetime. A non-zero vacuum energy is expected to contribute to the cosmological constant (probably dark energy) which causes the accelerating expansion of this present universe of ours.

The *'cosmological constant problem'* is, as we have seen, the incompatibility between the observed values of the 'vacuum energy density', the tiny but non-zero value of the cosmological constant (Λ), and the theoretically huge values predicted for it by QFT. The difference is as high as 120 orders of magnitude! And yes, Steven Weinberg called it "the worst theoretical prediction in the history of physics." How shall we understand this?

Other than tweaking Einstein's sacrosanct GRT to modify his gravity field equations, the *Anthropic Principle*, both weak and strong versions, has gained increasing acceptance. Anthropic arguments posit that only spacetime regions of small vacuum energy, like ours, are capable of supporting intelligent life who may then venture such impudent questions. That said, 'modified gravity' solutions (e.g. Milgrom's MOND) to the prodigious cosmological constant problem are still considered promising alternatives.

The 'QED quantum vacuum' describes the electromagnetic interactions between electrons and photons in the ZPE field. The Quantum Chromodynamics (QCD) vacuum describes interactions between quarks and gluons in the ZPE field. Elementary particles are therefore 'excited states' of the all-pervading ZPE quantum vacuum field state itself, and all matter-energy qualities and properties are but 'vacuum fluctuations' spontaneously arising and instantly exiting interactions with the basal quantum Zero Point Energy Field. From such fleeting proto-physical quantum vacuum fluctuations—arising in the primordial ground of everything—arise our beloved all too impermanent spacetime realities. Thus for Middle Way Buddhists ultimate *shunyata*/emptiness (of intrinsic existence) and impermanence (*anitya*) (of relative conventional existence) are the two foundational principles of appearing spacetime reality.

For quantum cosmology ZPE is considered to be the probable explanation for the 'cosmological constant *Λ*' and thus of the mysterious dark energy that accelerates our expanding receding universe. And that creates the 'cosmological constant problem', as we have just seen.

How much energy is actually contained in this mysterious "sea of energy" (Dirac) that is the ZPE unified quantum vacuum field state? Yes. QFT requires it to be very large. The Heisenberg Uncertainty Principle allows the vacuum energy to be large enough to produce quantum field interactions, yet small enough to remain consistent with the equations of Einstein's General Relativity (GRT) which describes a cosmological 'flat space'. Here QFT, the physics of the very small, and GRT, the physics of the very large, remain mathematically incommensurable theories of physical field quantum interactions. And yes, we need a quantum gravity theory (QGT) to quantize Einstein's gravity and unify these two prodigious pillars of recent theoretical physics. And we need a 'post-empirical', post-quantum Noetic Quantum Ontology for that.

Foundational Quantum Ontologies in Review

Mereological Prelude. Well, what is the foundational formless, timeless *ultimate* ground of the fundamental *relative* proto-physical Unified Quantum Vacuum zero point energy field (ZPE)? The physical, quantitative, cosmic spacetime *physical ground* that is the universal quantum wave function (*Ψ*) arising from its non-empty cosmic ZPE vacuum field are perforce *ultimately* grounded in a subtler, all subsuming, post-quantitative, trans-rational, post-empirical, formless primordial awareness emptiness *kosmic* 'groundless ground'—boundless "implicate order of the vast unbroken whole" (Bohm) in which, or in whom they continuously physically manifest as our cosmic spacetime particulars.

This noetic aboriginal ground is a mereological (part-whole relations) necessity, as we have seen. Parts perforce require more inclusive wholes that include them. Wholes require their participating instantiating parts. Mereology thus explains the urgent 'grounding relation' of we unfolding participating parts to our all embracing enfolded whole, by whatever grand name and form (*namarupa*). We should feel better already. Let's now rest for a few happy moments in that urgent lucent truth. [*Appendix A*]

Recognizing, then realizing this great truth requires noetic *contemplative technologies* and research methodologies. [Boaz 2023] Such Contemplative Science utilizes both quantitative objective, paraconsistent third person data sets, and the qualitative, though still objective data sets of personal, subjective, introspective, even contemplative first person reports of highly experienced meditation practitioners and the very subtle minds of their amazing masters. [Wallace 2009; Begley 2007; Boaz 2022] These remarkable beings naturally weave their nondual primordial love-wisdom mindstream into the splendent fabric of sociocultural space and time—for the benefit of us all.

Thus are the Mahayana/Tibetan Vajrayana Middle Way Madhyamaka Two Truths—spacetime Relative Truth, and post-empirical, all-pervading nondual Ultimate Truth that pervades it—unified in the Buddhist nondual *Perfect Sphere of Dzogchen.* This *kosmic* gift of spacious, empty (*shunya*) unbounded whole (Basic Space of *chöying/dharmadhatu*), nondual *ultimate* reality itself (*dharmakaya, kadag, Tao*) is the formless, timeless, selfless perfectly subjective *kosmos* primordial ground state in whom this relative cosmos and our theories about it arises and plays.

Again, mereologically, the multiplicity of the reality dimension of physical and mental form—the particular cosmic parts—are perforce subsumed by the greater primordial, all embracing boundless *kosmic* awareness-consciousness whole itself. So many words for That that cannot be told in words.

What is the current physics explanation for this ultimate ground of the penultimate quantum ZPE vacuum energy field? Spoiler. It's not all that ultimate. It still remains in the classical, para-physical reality dimension. As if there were nothing more inclusive or profound than Big Science's well considered cosmology conjectures. We are slowly growing beyond that bygone classical cognitive limit. And yes, the quantum ontologies now on offer have failed to probe beyond toward a 'post-empirical' holistic integral Noetic Quantum Ontology.

Foundations for a Noetic Quantum Ontology. Relativistic Quantum Field Theory (QFT/QED) is universally considered to be our most fundamental if still ontologically incomplete theory of the behavior of mass/matter/energy of the microphysical systems that constitute appearing macrophysical spacetime reality—trees, stars, and all of us. Its predictive successes have been spectacular. It has given us the computer, TV, smart phones, laser communications, and the nuclear bomb.

The subjective voice of QFT/QED as it arises in the 'superposed' pre-collapse 'universal quantum wave function' (the Ψ-wave) of Irwin Schrödinger has left that prodigious theory vulnerable to a panoply of objective incommensurate 'foundational interpretations'—quantum philosophical ontologies. What does the great Relativistic Quantum Field Theory (QED/QED) reveal about the *ultimate* nature of appearing spacetime reality whose microcosmic electron position and momentum it presumes to measure? Just what is the ontic ultimate nature of this physical reality that is described by reasonably obscure formalist quantum mathematics?

Such ontological philosophical speculation naturally arises from the recondite mathematical formalism of quantum descriptions of physical spacetime reality. It is here that we enter in the cognitive dimension that is all pervading formless *ultimate* primordial ground of this *relative* spacetime regime of physical and mental form—and our conceptual theories about it—to wit, our 'global web of belief' (Quine 1969). Thus do we continue our grail hunger for a providential, reasonable centrist integral *Noetic Quantum Ontology* whose elements I shall finally describe below.

A Very Brief History of Quantum Physics. As to the historical cognitive processional of 20th century physics, we have come from the smug naïve certainty of empyrean classical physics at the turn of the century; to the discovery of the 'quantum of action' in 1900 by Max Planck; then the Special Relativity of his friend Einstein in 1905 grounded as it is in Maxwell's Equations, 1862 mathematical foundation of classical electromagnetic theory. Then on to the Rutherford-Bohr model of the atom in 1913; and to the macrocosmic matter waves of Prince Louis de Broglie in 1924 which showed that microscopic quantum physics must apply to macrocosmic matter. Now the 1927-1928 contribution of Warner Heisenberg's Principle of Uncertainty, and Niels Bohr's 1928 Principle of Complementarity, just after the wondrous 1927 Ψ-wave equation of Irwin Schrödinger; then on to Paul Dirac's 1927 'transformation theory' that demonstrated the consistency of QFT with Einstein's Special Relativity Theory, and finally to Richard Feynman's 1959-1961 enhancement of Dirac's 1928 Quantum Electrodynamics (QED) that further 'relativised' QFT by taming the vexing 'problem of infinities' through an admittedly problematic but very practical 'remormalization' strategy.

From this noble history we observe the natural but halting evolution of the objectivist physicalist bias of classical physics on to the breakthrough that is this inchoate inherent subjectivity of 20th century quantum physics.

A proper much needed post-quantum Noetic Quantum Ontology is a bid to understand the *ultimate* aboriginal nature of the very ground of that history of *relative* conditional conceptual QFT/QED quantum understanding; the primordial all subsuming ground of the formalist quantum mathematical operators in which knowledge and knower together arise and participate.

Our monumental quantum story is far from complete. As we attempt to unify QFT/QED with the gravity of Einstein's General

Relativity Theory (GRT) we may expect some radical changes to both. Here we shall continue to enhance our holistic understanding as metaphysical qualitative quantum ontology illumines quantitative quantum and relativistic epistemology, methodology, and phenomenology.

Why not let well enough alone? Why worry about metaphysical ontology and just enjoy the formalist quantum mathematics that have bestowed upon humankind our precious smart phones, microwave ovens, and the rest? Perhaps we should "Shut up and just calculate". [Mermin] Well and good.

However, should we desire to know what this most sublime theory in the history of Science reveals about the actual *ultimate* nature of the human mind—human consciousness/experience, and the cognitive and behavioral causes of human happiness—then we must at last engage the inherent relationship of that *ultimate* metaphysical ground in which such a *relative* quantum theory arises. In this healthy ecumenical spirit let us then further engage Big Science's hitherto feeble attempts at foundational quantum ontology.

Quantum Ontologies Now on Offer. Of the twenty or so foundational 'interpretations of quantum mechanics' now abroad in the quantum mindscape we've seen that there are seven on offer that have commanded the most critical attention. None of them are 'post-empirical' integral noetic. None refer to the prodigious metaphysical primordial ground in which the *Ψ*-wave function spontaneously arises—the necessary 'grounding relation' of relative theory to its prior and present ultimate source condition. RQM comes closest.

We have seen that these well considered extant quantum ontologies include: 1) the original default 1927 antirealist 'proto-collapse' Copenhagen Interpretation of a collapse-ambivalent Bohr and Heisenberg; 2) nonlocal yet still realist 'hidden variables' models (e.g. the Bohm-deBroglie pilot wave theory) which attempt to explain away the inherent random subjectivity of the quantum wave function; 3) the von Neumann-Wigner "consciousness causes collapse" model in which a subjective human consciousness is necessary to interpret and therefore complete an objective quantum measurement, which may include Stephen Hawking's antirealist, or para-realist Model Dependent Realism (MDR) view (*The Grand Design* 2010); 4) local realist GRW and other local 'collapse models'; 5) stochastic epistemic Quantum Bayesianism (QBism); 6) the radical, hyper-realist 'purely mechanistic' Many Worlds Interpretation (MWI) which I have mercilessly criticized above. 7) We have in some detail explored what is perhaps the most fluent of the bunch—Carlo Rovelli's Relational Quantum Mechanics (RQM). [We shall once again revisit these foundational ontologies below.]

Each of these ontic 'foundational interpretations of quantum mechanics' is an attempt to explain or explain away Irwin Schrödinger's 1927 ineffable 'universal quantum Ψ-wave function' of the 1928 Quantum Field Theory (QFT) of Dirac and Heisenberg.

QFT is here to stay. A settled foundational quantum ontology must be a centrist "noetic ontology" that engages both relative and ultimate dimensional voices of QFT. Such a 'final' quantum ontology will certainly utilize quantum quantities common to most of these seven, but must now at long last engage a qualitative 'post-empirical', postformal, paraconsistent holistic ontology that is prior to, yet includes and subsumes the best of both present and new revisions of the requisite formalist quantum mathematical operators; in short, a centrist middle way integral Noetic Quantum Ontology.

Thus do we begin to heal and unify objective quantitative cognition with subjective qualitative cognition. These two modalities of our unified human cognition have been hitherto torn asunder under sway of the prevailing classical, mechanistic materialist/physicalist biases and belief systems of Modern Science. And that deep background (subconscious) cultural 'global web of belief' (Quine 1969) includes the dualistic metaphysical 'scientific' theory laden dogmas of monistic Scientific Local Realism and monistic Greek Scientific Materialism/Physicalism that has beset and engulfed our Modern and Postmodern Western mind and its scientific and intellectual culture.

This unifying paradigmatic *'mind change'* in Big Science has proven exceedingly difficult given 400 years of European Enlightenment physics—'the idols of the tribe'—encaged as it is in the 'scientific' realist/materialist ideologies that are our paradigmatic noble Greek and Hebrew metaphysical legacy.

Ultimate Spirit Embraces Relative Quantum Field Theory. QFT may be seen as physics' nascent *relative* epistemic cognitive architecture for accomplishing an *ultimate* ontological understanding of the whole nature of appearing reality, both objective and subjective. Yes. The relative quantitative QFT/QED mathematical formalisms must now be integrated with qualitative ultimate all subsuming *kosmos,* boundless whole itself—nondual primordial awareness-consciousness ground whence relative spacetime stuff and our ever-evolving inherently incomplete theories about it emerge.

I have argued here and elsewhere that panpsychic (priority monistic cosmopsychic) acausal nondual Buddhist *Dzogchen* as it arises from its conceptual foundation in causal Buddhist Middle Way Prasangika Madhyamaka philosophy constitutes an *ultimate* foundation for the *relative* epistemic 'universal quantum wave function' mathematical formalisms. [Boaz 2022 *Ch. VII*] I have above outlined above such a 'post-empirical' (QFT itself is post-empirical) ontology, a centrist Middle Way Noetic Quantum Ontology that is a foundational interpretation of QFT with its prodigious universal quantum Ψ-wave function as it is instantiated in its timeless nondual noetic primordial awareness ground. I shall proceed with that view below.

The immeasurable challenge is this. That greatest of human intellectual achievements, the prodigious Standard Model of particles and forces, with its recent *lambda* Λ-CDM (cold dark matter) Standard Model Cosmology is known by all the players to be incomplete. It still clings to the classical, orthodox, waning paradigm dogmatic materialist metaphysic that is extreme objectivist Scientific Local Realism and the metaphysical Physicalism/Materialism of a bygone classical Galilean-Newtonian cosmos. Such an ontology sees only objectively 'real' purely physical objects (substance) existing observer-independently, permanently frozen in a substantialist real time (*t*), eternally in an absolute, objectively real, mechanistic purely physical Minkowski 4-D spacetime manifold.

Good news! A new syncretic paradigm is afoot in the recent physics and cosmology cognosphere. The old classical knowledge paradigm is now being integrated with the inherent 'objectively random' subjectivity of the QFT/QED quantum paradigm. That paradigmatic unification, with the counsel of causal Buddhist Middle Way Madhyamaka philosophy and its acausal *Ati Dzogchen* ground portends, as I have said, the advent of our rapidly evolving 21st century Noetic Revolution in objective Science and its perfectly subjective nondual Spirit ground.

In brief, a classical, purely physicalist, objective, local realist, observer-independent spacetime has now fallen on hard times. Physicists are at last beginning to hear Einstein on time: "Time—past, present, future—is an illusion; albeit a very persuasive one." With new work in quantum cosmology most physicists have thrown out the absolute reality of empty space as well. The ontic result leaves our beloved 4-D spacetime realities relatively real, yet not ultimately real. Sounds like Middle Way Madhyamaka Buddhist philosophy. [H.H. Dalai Lama 2004; Boaz 2020, *Ch. V*]

In any case, the notoriously perverse mathematical incommensurability of QFT/QED with Einstein's General Relativity Theory (GRT)—the formalist split between these two great pillars of modern physics—will continue unabated without an ideological softening of Modern Science's hyper-objectivist monistic Metaphysical Scientific Local Realism with its monistic Physicalism. QFT refuses to be mathematically crammed into a procrustean bed of classical observer-independent realist/materialist physics—try as we may. Sadly, most theoretical physicists remain fixed in this bygone classical ideology.

We have seen that paraconsistent logic and mathematics offers a cognitive respite, and a reasonable rejoinder to this ostensible epistemic paradox that is objectivist formalist QFT math coexisting with the postformal subjectivity of a 'superposed' pre-collapse quantum Ψ-wave function. Let us listen and hear both of these voices—objective and subjective—of the wondrous quantum theory that we may reap all of the benefits. There is no need to affirm mathematical quantum objectivity and deny the inherent subjectivity of the quantum ontology to which it points. Proper Science engages the data that appear to our senses and to our mind in all its objective and subjective splendor, while surrendering ideological bias, no matter how entrenched.

The waning classical scientific knowledge paradigm view is now considered by most philosophers of physics, and a few theoretical physicists, to be, in a final analysis, a failed ontology. How is this so? 1) It contradicts the inherent acausal 'objective randomness' and therefore inherent subjectivity of quantum theory, to wit, 'always correct' QFT/QED. 2) It fails to engage the nondual, basal, perfectly subjective primordial awareness-consciousness ground in which, or in whom our objective, conceptual and mathematical quantum worlds arise. We are seeing once again that Science must continue to surrender its ideological 'taboo of subjectivity' and explore *a priori* subjectivity and objective paraconsistent logical intuitionism in order to approach such a unified understanding. This shall require post-bias intellectual honesty, courage, and an open post-scientific Zen Mind-beginner's mind.

Post-Quantum Holism. Yes. We desperately need a unifying Quantum Gravity Theory (QGT) to heal this seeming epistemic split between the minute microcosmic realm of Max Planck's Planck Scale 'quantum of action' (this Planck constant of nature is $\hbar$ = 13.1Q), and the vast large scale macrocosmic dimension ruled by Einstein's wondrous gravity equations in his prodigious General Relativity Theory. For that we must engage a metaphysical ontology that reaches beyond the mathematical formalism of QFT/QED.

Some physicists (e.g. David Bohm), and most philosophers of physics, along with Mahayana Buddhist philosopher-practitioners know that there is no innate *ultimate* dimensional separation between our appearing microcosmic and macrocosmic phenomenal regimes. The apparent *relative* separation is semiotic and conceptual. Indeed, the whole of physical spacetime appearing reality with its monumental quantum Ψ-wave function, the consciousness of an observer/experimenter, and its decoherent measurement instruments is already unified and subsumed in the formless, timeless, boundless, indivisible, nondual primordial awareness-consciousness ground in which this whole shebang arises. Such a holistic metaphysical understanding must be integrated into an integral noetic metaphysic of the universal quantum wave function with its arcane mathematical formalisms, as we have seen. A propitious, post-classical integral Noetic Quantum Ontology imperfectly accomplishes this holistic aim.

Here's the rub. Contemplative blissing out in the perfectly subjective ground of post-quantum being is not enough. We must skillfully engage our objective cognitive capacity to conceptually and mathematically explicate that prior unity of objective quantum form and its illusive subjective nonlocal entangled quantum emptiness ground while remaining present to the prior trans-conceptual, nondual truth of the matter.

That is to say, we maintain an awareness of the present state of nondual unity of our perennial Two Truths—relative spacetime form, and its formless, selfless emptiness ground while doing the math. We construct our integral Middle Way Noetic Quantum Ontology, and any other such emerging quantum ontologies upon that prior and present unity. A bitter cognitive pill indeed for 21st century quantum physics still clinging as it does to a bygone Newtonian, realist/materialist classical physics paradigm.

This scientific and cultural Kuhnian knowledge 'paradigm shift' is well under way as we begin to surrender our absolute Scientific Local Realism, and its more recent issue—'Structural Realism'—along with 'old paradigm' Scientific Physicalism ideologies to the emerging theme of quantum holism. Let this absolutist Local Realism/Physicalism cognitive bias transform into a pragmatically useful relative-conventional practice.

Thus it is, a radical scientific and cultural Kuhnian paradigm shift in perceiving and thinking is now upon us as the descending objectivist classical relativistic physics paradigm is gradually surrendered to the ascending subjectivity of the quantum physics paradigm with a postformal 'foundational quantum ontology'. So yes, we still require a multidimensional integral Noetic Quantum Ontology to reveal the more subtle benefits of that peerless quantum theory. This Quantum Revolution has precipitated the next global revolution in science, culture, and religion/spirituality. I have come to call it *The Noetic Revolution in Matter, Mind, and Spirit.* [Boaz 2023]

On the accord of philosopher of science Thomas Kuhn, such a new science paradigm 'gestalt shift' requires two or more generations to become settled dogmatic orthodoxy. As the old paradigm tenured acolytes expire, new paradigmatic blood enters the hallowed halls of academic learning—but the blink of an eye in the fullness of time.

Still, I'm impatient. May this lugubrious process be somehow expedited? With a modicum of paraconsistent logical 'special pleading' we shall see that it may. Indeed, as we have seen, the random, acausal and antirealist subjectivity of the global quantum metaphysic has now nearly dethroned the prevailing Scientific Local Realism and prosaic 'common sense' metaphysic of Newton's and Einstein's observer-independent absolutely real world out there (RWOT). Perhaps we need only to dump the ideology that is Scientific Local Realism altogether; or at least commit it to the outer darkness of some 'spooky' gossamer antirealist metaphysic, as did Niels Bohr. It is after all still essential for calculating the position and momentum of objectively 'real' electrons after they have 'collapsed' from the surreal subjectivity of an infinity of 'quantum superpositions.

In any case, we have seen that the next step in this urgent process of doctrinal, even ideological scientific change must be a reasonable, imperfect centrist and integral (containing all the parts that make the whole complete) Noetic Quantum Ontology metaphysic that unifies the realist, conceptual, objective dimension of quantum formalism (and the human consciousness that experiences the mathematics), with the subjective dimension of formless enfoldment in all subsuming primordial awareness ground in which (or in whom) this all unfolds for us and emerges in space and time. It is this unifying 'grounding relation'—grounding by subsumption—to which the inherent subjectivity of the superposed universal quantum wave function points. Such a providential Noetic Quantum Ontology shall facilitate our emerging inchoate scientific and cultural paradigm shift; to wit, *The Noetic Revolution* in matter, mind and spirit that is now abroad in our global human cognosphere. [Boaz 2023]

Quantum Nonlocality, Quantum Ontology and the Problem of Consciousness

Relativistic Quantum Field Theory: Variations on a Theme of Wholeness. Let us now further engage the inherent subjectivity of nonlocal entangled ZPE quantum emptiness and the emergence of objectively 'real' spacetime form.

We've seen that quantum mechanics as it has evolved into Quantum Field Theory (QFT) and then relativistic QFT/QED has engendered an incipient global scientific and indeed cultural epistemic and ontological revolution. Objective things are not at all as they appear! The illustrious history of both Western and Eastern philosophy and religion has pondered this urgent truth since its beginning in prehistory. Now, with the advent of the inscrutable quantum theory the comfy certainty of 'common sense' Metaphysical Scientific Local Realism has become engulfed in a scary web of Niels Bohr's 'quantum uncertainty'. What hath God wrought!?

QFT/QED is the most precise, predictive, unifying, well tested scientific theory ever conceived in the 400 years of Modern Science. Yet an interpretive explanation of its meaning and significance to the *ultimate* ontic nature of *relative* quantum spacetime reality remains still an utter conceptual mystery. Yes, that conundrum has become known as the "quantum enigma", or the "quantum mystery".

We've seen that disentangling quantum nonlocal entanglement is the sector of physics known as 'Foundations (or Interpretations) of Quantum Mechanics'. Quantum mechanical predictions are always stochastically correct. But how shall we choose between these proliferating competing quantum ontological interpretations? What does relativistic QFT reveal about the ultimate nature of appearing relative conditional quantum reality? Some have said, "Forget philosophy, just do the calculations; that's enough". We shall have our computers and smart phones. Still we need to know what QFT means for our understanding of the actual ultimate nature of appearing relative quantum space and time.

[Broadly construed, *Ontology* (Gr. *ontos*) inquires as to the ultimate nature of being. *Epistemology* inquires as to how we may know it. *Phenomenology* inquires into our human experience of it. Objective *Science* gives us a mathematical and physical epistemology. Nondual perfectly subjective *Spirit* is the ultimate ground or boundless whole in whom this wondrous *kosmic* process arises and is instantiated.]

Quantum mathematical formalism is by default an instrumentalist ("Just do the math and forget the philosophy") antirealist ontic view. The formal truths of mathematics—1+1=2; the proofs; 3 is a prime number, and the rest—are *a priori* 'necessary truths', necessarily true in all universes of discourse, anywhere in the cosmos. But what is the ontological *ultimate truth* foundation of such relative, conditional formal mathematical objects? Are they human observer-independently real, existing 'out there' somewhere for smart mathematicians to discover (mathematical, usually Platonic Realism), or are they cognitive inventions of the human mind, a precision mental game of numbers that posits no ultimately existing mathematical objects (Mathematical Antirealism)?

Perhaps there is a centrist 'middle path' wherein Mathematical Realism with its abstract math 'Platonic ideal forms' existing relatively, conventionally—the relative truth of Mathematical Realism—yet not absolutely or ultimately, which is the relative truth of

Mathematical Antirealism. After all, relatively real spacetime forms and their ultimate primordial ground are perforce an indivisible prior ontological unity. As Buddha told, "Form is empty; emptiness is form." In such a centrist view mathematical, physical and mental objects are relatively real, but not ultimately real; as in the Buddha's Two Truths, relative form and its ultimate emptiness ground.

Quantum Field Theory has demonstrated that its foundational microcosmic wave-particle fields that comprise all *relative* physical reality admit of no *ultimate,* intrinsic, observer-independent existence! Einstein hated it. Bohr loved it.

By most of the above seven 'foundational interpretations of quantum mechanics' the properties of physical matter-energy particle-waves are 'uncertain' (Heisenberg's fundamental Principle of Uncertainty) as to their intrinsic existence prior to an observation/ measurement of their properties of position and momentum by a sentient observer consciousness interpreting a measurement instrument. Prior to such experience (*a priori*) the stuff of reality abides in a diaphanous metaphysical (beyond physical), pre-collapse Neverland of nonexistence called a 'quantum superposition' of an infinity of all possible emergent matter-energy particle-fields. So arises the 'quantum measurement problem' and the spooky 'problem of consciousness'. Let us then briefly engage that.

The 'Hard Problem of Consciousness'. David Chalmers (1996) has explained the "hard problem" as the 'explanatory gap' (Heisenberg's *schnitt*) between our subjective, personal first person direct experience, and objective brain structure and function. How is it that our inner subjective experience (*qualia*) arises from purely objective brain matter? "How does the water of physical brain become the wine of conscious awareness?" [Colin McGinn] That is our perennial metaphysical 'Mind-Body Problem'.

Is our phenomenal experience more than the sum of its physical parts? Is human consciousness, human experience, reducible to mere objective purely physical brain matter? Perhaps after all, human consciousness does not ultimately arise from physical brain, although its physical 'neural correlates' surely do. Perhaps human consciousness arises from/in the cognizant 'implicate unbroken whole' of nondual primordial awareness-consciousness ground itself, and is thus ontologically prior to mere physical brain structure and function. That is the present dilemma.

In any case, our question remains. How is it that after we have answered the 'easy problems' of human cognitive conceptual and behavioral functions, the hard problem of consciousness remains unanswered? How and why do these quantitative objective physical brain functions produce qualitative subjective non-physical experience? That is the hard problem of human consciousness. Is the human concept-mind capable of resolving the 'hard problem of consciousness'? Physics and neuroscience have provided no answers beyond the doctrinaire dogma that is naïve 'scientific reductionism'—conceptually reducing *all* human experience to purely physical brain matter.

The not so hard problem is that our *relative* human consciousness perforce mereologically arises in its *ultimate* all subsuming primordial awareness-consciousness ground, the vast unbroken boundless whole itself. *Real clear mereology: the whole includes* its *parts; parts participate in their greater whole.* Not so hard to grasp.

No surprises here. Human subjective experience is *ipso facto* dimensionally prior to or beyond the epistemic reach of pure objective and mathematical inquiry. To insist that it is not constitutes some species of 'category mistake', an informal logical fallacy. Our inner, private, subjective mental life remains mostly hidden from scientific objective observation, let alone mathematical description. Astoundingly, the prodigious methods of Science have utterly failed to penetrate the nature of human experience, the very basis of our human being! What's missing in hard, objective Big Science physics?

Good news! The paradigmatic marriage of objective Science and its perfectly subjective nondual Spirit ground is inherent in our emerging Noetic Revolution. As this revolutionary 'gestalt switch' descends upon physics and neuroscience, imperious Big Science has been compelled to explore the art and science of subjective modes and methodologies for knowing, even—Yikes!—contemplative technologies. [*Appendix A*] New emerging scientific paradigm indeed. [More on the 'hard problem' in Boaz 2023 *Ch. VI*]

Thus have the conundrums wrought by QFT created a Kuhnian 'scientific crisis' that portends the present 'scientific revolution' with its accompanying knowledge 'paradigm shift' that results in an emerging radically new 'scientific paradigm', to wit, the Noetic Revolution in matter, mind and spirit that is now upon us.

Thus is Big Science physics now forced, haltingly, to consult philosophy of mind, consciousness studies, and the emerging academic discipline of Contemplative Science.

Buddhist contemplative philosophy has engaged the 'hard problem of consciousness' for 25 centuries, and has offered profound revelations. How is this so?

We've seen that "Form is empty; emptiness is form." The Buddha's Two Truth dimensions: the prior and present unity of relative spacetime form, and its ultimate boundless emptiness *dharmakaya* ground, infinite unbounded "implicate order of the vast unbroken whole". [David Bohm] We've also seen that the world of Big Science with its 'universal quantum wave function' may be seen as the dualistic dimension of relative, objectively appearing spacetime form continuously arising from its noetic nondual primordial 'groundless ground'.

[The basal formless emptiness ground of all appearing reality is 'groundless' because even That is *ultimately* nonexistent; "empty of any shred" of ultimate intrinsic existence, though it still appears to sentient consciousness as *relatively* really 'real' spacetime existence. Mahayana Buddhism calls this "the emptiness of emptiness".]

The realm of Spirit is then the nondual primordial dimension of *kosmic* 'Basic Space' (*chöying, dharmadhatu*), the vast selfless, formless ultimate reality awareness-consciousness *dharmakaya* ground that subsumes and pervades spacetime form, and in whom it arises and is physically instantiated. Told Gautama the Buddha of this present age, "Form is not other than emptiness; emptiness is not other than form."

It bears repeating. Broadly construed, the province of objective Science (physics) with its matter-energy-light form ($E=mc^2$) abides in a causal interdependent relationship (*pratitya samutpada*) with acausal perfectly subjective nondual Spirit that is its formless primordial awareness-consciousness all-embracing ground. Viewed ultimately, that relation is one of identity. In short, our *relative* dimension of form/matter and human consciousness arises and abides in *ultimate* noetic nondual primordial consciousness Spirit ground, unbounded all subsuming whole itself. These paradigmatic worlds of being are an ontologically prior yet phenomenally present unity. Yes. "Form is empty; emptiness is form." Quantum nonlocal entangled form is empty; quantum emptiness is form.

Toward a Centrist Integral Noetic Quantum Ontology

We have explored in some detail the urgent epistemic need for a centrist integral Noetic Quantum Ontology. We have established a rather protracted foundation for it. I must now complete the development of that ontology as best I can.

First Consider the Primordial Reality Ground. I have argued here and elsewhere that the epistemic evolution of the physics revolution that is Relativistic Quantum Field Theory (QFT/QED) with the mysterious subjectivity of its superposed 'universal quantum Ψ-wave function' is an inchoate expression of that nondual noetic perennial wisdom mind ontology presenting here in its exoteric, relative, conceptual cognitive modality. The universal quantum Ψ-wave function is thus a dualistic, conceptual, formalist mathematical expression of the emergence of physical *relative* spacetime, mereologically always already abiding within its prior nondual fundamental ontological ground, *ultimate* primordial awareness-consciousness itself, that vast boundless whole, by whatever grand name.

In short, relative quantum reality and its physics is necessarily subsumed in the ultimate reality that is the metaphysical natural timeless, formless, selfless nondual emptiness ground of all appearing Reality Being Itself.

From that primordial ground arises our incipient foundational Middle Way integral Noetic Quantum Ontology.

We have seen that thus far physicists and philosophers of physics have failed to produce a settled centrist 'middle path' (between the metaphysical extremes of absolute physical existence and absolute idealist nonexistence) Noetic Quantum Ontology in their various philosophical interpretations of QFT that serves as an ultimate *grounding relation* for their relative mathematical formalisms. This failure is as I have suggested, the result of modern physic's refusal to venture beyond the limits of objective classical physics and engage the spooky quantum nonlocal entangled subjectivity of human consciousness as it arises in its primordial all subsuming awareness-consciousness nondual Spirit ground.

It is through the disciplines of philosophy of physics, philosophy of mind, the East-West Science of Consciousness, and its Contemplative Science ('mindfulness of breathing', *shamatha, vipashyana*) that we come to recognize, then realize that great *one truth unity* of our perennial Two Truths—relative form and ultimate emptiness.

Ontology, the inquiry into the nature of ultimate being itself, is by its very nature metaphysics, literally 'beyond physics'. And yes, the inherent subjectivity of metaphysical ontology has from the beginning been taboo in objective physics generally, and ironically, in subjective quantum physics particularly. The formidable fundamental subjectivity and 'quantum uncertainty' (Heisenberg) of the non-causal 'objective randomness' of quantum mechanics—the basal ZPE utterly random quantum vacuum fluctuations—has forced an unbidden confrontation with the innate subjectivity of human consciousness, and therefore with the primordial awareness-consciousness ground of That (*tat, sat*). This ontic and epistemic conundrum has become known as the 'post-empirical' 'quantum mystery', or the 'quantum enigma', the "lucid mysticism" of quantum pioneer and Nobel laureate Wolfgang Pauli.

The *observer-dependent* antirealist freedom of the quantum view is indeed an unforeseen revolution in a hitherto physicalist/realist *observer-independent* 'classical' physics universe of discourse. QFT follows our human wisdom tradition of a spacetime reality that is essentially dependent upon mind, the consciousness of a relative sentient observer arising in and not separate from its ultimate primordial awareness ground; Suzuki Roshi's relative conventional 'Small Mind' participating in its prior all-pervading 'Big Mind' in whom it arises, participates, and is instantiated.

These are the 'Two Truth' dimensions—relative and ultimate—of the all embracing "implicate order of the vast unbroken whole itself", as David Bohm told it. It is the noetic monistic panpsychic cosmopsychic ontology of this nondual (*advaya*, "not two, not one, but nondual") aboriginal source or ground that is fundamental. That fundamental, ultimate, perfectly subjective primordial reality base—'Basic Space' (*dharmadhatu, chöying*), Buddha's one truth unity (*dzog*) of his Two Truths, ultimate and relative—transcends our dualistic concepts and beliefs about it. Yet, it may be experienced directly (*yogi pratyaksa*), trans-conceptually, contemplatively by way of yogic 'penetrating insight practice' (*shamatha/vipashyana*) under the guidance of the very subtle mind of a qualified meditation master. Such practice makes real and contemplatively certain the perfect subjectivity of the noetic nondual primordial 'groundless ground'.

We've seen that such a noetic metaphysical ontology is conspicuously absent in the recent metaphysical 'foundational interpretations' of the wondrous, 'always correct' quantum theory.

If we are to construct a bridge between the human cognitive dimensions of relative, dualistic objective Science, and ultimate nondual perfectly subjective Spirit in whom it arises, we require such a unifying centrist (between absolute existence and absolute nonexistence) integral (containing all the essential parts that complete the whole) Noetic Quantum Ontology.

Moreover, we've seen that such a noetic metaphysic is required for a mathematically, or a para-mathematically consistent Quantum

Gravity Theory (QGT) that quantizes Einstein's entropic geometrical gravity—General Relativity Theory (GRT)—thereby unifying these two foundational pillars of physics into a mathematically commensurable GRT and QFT unity. This is, as Hamlet would tell were he here today, the monumental physics "consummation devoutly to be wished".

In the alternative, instead of quantizing gravity perhaps we should be 'gravitizing' Max Planck's 'quantum of action'. Is the secret of entropic great gravity sequestered somewhere in the quantum mechanics mathematical formalism; Einstein's "something deeply hidden" in the nonlocal entangled universal quantum wave function? Or, are there sunny quantum fields out there, or in here, merrily propagating in the dark recesses of Einstein's gravity field equations? Should we modify Einstein's gravity, which has been found wanting and inadequate at Planck Scale tiny quantum and vast intergalactic distances and energies? Or shall we modify QFT/QED/QCD to embrace gravity? Physicists are exploring both of these alternatives. Perhaps the solution lies in a 'middle path' that accomplishes a cognitive *qbit* of both.

The Seven Main Interpretations of Quantum Theory Revisited. We saw above that these ontic 'foundations' are: 1) the default antirealist quantum Ψ-wave proto-collapse of the original Copenhagen Interpretation (Bohr was never comfortable with 'Ψ-wave collapse'); 2) dynamical collapse models (GRW theory); 3) the 'Consciousness Causes Collapse' of von Neumann and Wigner; 4) hidden variables models (non-collapse, nonlocal Bohmian mechanics of the deBroglie-Bohm pilot wave theory); 5) Quantum Bayesianism or QBism, the non-objective, non-ontic probabilistic epistemic approach as to our human perspectival 'degrees of belief' (Fuchs, Caves and Mermin); 6) quantum decoherence or the 'purely mechanistic' branching worlds, multiple, alternate universes theory called the Many Worlds Interpretation (MWI) of Hugh Everett, Bryce DeWitt, David Deutsch, and recently, of Sean Carroll. [For a biased but informative survey of these foundations of quantum mechanics see

Carroll 2019.] 7) We have also examined perhaps the most promising extant quantum ontology—the Relational Quantum Mechanics (RQM) of Carlo Rovelli.

Once again, none of these 'foundational quantum interpretations' venture beyond the mundane limit of the classical banal Scientific Local Realism/Physicalism ontology.

All of these ontic philosophical 'foundational interpretations of quantum mechanics' correctly apply objective, conceptual, *relative* human consciousness to the quantum ontology problem, but all fail to engage—objectively conceptually or subjectively/contemplatively—the 'deeply hidden' prior *ultimate* ontological primordial awareness-consciousness ground in which the Ψ-wave mathematics, and its mathematicians arise. That our human spacetime physical and mental life-world arises in, is grounded in, and participates in a greater all-pervading primordial whole is, astoundingly, never considered in these tiresome standard realist, objectivist, physicalist quantum ontologies. It's still taboo and just too 'spooky' (Einstein's *spukhaft*).

All of these 'interpretations' are *relative* conceptual objectivist attempts at grasping a necessary trans-conceptual *ultimate* ontology that grounds the conceptual quantum formalism of the universal quantum wave function. None have succeeded. I have elsewhere referred to this 'post-empirical' quantum predicament as "the problem of quantum ontology"—the hitherto largely ignored challenge of addressing the ultimate essence, nature, ground and behavior of Schrödinger's global, universal quantum Ψ-wave function—the very ontological heart of Quantum Field Theory; not to mention the ground of Schrödinger himself. Any quantum mechanics interpretation must include such a foundational centrist, integral Noetic Quantum Ontology in some modality if we are to explicate "that which is deeply hidden" (Einstein) within the abstruse quantum formalist mathematical operators of the wondrous Ψ-wave function.

It seems to me that real clarity in any physical theory requires a cognitive amalgam, a centrist middle way that engages both voices of our human noetic (nondual, body, mind, spirit subject-object unity) cognitive doublet—objective conceptual mathematical cognition; and subjective, intuitive, contemplative cognition.

Therefore, we shall herein attempt to discover how this revolutionary quantum worldview may enable us to bridge the 'consciousness explanatory gap'—Heisenberg's knowledge *schnitt*—between relative objective conceptual experience and its 'neural correlates' in brain, and the subjective fundamental ultimate ground of reality itself; between objective Science and its perfectly subjective Spirit ground, by whatever grand name or concept, in whom this all arises and participates.

Quantum Emptiness and Buddhist Emptiness: Variations on a Theme of Wholeness. Middle Way Buddhist philosophy points to appearing spacetime forms that are relatively, objectively, scientifically real; but are subjectively, ultimately immaterial nondual Spirit itself—formless emptiness/*shunyata*, essence and nature of Ultimate Truth 'Basic Space' (*chöying*), Samanthabhadra, *dharmakaya* Buddha of the vast empty boundless whole itself.

So, form is relatively, conditionally, conventionally, objectively really real! Yet, the Buddhist absence/emptiness of form's *ultimate* existence is still an absence of *something*. Buddhist emptiness *shunyata* is *ipso facto* an emptiness of something. So this absence/emptiness exists. The absence of an elephant in the room logically implies the existence of at least one elephant somewhere. So, both nonlocal quantum entangled emptiness, and Buddhist emptiness of intrinsic ultimate existence do exist!

Therefore, this Buddhist view of the prior unity of the Two Truths that are ultimate formless emptiness/*dharmakaya* and its relatively arising spacetime form is not a philosophically idealist nihilistic denial of objective spacetime form altogether. Indeed, it is an affirmation of the physical and mental form that fills the worlds of space and time. If human beings in embodied form did not objectively exist who is it that enters in and practices the Buddha's Eightfold Path to liberation from suffering? Who is it that practices compassion toward living beings? Who is it that ponders quantum ontology? Who is it that spiritually realizes this selfless, formless, *buddic* noetic ultimate emptiness ground that is instantiated by spacetime form? Indwelling primordial Presence of That (*tathata*).

Emptiness/*shunyata* represents a profound aspirational mean between the false dichotomy of relative absolute permanent substantial existence (Metaphysical Scientific Local Realism/Physicalism) and the ultimate nonexistence (Absolute Idealism) of emerging spacetime form.

The Relative Truth of Science, objective quantum spacetime form, and our scientific theories about it arise, appear, and are instantiated within the Ultimate Truth of nondual Spirit, perfectly subjective all subsuming ultimate quantum emptiness ground. An inchoate fundamental one truth unity of the 'Two Truths' that constitutes our integral Middle Way Noetic Quantum Ontology!

We have seen many times in these pages that the Two Truths, relative and ultimate, of the Buddhist centrist 'middle path' must be viewed as an ontologically prior yet phenomenally present one truth unity. Our perennial Two Truths—Ultimate Truth (*paramartha satya*), formless, timeless, selfless, perfectly subjective all-inclusive primordial awareness-consciousness ground of everything—and Relative Truth (*samvriti satya*), objective physical and mental emerging objective quantum spacetime matter/energy/light form ($E=mc^2$) continuously arises in/as this perfectly subjective ultimate 'supreme source' ground of everything. These two reality dimensions are always already a nondual primordial unified *one truth unity (dzog),* invariant throughout all spacetime matter-energy motion human cognitive reference frames—indivisible and interdependent. Our fundamental Noetic Quantum Ontology perforce expresses such a holistic unified view.

And yes, the present state of quantum ontology—the foundational 'quantum mechanics interpretations' on offer—have failed to even consider such a post-classical unified ultimate metaphysical ontology. Quantum ontology is stuck in classical physics.

Still, the Buddhist centrist 'middle path' ontology parallels some of our objectivist realist/materialist 'quantum interpretation' views. But a viable settled quantum ontology must balance the inherent ontological subjectivity of the nonlocal nondual interconnectedness of the global 'universal quantum wave function'—nonlocal entangled ZPE quantum emptiness—with the objective reality of the seemingly separate infinitely abundant mathematical wave functions of macrocosmic cosmopsychic local spacetime form with all of its 'uncertain' subatomic microcosmic micropsychic particle-fields.

A robust quantum epistemology with its mathematical formalisms must build upon such a foundational ontological/metaphysical base. It must then transcend and embrace its own inherently dualistic mathematical formalisms in a centrist theory that describes the interdependent relationship of the ontological identity of the nonlocal nondual *ultimate* noetic subjectivity of its superposed universal quantum wave function with its many objective physical and mental wave functional entities. These infinitely many superposed Ψ-waves emerge in real *relative* macroscopic spacetime existence upon each microscopic quantum measurement, or macrocosmic 'observer-dependent' reality constituting human perception. Ontology is prior to, but not separate from epistemology which is its natural cognitive extension in real space and time. We can no longer split, ignore, or deny the natural interdependence of these two cognitive dimensional *relative* functions that comprise *ultimate* reality itself.

We've often seen in these pages that until a settled nonlocal quantum ontology emerges, there shall be no Quantum Gravity Theory (QGT)—the great mathematical consummation that quantizes gravity, finally unifying the hitherto incommensurable two great theoretical pillars of Modern physics, namely, Albert Einstein's General Relativity Theory (GRT) and Dirac's and Feynman's Quantum Electrodynamics (QFT≈QED); if such a purely formal mathematical intention is indeed logically possible at all. And if not we shall turn to alternative paraconsistent 'intuitionist' math and logic, as we have seen above.

Recall that there is at work here in the process of the arcane discipline of physics a rather humorous cosmic irony. The imperious laws of physics work perfectly well in practice. Spacetime reality always spontaneously shows up for our laws of physics to measure and predict. No problem whatsoever. Now if only we could make these laws work consistently in theory! And that shall require a paradigmatic expansion of our scientific theory beyond the present limits of Metaphysical Local Realism/ Physicalism.

Well, is the *ultimate* nature of appearing reality local, observer-independent, existing in an objective and physical "real world out there" (RWOT); or is it nonlocal, observer-dependent, theory-dependent, measure-dependent, subjective and immaterial? Such a false dichotomy has now become cringe-worthy. So how about a nice centrist middle way between absolute existence, and absolute nonexistence? Mahayana Prasangika Madhyamaka Buddhists have done a good job with it. Let's use it in physics.

In short, the real unity of objective Science and perfectly subjective Spirit—the practice of that unity—will engage a subjective, ultimate Noetic Quantum Ontology such as the one I have suggested in this section; an ontology that transcends yet embraces objective, relativistic quantum formalist mathematics. Schrödinger's protophysical 'universal quantum wave function' perforce emerges from a more inclusive, formless, timeless all pervading fundamental ontic ground; name it as you will. Indeed, spacetime itself arises from and is mereologically instantiated in that basal 'Basic Space' (*dharmadhatu*) of our primordial awareness-consciousness ground (*dharmakaya, chittadhatu, jnanaprana*).

All of it, and all of us are interdependently interconnected in that vast infinite whole. We are not separate and alone in a vast meaningless *kosmos.* We *are* that noetic *kosmos!* Bright indwelling Presence of That. Our mereological participation necessarily makes it so. To borrow an ancient Hindu aphorism—*Tat Tvam Asi*—That I Am. That is, on the accord of the wise, the true nature of our mind and its human consciousness experience. That (*tat, sat*) is our 'supreme identity' of that 'supreme source ground'. Perfect reality process. Luminous clarity. *Mahasukha!* Great joy!

Such a holistic view may remind us that this noble aspiration to unity, and the wakeful compassionate wisdom of happiness of that, is 'always already accomplished', deep within us! Objective Science (form), and perfectly subjective Spirit (emptiness)—Buddha's Two Truths—are always an ontological prior and phenomenally present unity. All the masters, saints, sages, and *mahasiddhas* have told it. Jesus said, "That happiness you seek, the Kingdom of God, is already present within you ... and it is spread upon the face of the world ... but you do not see it." [*Luke 17*] Now we can see it. "The Tao that can be named is not the primordial Tao." "Wonder of wonders all beings are Buddha."

From an ultimate view the primordial Two Truth reality dimensions—relative spacetime form and its nondual ultimate spacious emptiness ground—are utterly indivisible. And now we can clearly see it (*samadhi, dhyana, vipashyana, moksha, satori*). It is in this liminal transitional vivid open cognitive space, that cognizant 'explanatory gap'—conscious awareness portal into our infinite 'supreme source' ground—that the real work of realizing the unity of Science and Spirit begins, and ends. It is That to which we awaken through the assiduous practice of the psycho-spiritual path, under the compassionate guidance of the enlightened mind of the meditation master.

A centrist integral Noetic Quantum Ontology understands and spontaneously expresses that prior and present unity of objective Science grounded in its perfectly subjective Spirit whole.

The Prior Unity of Science and Spirit

The Really Real, and the Not So Real. We have already seen that beginning with John Stewart Bell's monumental 1964 Bell's Theorem, and 50 years of confirming nonlocality physics 'Bell Test' loop hole canceling experiments (18 of them through 2022) 'spooky' quantum nonlocality/entanglement is now considered by the recent protagonists of this 100 year scientific drama to be 'scientifically' proven. Einstein and Bohm with their last gasp conjecture for a 'nonlocal hidden variable parameter' to 'save the appearances' of Scientific Local Realism stands refuted by Bell's Theorem for most theoretical physicists, and nearly all philosophers of physics who stop calculating long enough to consider the deeper ontic meaning of our wondrous quantum theory.

Well, has John Stewart Bell had the last word? What is the seed of common sense relative empirical truth in Scientific Local Realism/Physicalism that cannot be credibly denied? Whether spacetime stuff is *ultimately* real—Middle Way Buddhists deny that it is—appearing objective reality must be at least *relatively* really real! After all, here we are, along with real trees and stars. Mahayana Buddhists agree. Whether or not we reify quantum mathematical abstractions into a real objective existence, spacetime stuff and beings are everywhere! What is the existential reality status of a conscious embodied mind that denies its own existence? We call such a one bad names, like 'solipsist', or 'nihilist', 'schizoid', or even 'psychotic'. The yoke of the burden of rejoinder for antirealist and idealist skeptics of an observer-independent RWOT is heavy indeed. Is there a 'middle path between such fallacious metaphysical extremes?

Yes, of course. We must avoid the epistemic false dichotomy that insists that the nature of appearing reality be *either* ultimately real (Metaphysical Scientific Realism/Materialism), *or* ultimately illusory (antirealist Metaphysical Idealism). And yes, there is a centrist 'middle path' between these metaphysical extremes of absolute existence and absolute nonexistence. Such a path was elaborated by Gautama the Buddha of this present age 26 centuries past. We might well consider this noetic wisdom of the ages when constructing our ambitious but much needed centrist integral Quantum Noetic Ontology.

Recall that from the metaphysical ontology you choose arises the karmic cause and effect reality you deserve. Often a difficult choice. Perhaps it is better to err on the side of ontic and epistemic holism; and the altruistic, compassionate *bodhicitta* happiness inducing conduct that arises herein.

The history of religion and philosophy, both West and East, might be seen as a dispirited cognitive program of such a false dichotomy that is an absolute distinction between objective, monistic 'scientific' Metaphysical Realism/Physicalism; *or* a subjective, antirealist monistic Metaphysical Absolute Idealism. But this bedeviled history of our species' quest for absolute objective certainty, for some absolute truth that will bestow endless comfy human knowledge and happiness has been necessary in order to arrive, individually and collectively, at our present liminal developmental awareness-consciousness juncture—the 'turning point' toward scientific and spiritual wholeness—and the authentic human happiness that always already abides here.

Clearly, we need a centrist middle way between these ontological metaphysical extremes that present to dualistic thinking mind as *either* absolute objective existence *or* absolute subjective nonexistence. "To be or not to be." Perhaps the truth of the matter is "To be *and* not to be": the paraconsistent truth of our perennial wisdom Two Truths trope.

Recall that human objective conceptual cognition is inherently dualistic. The semiotic logical syntax of language is truth functional or two valued—either true or false. Conceptual dualistic mind construes things as separate, as *either* this, *or* that. In Aristotle's logic it is either 'A' or 'not-A', but not the unity of the two. So language—semiotics (logical syntax, semantics/meaning, pragmatics/speech acts)—has evolved as the destructive duality of a knowing subject 'I' and an 'other' essentially separate object. 'Me', and everything else. Self (*atman*), and 'noself' (*anatman*). We have seen that Mahayana Prasangika Madhyamaka Buddhist philosophy and practice has profoundly accomplished a logically consistent pragmatic Middle Way between such dualistic thinking and its resultant, often harmful self-centered conduct—primary cause of human suffering—fear, anger, ethnic hatred, endless war and despair. Less emphasis on self-ego-I, and enhanced emphasis on compassionate selfless noself represents the relative antidote of both Gautama Shakyamuni the Buddha (selfless *anatman*) and Jesus the Christ (selfless *kenosis*).

Quantum Field Theory has unwittingly pointed to such a centrist middle way ontology through its distinction between the spooky *subjective* inherently entangled ZPE nonlocality of the superposed universal quantum wave function (Ψ) prior to its often presumed 'collapse' via a conscious observation—'consciousness causes collapse' (Wigner)—into an apocryphal safe and sane local *objective* reality, our beloved really 'real world out there' (RWOT). Here our objective observer-independent realities are neatly objectified/reified via an inherently subjective process of an observer-dependent human consciousness—human mind creating, imputing, and reifying its appearing realities. David Finkelstein's 'Universal ontological relativity' indeed.

Perhaps our much valorized yet incomplete Quantum Field Theory is, as was Einstein's Relativity before it, an ideological interregnum (between ruling kings) awaiting that next more inclusive, yet always incomplete scientific world view. We require a scientific view that awakens us to our already present connection to the aboriginal *ultimate* dimension in whom this *relative* spacetime dimension arises, that we may directly experience, and know and feel who we actually are, our 'supreme identity', bright Presence of That, prior to any 'scientific' ideological judgments and beliefs about it.

Then, through this joyous 'primordial wisdom' (*jnana, yeshe,* gnosis) we conceptually unpack the nondual ultimate truth of the matter in order to benefit living beings—including our Mother Earth—being here in quantum space and time. Such kind compassionate ethical conduct (*bodhicitta, karuna, ahimsa, daya,* altruism, *pathos, rahmah, riham*) is considered in many of our noetic wisdom traditions to be the very secret of human happiness. Compassionate love-wisdom mind may be considered the *action/conduct* of our always already present indwelling Presence of the primordial ground in whom these spacetime realities of ours all arise and participate.

Such is the great truth of the prior and present unity of relative, dualistic, objective Science and its ultimate, nondual, perfectly subjective Spirit ground—this Two Truths union that is the one truth of nondual ultimate reality itself. The Voice of all beings—songbirds, cats, people, buddhas—bespeaks this sublime one truth unity (*dzog*). Verily, "It is already accomplished from the very beginning" (Nagarjuna), deep within us. Primordial Presence of That. That is the real. On the accord of the noetic wisdom traditions of our species that is the clear and simple truth of our being here to which we now awaken, breath by mindful breath. [*Appendix A*]

The Wisdom of Non-Seeking. The easeful love-wisdom of compassionate non-goal directed human action is known in the Chinese Taoist noetic wisdom tradition as *Wu-Wei*. *Wu-Wei* seeks nothing at all. Not even *Wu-Wei*. "The *Wu-Wei* that seeks *Wu-Wei* is not *Wu-Wei*." [Chuang Tzu]

"We are betrayed by destinations." [Dylan Thomas] Our East-West human noetic wisdom tradition has—for at least 10,000 years—attempted to both conceptually and contemplatively understand the prior, already present unity of these two all too human cognitive modalities, our *noetic cognitive doublet* that is both objective conceptual, and subjective contemplative human cognition.

To be sure, objective Science—physics, cosmology, neurobiology—and subjective contemplative Spirit/spirituality require a providential Two Truths ontology, objective relative and nondual ultimate that is the 'grounding relation' of such an objectively fluent holistic epistemology. We require the whole subject-object picture in order to see its prior and present unity.

Thus is Heisenberg's liminal transitional *schnitt* or 'explanatory gap' between objective local phenomenal experience and subjective nonlocal, nondual uncertain experience bridged, at least conceptually. And it is unified non-conceptually through the direct contemplative experience (*yogi pratyaksa*) that is the subject of the emerging knowledge discipline known as Contemplative Science. [Boaz 2022] Quantum pioneers Bohr, Heisenberg, and Schrödinger were all fluent in Eastern contemplative philosophy, if not actual mindfulness meditation practice.

Perhaps in this hapless monumental grail quest for human absolute objective certainty we shall discover a unifying, if ever incomplete theory that unites the objective classical relativity of Einstein's local GRT with the inherent subjectivity of nonlocal QFT/QED. I have herein argued that the cognitive architecture, a noetic transept for such an ontic/epistemic project must include a 'post-empirical' fundamental Noetic Quantum Ontology. I have offered the fundaments of such an ontology above.

Moreover, we seek certainty in all the wrong places. Please don't believe this, but consider it: certainty does not abide in objective conceptual cognition. Yet non-conceptual nondual certainty abounds in contemplative 'mindfulness of breathing'!

No doubt that a paraconsistent mathematical consummation of QFT shall add greatly to the relative human happiness of theoretical physicists. And a Noetic Quantum Ontology shall add confidence to the quest of scientific minded contemplatives for ultimately subjective nondual certainty, which on the accord of the Buddhas and the *mahasiddhas* of our noetic wisdom traditions is "already accomplished from the very beginning", *buddic* wisdom mind within us, prior to our adventitious happiness seeking strategies. Indeed, it is such a love-wisdom mind noetic imperative to which we gradually, then suddenly awaken upon this confusing, difficult, joyous life Path to wholeness.

The Two Truths—real relative and super-real, even *fantasque* surreal empty ultimate—beget these two corresponding and interdependent quests for certainty—objective scientific and subjective spiritual—until that is, the seeking subject, the self-ego-I, surrenders the eternal quest for the objects of its quest.

Now, in the ultimate realization and peace of buddic noself/anatman dawns the Wu-Wei wisdom of non-seeking—the 'wisdom of uncertainty'. That heart seed of enlightenment that is our 'innermost original wakefulness' is always present numinous Buddha mind Presence that we already are now. We begin to awaken and spontaneously express that great indwelling love and wisdom truth through compassionate practice of the psycho-emotional-spiritual Path under the gentle guidance of the meditation master. Please consider this well.

Please permit me once again to offer this great teaching of Jesus the Christ: "That which you seek ... the Kingdom of God, is already present within you ... and it is spread upon the face of the world, but you do not see it." [*Luke 17*] Buddha told, "That which you seek is already present from the very beginning.... Wonder of wonders all beings are buddha." The not seeing this great truth, this always present primordial wisdom is primal egocentric ignorance (*hamartia*/sin, *avidya, marigpa, ajnana*), primary cause of human suffering and human evil—fear, anger, hatred, despotism, and endless war.

As to that "something deeply hidden" (Einstein), this interdependent centrist relationship between relative local causality in Science and acausal nondual ultimate Spirit is the prior and present unity of our perennial Two Truths. John Stewart Bell, arguably the most profound quantum physicist since Irwin Schrödinger, has revealed the 'post-empirical' trans-rational truth of the primordial universal quantum wave function that abides at the nondual noetic heart of the quantum enigma. [*Speakable and Unspeakable in Quantum Mechanics,* Bell 2004]

Knowledge of the prior and present unity of dualistic objective Science and its perfectly subjective nondual Spirit ground—primary aim in our quest for a Noetic Quantum Ontology—requires that we identify not only the objective science and epistemic and contemplative methods of what is to be known, and how it is known, but the subjective inherent noetic nature of the knower, the one who knows.

A Noetic Quantum Ontology shall become useful to physics and contemplative studies only when we have discovered the inherent nature and wisdom capacity of the knower, that one who knows. Let us then more deeply explore just who it is that discovers this happy unity of our perennial Two Truth dimensions that are relative objective Science, and perfectly subjective nondual Spirit in whom it arises

Who Is It That Knows?

This embodied, selfless reality-constituting continuum of life-giving *prana* spirit wind (*lung, hsing-ch'i, pneuma, ruakh, breath of life, bioenergy*) upon the quiescent mindful breath—that is the one who knows. "Mindfulness of breathing" as it was taught by Gautama the Buddha and Jesus the Christ animates our human physical and mental form via the all pervading subtle wisdom *jnanaprana* spirit energy life force arising within the primordial ground of embodied form—radiant indwelling Presence (*vidya, rigpa, christos*) of That. 'Spirit' translates 'breath' and breath translates nondual Spirit in the great noetic Primordial Wisdom Tradition of humankind.

A Mereological Proof for the Existence of God? Voila! We have now discovered a directly experiential (*yogi pratyaksa*) existential 'proof' for the existence of a post-theistic, non-Creator, non-anthropomorphic, not separate, trans-conceptual, all embracing ground that is noetic nondual God! That *ultimate* perfectly subjective vast boundless whole perforce subsumes its *relative* participating parts, including all of us. Just so, we the gently embraced parts are an inherently unified participating aspect or selfless Presence of that formless, timeless, nondual 'groundless ground'.

That always already present luminous Presence of nondual Godhead is the primordial Spirit ground of all arising spacetime particulars, vast infinite all subsuming unbounded whole of reality itself in whom we as the relative participating parts perforce arise and participate, recognized, then realized upon our intimately present *prana* spirit breath (*lung, ch'i, pneuma*-Holy Spirit). Nondual Godhead *is* spirit-breath. Our *jnanaprana* wisdom breath is our always present immediate connection to That. That God breathes us!

Here now upon the breath in the belly is the always already present Presence of existence of selfless, formless, timeless directly experienced (*yogi pratyaksa*), nondual perfectly subjective Godhead, above and beyond our concepts and beliefs about it; by whatever glorious name. Numinous innermost bright Presence of That! We should feel better already.

Be all That as it may, such a proto-mystical, trans-rational explanation—particularly if it contains the epithet 'God'—is not at all agreeable to our dualistic conceptual 'scientific' sociocultural 'global web of belief' that is our collective self-ego-I. That ultimately non-existent but relatively all too real atman-self entity—steeped as it is in the prevailing secular deep cultural background Greek Metaphysical Local Realism ideology—has become the prevailing metaphysic of Western culture, namely, Scientific Materialism/ Physicalism. This hyper-objective 'Scientific Local Realism' metaphysic has plundered the Western heart and mind of its indwelling, innate subjective love-wisdom nature. And we are scarcely aware of it. Dispiriting metaphysic indeed.

Yes. We transcend such dualistic conceptual belief systems in 'post-empirical' nondual Godhead via our contemplative transpersonal mindful mantra breath, quite beyond the realm of mere concept and belief. Here, along with the great foundational quiescent mantra *OM AH HUM*, one of our idiomatic noself help mantras might be Suzuki Roshi's "No time, no self, no problem." Or the short version of the beautiful Jesus Prayer.

In short, the directly experienced (*yogi pratyaksa*) nondual unity of self-ego-I (*atman*), and its aboriginal primeval selfless noself (*anatman*) Spirit ground remains potentially present to our awareness continuum, between distractions, more or less constantly.

We *connect* to that post-theistic, non-anthropocentric, nondual Godhead via our *choice* of 'awareness management'—the 'placement of awareness-attention' upon our perfectly subjective always already present Presence at the heart, and through the yogic breath in the belly. [*Appendix A*] As Buddha told so long ago, "Noself is the true refuge of self." Our "mindfulness of breathing" makes it so.

*Primal Ignorance (avidya, marigpa, ajnana, hamartia/*sin*).* Following this imperfect continuity of trans-conceptual quiescent love-wisdom mind 'noself' experience, self-ego-I too often bemuses its unrealized subtle noself wisdom potential by conceptually, dualistically reducing the perfect subjectivity of this singular whole shebang to mere objective dualistic conceptual cognition. Here 'self' and 'other' of the pernicious 'subject-object split' remain separate. The result is that this sublime prior unity of our two human reality dimensions, our *noetic cognitive doublet*—1) relative, conceptual, objective, physical, and 2) subjective, intuitive, emotional, spiritual—becomes almost hopelessly obscured. That unhappy result is human suffering—negative afflictive emotions of fear/anger, ethnic hatred, grasping desire, greed and pride that cause despotism, despair, and endless war.

'Scientific reductionism'—reducing ultimate selfless noself love and wisdom to mere relative knowledge of a conceptualizing self—facilitates this painful process of separation from our primordial love-wisdom mind home. We exhaust our brief lives seeking happiness that is already present within. Seeking our already present happiness is a kind of unhappiness; is it not? Therefore, we cannot *become* happy in some exalted future happiness mind state; but we can *be* happy here and now. That always present 'be here now' happiness is like coming home. Mindful 'awareness management' realizes it.

Some of our less enlightened internal ego identities desire to cram such lucent nondual wisdom into a stunted procrustean corpus of materialist "common sense" (Bertrand Russell's "metaphysics of the stone age") notions of a purely physical spacetime reality peopled with objectively embodied, inherently separate and competing egocentric thinking automatons. Ominous worldview indeed.

We shall see that the antidote to such ego-I duplicity is relative self surrender to ultimate selfless noself Presence of the ground, that they may work together as a noetic wisdom team. This contemplative dialog is engagement of the union of *shamatha* mindfulness practice and *vipashyana* analytic insight practice, the direct seeing (*samadhi, dhyana, Zen*) of revealing, penetrating insight meditation. We have seen in Chapter I that such a process requires a secure, flexible, intelligent, courageous and contemplatively trained self-ego-I; and a qualified 'spiritual' mentor to guide the awakening process.

Sadly, this cup of wisdom tea is perhaps several generations hence for habitually realist-materialist-reductionist Western physics and cosmology practitioners. This in spite of the antirealist construal of the wondrous inherent subjectivity to which Schrödinger's quantum Ψ-wave function inexorably points.

Thus it is, our monolithic, one-dimensional materialist individual and collective common sense self-ego-I is rarely so intelligent as to utilize the contemplative cognitive technologies of recent Contemplative Science. Narcissistic self is disinclined to admit of our trans-conceptual noself Presence for all too real illusory entity denies this spooky ultimate love-wisdom mind dimension by fearfully, habitually, conceptually construing or reducing it to the mere gross physical dimension, namely, relative purely physical-chemical brain structure and function. This untidy bit of 'functionalist' philosophical 'Small Mind' conjuring is known to the philosophy trade as 'scientific reductionism', and to the construal of this very ordinary 'Small Mind' as 'common sense'.

'Cognitive Construal'. We human beings—scientists, philosophers, teachers, and regular folks—*construe* our realities not as they are, but as we (self-ego-I) desire and hope them to be. The antidote to such grasping biased egocentric thinking is selfless quiescent mindfulness meditation, 'mindfulness of breathing'. That connects us to the 'one who knows', already present bright lucent Presence of That (*tathata*).

Sadly, course relative conditional untrained mind has little or no conscious awareness of that fundamental liberating cognitive process. We have hitherto labeled that difficult process 'primal ignorance' (*avidya, marigpa, ajnana, hamartia*/sin).

Now we can see that whole 'perspectival' process clearly. So *feel* that indwelling primordial *buddic* love-wisdom mind for a few moments upon the mindful breath in the belly—life force *prana* spirit wind entering in upon each and every breath—whether or not such a yoga is now a part of your present 'global web of belief'. Go ahead and do it now.

Who is it that knows? We now know who it is that knows, feels and realizes the ultimate truth of foundational selfless compassionate 'noself' in an illusory 'no time' arising in nonlocal nondual noetic Buddhist emptiness (*shunyata*), and the inherent subjectivity of quantum emptiness that emerges as this all too real spacetime reality of ours. Let us then revisit in a new light just who it is this noetic primordial awareness Presence That I Am.

Who Am I? What is My Mind?

What is your mind? Please consider this bit of wisdom from the ultimate view of the inherent noetic Primordial Wisdom Tradition of our species. *I Am That I Am*—'innermost esoteric' love-wisdom identity Presence of the vast boundless whole that is our Heart's desire—that human happiness which we constantly seek. That is the *'I Am Presence'* of Jesus and the Hebrew prophets, and of Gautama the Buddha. It is That (*tathata*) to which we aspire; That to which we awaken each moment; each mindful breath, whether we know it, or believe it or not. Spooky indeed to the concept-mind of our collective self-ego-I, and to our deep cultural background (mostly unconscious) objectivist, materialist, physicalist 'global web of belief'. [Quine 1969] Indeed, there are more things in heaven and earth than are dreamt of in our conceptual materialist philosophy. What to do?

Well, how about some good news? Self-ego-I, in due course and by grace, and with a *qbit* of persistent contemplative practice under the guidance of an authentic meditation master ceases to be a deceiver and obstructer to our psycho-emotional-spiritual growth. It becomes instead a self-interested but mostly non-judgmental ally in this cognitive bias-busting great process of awakening to our always present instant, open primordial Buddha mind Christ mind Presence (*vidya, rigpa, christos, buddhajnana, buddhadhatu*). It is that trans-conceptual primordial awareness-consciousness itself that is the very Nature of Mind and all its experience. What is your mind? On the accord of the subtlest or 'highest' teaching of our great wisdom traditions—Hindu Advaita Vedanta, Buddhist *Dzogchen,* Taoist

Tai-chia, and Abrahamic esoteric monotheism—*That* is your mind! By whatever grand name or concept.

So be kind to this strange guest of your phenomenal world that pretends to be only your self-ego-I. Relate to it and love it as the mother loves and gently corrects her wayward child. You are that child. And you are that mother.

I Am That I Am. Recall that exoteric relative ego-self cognition and esoteric selfless ultimate *bodhi* mind cognition are not ultimately separate. Relative self and ultimate 'noself' are indivisible and not separate. Remember that selfless noself Presence is the 'ultimate refuge' of relative self-ego-I. Thus do we 'take refuge and actualize *bodhichitta*', the not always obvious altruistic secret of human happiness. Many of my readers are very much aware of this happiness fact. If it's news, please consider it well.

Please recall also our exoteric/outer knowledge and esoteric/inner noetic nondual wisdom—our perennial Two Truths, Relative Truth (embodied self and its experience in time) and Ultimate Truth (timeless selfless boundless whole in whom this all arises)—are always already a non-conceptual, nondual, utterly interdependent ontic prior yet always present *one truth unity*. Our relative self stands always at the liminal threshold of recognition and realization of all pervading formless, timeless ultimate noself love-wisdom mind. Our healing love-wisdom mind already knows this. Good news indeed!

This unified Two Truths—relative and ultimate—the 'ontologically relative' (we establish our realities via our concepts and beliefs), all-embracing *one truth unity* is invariant through all human experience. So, as Buddha told, "Rest your weary mind and let it be as it is; all things are perfect exactly as they are." That ultimate truth dimension obtains even as the natural adversity of this Relative Truth dimension ravishes us.

Thus does primordial awareness wisdom express the selfless one truth unity (*dzog*) of our indwelling Buddha mind (*buddhajnana*) Christ nature (*christos*) that endlessly embraces the dimensions of the Two Truths of our being here in time. The Buddha and the Christ both taught compassionate love guided by wisdom. Love and wisdom are the two limbs of Buddhism and Christianity.

Awareness Management. That profound but simple subtle teaching rides each mindful breath. Place your attentional awareness upon the gentle *prana* wind of your breath in the belly, and at the spiritual Heart (*hridaym*), and at the 'third eye' (medial prefrontal cortex) just behind the forehead—and see for yourself. [*Appendix A*] Such an awareness management 'placement of attention' will immediately produce alpha, 'waking theta', and for experienced meditators, high frequency gamma brain rhythms—the 'relaxation response'. [*Appendix B*, "The Neuroscience of Mindfulness Meditation"] Go ahead and do it now for a whopping two minutes. Longer if you find it like unto a blissful yogic swoon.

We have seen again and again that this freeing state of knowing-feeling love-wisdom mind Presence is inherently always present each moment now at the numinous spiritual Heart (*hridyam*) of the human being. Mindful mantra breath makes it so. *OM AH HUM* instantly connects us to the *jnanaprana* spirit wind in the belly. All the masters and mahasiddhas of the 'Three Times'—past, present, future—have told it. This is the acausal nondual "fruitional view" that embraces and subsumes the duality of the causal view—cause and effect—be good and practice now in order to accomplish a happy result in some future mind state.

The outer, exoteric understanding of our great Primordial Wisdom Tradition is primarily causal. This view is represented in the Buddhist tradition by the *Pali Canon* of Theravada, and by the Two Truths trope motif of the Mahayana Causal Vehicle. The 'innermost secret' understanding is present in acausal nondual *Dzogchen* View and praxis, the Great Completion of the causal Mahayana path.

Well, does Buddhahood have a cause? Relatively, a big yes. Ultimately, not necessarily. We find the resolution of the duality of the noble Buddhist Causal Vehicle in the nondual fruitional teaching of Buddhist *Dzogchen*, Essence *Mahamudra*, Definitive Madhyamaka, and *Saijojo* Zen. [Boaz 2020] Here, in timeless now, the fruition—the enlightened ultimate happiness result—"Is already accomplished from the very beginning". [*Dzogchen* founder Garab Dorje] It is always 'primordially present' here and now upon each *jnanaprana* love-wisdom breath. Ultimately, we are always already That! That is the actual Nature of Mind. That is who it is that I AM! Our 'supreme identity' of the 'supreme source' that is ultimate reality Being Itself. That is the teaching. Mindfulness meditation awakens us to that great nondual (subject-object unity) truth.

Hence, it is our indwelling nondual, non-causal, transpersonal, trans-conceptual love-wisdom mind—numinous, luminous Presence of That—to which we awaken, breath by mindful breath, upon life force subtle *prana* wind (*pneuma*-Holy Spirit, *lung, ch'i*) energy entering in upon each mindful breath. Solid self-ego-I here dissolves into primordial 'noself' Presence (*rigpa*) from which it has never departed. The *Prajnaparamita* (Perfection of Wisdom) Mantra reveals: *"Gate gate paragate para samgate bodhi svaha."* "Gone gone beyond, gone utterly beyond; now perfect wisdom." *Prajnaparamita* is the Divine Wisdom Mother of all the Buddhas of the Three Times. Good company (*satsang*) indeed.

Brief Review of the View. The splendid irony and paradox here is that in the fruitional view primordial wisdom mind is, as we have so often seen in these pages, always already present, deep within us. "That which you seek is already present within you." [Jesus] "Wonder of wonders all beings are buddha." [Buddha] Not seeing it is atavistic primal human ignorance (*avidya, marigpa, hamartia/* sin). The profound remedy is ultimate primordial awareness wisdom (*jnana, yeshe,* gnosis).

How shall we connect? *Awareness management*: conscious 'placement of attention/awareness' upon your *prana* breath in the belly, your heart chakra, your 'third eye' forebrain, and your "great bliss crown wheel" as you open your crown chakra to bright Presence (*vidya, rigpa, christos*) of the primordial ground of everything. So, the methodology is mindfulness meditation (*shamatha/vipashyana*), and 'deity practice' under the enlightened guidance of a qualified spiritual mentor. And yes, it's like coming home after a long and difficult journey.

Thus is mindfulness meditation a conscious finite awareness portal, a spacious conceptual 'explanatory gap' into infinite formless, selfless, timeless primordial awareness-consciousness ground of all arising and appearing phenomenal spacetime reality. Who am I? What is my mind? That is who you are. That is your mind! Human Happiness Itself.

This great, universal awareness *process*—and the numinous I Am Presence of the ground that feels and knows it—has been conceptually elaborated by our wisdom traditions, and even within traditions, in many different ways. The centrist Middle Way Madhyamaka Buddhist *Dzogchen* view and practice is both origin and aim of all our happiness seeking strategies. As H.H. Dalai Lama told, "Just open the door."

The Four *Dzogchen* Yogas: Practical Yogic Meaning in Highest Buddhist Practice

Primordial Wisdom Prelude. Clearly, we require an exoteric conceptual understanding of this open secret of human happiness, both relative human flourishing (*eudiamonia, felicitas*), and ultimate liberation from suffering (*paramananda, mahasuka, beatitudo*). And we need as well an understanding beyond our mere concepts, beliefs, and cognitive biases—of whatever Path we presume to practice. Indeed, that is the main purpose of this book. I pray that the following, very brief Buddhist wisdom semiotic construction shall prove helpful for my reader who wishes to further engage Buddhist view and practice as a means to understand the unity of modern quantum Science and premodern noetic Spirit. Or perhaps as a means to establish or to continue an established Buddhist or other spiritual practice. Some knowledge of the seemingly obscure discursive terminology of Indian and Tibetan Buddhism, and of esoteric Judaic-Christian mysticism is desirable and useful.

Yet, caution is advised. Let us come to know these terms conceptually yes, but without distracting too much from the non-conceptual 'innermost secret' nondual direct spiritual experience (*yogi pratyaksa*) of the teaching. Once again, we must use both modes of our *noetic cognitive doublet*—objective conceptual, and subjective contemplative spiritual. How do these interdependent terms relate to one another? How do they facilitate our compassionate ethical conduct and thus our human happiness? How does the practice that is described by these conceptual terms result in awakening to our indwelling love-wisdom mind—"the peace that passes all understanding"?

In the beginning *shamatha* or 'mindfulness of breathing' begets *sati, smrti, shiné, bhavana, dranpa*, the mindful quiescent peace of awareness/remembrance that generates the *direct seeing* of *vipashyana* (*samadhi, dhyana, ch'an/zen, moksha*) that may in due course become the '*samadhi of certainty*' that realizes nondual unity of that all-embracing timeless, formless, selfless Ultimate Truth dimension—primordial emptiness Spirit 'groundless ground' itself (*dharmakaya, kadag, dharmadhatu, shunyata*) in whom the spacetime dimension of Relative Truth form arises, participates, and is instantiated. In short, the one truth unity.

I have told of the naturally arising luminous selfless clear light Spontaneous Presence (*vidya, rigpa, lhundrub*) of the nondual primordial ultimate ground is always already present at the spiritual Heart (*hridyam*) of the human being. That is our actual authentic primordial love-wisdom mind/nature—Buddha mind (*buddhajnana*), Buddha nature (*Buddhata, tathagatagarbha, sugatagarbha*), Buddha essence (*dharmadhatujnana*)—that recognizes, then permanently realizes the great truth of the aboriginal awareness-consciousness Nature of Mind (*sems nyid, cittata*) in whom this all arises and shines for us.

For a time one does not directly see it. At the beginning of the spiritual path this great noetic nondual truth is mostly, but certainly not entirely, mere concept and belief. But soon, as practice begins to stabilize the scattered 'wild horse of the mind' noetic nondual experience arises, the non-conceptual direct '*samadhi of certainty*' of this bright primordial Presence (*rigpa, vidya*) of the luminous 'supreme source' ground that *you* actually are—your 'supreme identity'—utterly beyond the concepts, beliefs, and biases of a narcissistic self-ego-I. Many of my readers have already accomplished that.

Nondual 'spiritual' realization of That (*tathata*) primordial ground, the boundless all embracing whole, is the union of *relative* discriminating wisdom (*prajna, sherab*), and *ultimate* nondual primordial wisdom (*jnana, yeshe,* gnosis). Such a knowing-feeling love-wisdom union abides utterly beyond any trace of dualistic separation between a perceiving knowing subject and its objects of perception—the odious subject-object split. Yes, it is this apparently real split between relative self-ego-I (*atman*) and the ultimate love-wisdom of prior 'noself' (*anatman*), this failure of recognition of the numinous Presence of our innate Christ-Buddha nature that is the root cause of primal ignorance (*avidya, marigpa, ajnana, hamartia*/sin), primary cause of human suffering—*dukkha* (Pali)*; duhkha* (Skt). How then shall we 'cut through', uplift and heal this adventitious ignorance?

The Four Yogas of the Tibetan Buddhist Great Perfection. The *Ati Dzogchen semde* (mind) teaching cycle expresses this prodigious wisdom practice processional as: 1) *Shamatha,* quiescent 'mindfulness of breathing'; 2) *Vipashyana* (*dhyana, samadhi, clarity*), the clear light direct seeing that is penetrating insight meditation; 3) *Kadag,* noetic nondual (*advaya, gnis-med*) cognition/apperception of the 'primordially pure' awareness ground/base (*gzhi rigpa*) of all appearing phenomena; and 4) Spontaneous Presence/*lhundrub* (*anabogha)* of the primordial *kadag* ground, direct always present Presence of that ultimate aboriginal ground abiding already as 'innermost secret supreme identity' of we human beings being here as esteemed guests of this our relative phenomenal world of space and time. That is the supreme gift of our lives—a little *relative* time in which to awaken to the perfectly subjective *ultimate* invariant one truth unity (*dzog*) of the matter. [Boaz 2020]

These *Four Yogas* of the *Dzogchen semde* (mind) teaching cycle—*shamatha, vipashyana, kadag ground,* and *lhundrub* Presence of that ground—are the human cognitive mind states and life stages that provide the contemplative love-wisdom foundation of the Indian Mahayana and Tibetan Vajrayana Buddhist spiritual Path. Together they open the heart and mind to receive the archetypal mythopoetic Great Love, entropic gravitas of the formless primordial awareness ground that spontaneously expresses itself in human form, and in our behavior and conduct as skillful, kind, compassionate *bodhichitta* practice. That practice is the thought, intention, and action for the benefit of living beings, the very secret and primary cause of our human happiness—both relative conventional, and Happiness Itself—ultimate harmless happiness that cannot be lost. That *bodhichitta* (mind of enlightenment) is grounded in the precious Buddhist Three Jewels: the Lama/Guru as the Buddha, the *dharma* teaching of the Buddha, and the crucible of learning that is the precious *Sangha* spiritual community. We connect—again and again, 'brief moments many times'—by 'taking refuge' in these three spiritual gems.

Thus does the unity of the two wisdoms, relative *prajna/sherab* and ultimate *jnana/yeshe/*gnosis, with compassionate loving-kindness of the *bodhichitta* of engaged action, represent the two limbs of the Buddha's teaching—compassionate love and skillful means/method of engaged wisdom.

The Four Dzogchen Yogas are inherently indivisible, an ontic prior and phenomenally present one truth unity, invariant through all human cognitive frames of reference. Quiescent *shamatha* is the foundation for penetrating direct *vipashyana samadhi* insight. *Vipashyana* naturally arises on the basis of mindful *shamatha.* From this twofold meditation foundation then arises our great love-wisdom doublet, nondual 'wisdom of emptiness' of the vast *trekchö kadag* whole or ground, and its primordial *tögal lhundrub* Presence arising therein, and spontaneously expressing itself as altruistic compassionate *bodhichitta* for the benefit of all living beings—including our precious Mother Earth.

It is from this spontaneous *buddic* love-wisdom mind *rigpa/vidya* Presence that the thought, intention and engaged human action for the benefit of beings manifests here now in space and time. Selfless 'noself' Presence lifts and heals our egocentric difficult human lot. Mindfulness meditation, 'mindfulness of breathing' as Buddha called it, with *vipashyana* immediate direct penetrating insight is the skillful means/method that accomplishes that perfectly natural miracle of being here on this pretty blue planet in vast nearly endless time and space.

So many words for something so simple. Please consider this well through the practice of both modalities of our human *noetic cognitive doublet*—exoteric objective conceptual Science, and its esoteric subjective contemplative nondual Spirit ground. Our two hitherto incommensurable knowledge paradigms now unified. *Mahasukha*! Great joy!

Selfless Noself Help

No Self, No Time, No Problem At All. Self-ego-I lives mainly in the past and in the future. This present moment *now* is difficult for us. Yet now is when and where everything happens! Throw in some self-aggrandizing, or regretful past/future fantasy reverie, and we have a bunch of dysfunctional human minds. As good a definition of our painful 'human condition' as any. Unhappiness arising as fear, anger, hatred, despotism, war are the inevitable result. We now know the way out of this futile self-created cognitive cage. As recent Zen Master Suzuki Roshi told, "No self, no time, no problem." Let's conceptually unpack this a bit.

The truth of the matter, on the accord of the subtlest nondual teaching of our noetic wisdom traditions, is that relative self-ego-I and ultimate selfless 'noself' love-wisdom mind are a prior and present nondual subject-object unity, a noetic body, mind, spirit unity. How may we know and feel this?

Through the power of focused attention/awareness upon the breath in the belly rest mindfully and naturally in that peaceful noetic space, upon the *prana* wind of the mantra breath, until this life force energy arises spontaneously as the selfless thought, intention, and action for the benefit of all suffering sentient beings. We have so often seen that this altruistic precious *bodhicitta* is the primary cause—arising from 'mindfulness of breathing' meditation—of harmless human happiness. All of the wisdom masters have told it. Do we not already know this? Authentic and more or less selfless help for another results in our own happiness. Let us accept full responsibility for it now.

I have come to call this continuous happiness choice *awareness management*—conscious placement of our moment to moment attention/awareness upon our indwelling, always already present Presence of the noetic nondual primordial Spirit ground, vast boundless whole in whom we all arise and participate.

We've also seen that our usual seeking quest strategies for human happiness are based in wild, mindless self, the narcissistic separate self-ego-I and its 'I-Me-Mine' effort to acquire much stuff, and to control everything, and to gain power over others. I have told that this atavistic, narcissistic activity of self is known as primal ignorance (*avidya, ajnana, marigpa, hamartia*/sin). It is the root cause of human suffering; not to mention the suffering of nonhuman beings, including our precious Mother Earth. The antidote to such body-mind toxicity? Our always present indwelling love-wisdom mind Presence, of course. And how do we accomplish that? Mindfulness meditation, of course. How do we learn to practice that? We listen to and practice the wise, loving injunctions of those who know, of course. *Mindfulness meditation is an act of love.* It benefits both self and others.

The great avatars who have come to earth to save us from such adventitious primal ignorance have told it well. Said Jesus, "Forget thy self." Such *kenosis* is the 'self-emptying' that the great exemplar accomplished in order to light the way for human beings. Islam literally means "surrender of self". For Buddha, "There is no permanent self….All *dharmas*/phenomena are ultimately selfless *anatman*/noself." This 'crazy wisdom' (*yeshe chölwa*) is indeed the truth of the matter—when we open our heart and mind to see it, and then actually practice it under the selfless guidance of the subtle mind of the master.

Self and Noself Together at Last. For Buddha there is indeed a relative conditional self-ego-I to whom such phenomena continue to arise. So, we tread a skillful 'middle path' between narcissistic self and primordial luminous numinous 'noself'. We need our self. Noself cannot show up for work, nor buy groceries, not to mention practice the buddhadharma. And who is it that *chooses* to meditate? Is it not this problematic self-ego-I? For most folks, and even for yogis and yoginis, the self-sense is not going to vanish into a puff of primordially pure fairy dust. And if it did, *who is it* then that is happy and liberated? We need our self-ego-I; do we not? And we need to surrender this self to primordial noself—almost moment to moment—while still acting skillfully and compassionately in the world. And yes, this requires a bit of peaceful quiescent mindfulness meditation guided by a qualified meditation master.

Thus do we imbue this course, grasping contrivance that appears here in an illusory time as narcissistic self-ego-I with the all embracing great love of our always present selfless noself love-wisdom mind—like the mother's love that so gently corrects the selfish narcissism of the child. "The Child knows the Mother." Yes. As Buddha told so long ago: "Noself is the true refuge of self."

We learn to "let it be as it is", imperfectly, more or less moment to moment, between endless distractions, upon each mindful breath, and with each mantra prayer, and with each bit of sublime if dualistic liturgy. It is unruly self-ego-I that seeks and motivates liberation from suffering—harmless Happiness Itself—that is, most ironically, "already present from the very beginning" as our essential Christ-nature Buddha mind Presence—by whatever noble name, concept, or belief.

"What's in a name? A rose by any other name would smell as sweet." [Juliet Capulet] Those who know sometimes tell that our noetic love-wisdom mind Presence has a very subtle scent of roses.

Therefore, for many years upon this difficult, joyous Path we need the unruly cognitive doublet that is the imperfect unity of an objective self and our perfect noself working together in an imperfect harmony. No need to deny self; no need to try to transcend it. Don't beat it up; but don't pretend that you can bargain with it for the control over you, others, and all the random stuff that it desires. Buddha told, "Make yourself an ally." Your innate love-wisdom mind Presence is your constant inner guide. Your outer Lama, Ajahn, Roshi, Rishi, Shaman mirrors that Presence of the 'supreme source' that you always already are; your 'supreme identity'. "Turning yourself over to some Guru" is ill advised. Far from it. An authentic, qualified meditation mentor/master gently and subtly guides this rather delicate process—sometimes from far across the sea. It is your self who *chooses* to do the practice, with a bare minimum of ego judgments, and thereby reap the many benefits.

> It is already accomplished from the very beginning…
> To rest here without seeking more; that is the meditation.
> —*Dzogchen* Founder Garab Dorje (55 CE)

Verily, it is that liminal great awakening that is awakening now within you upon each mindful *jnanaprana* love-wisdom breath, and in each mantra prayer. *OM AH HUM.* Thus do we 'Keep the View', even in moments of most intense fear/anger distractions. Primordial Presence is always already present. Mindful practice is the ongoing continuity of remembrance of That (*tathata*); especially when you have forgotten.

Who is it that I am? *Tat Tvam Asi*: That I Am, without a single exception. So, "Rest your weary mind and let it be as it is; all things are perfect, exactly as they are." [Gautama Shakyamuni the Buddha] Luminous outshining always present vivid Presence of That.

Perhaps, some sunny day, we shall all shine together in that light. No self, no time, no problem at all. Now that you know, arise upon your mindful mantra *jnanaprana* spirit wind breath and do some good. Monitor your spiritual practice that it not become all about your self. The Path can become narcissistic. Become a good practitioner. Volunteer to help somewhere. Find a way. Your way. It will make your precious life happy, not tomorrow, but today—here and now.

Buddha told it well, "Noself is the true refuge of self."

Conclusion: From the Prior Unity of Science and Spirit Arises a Panpsychic Noetic Quantum Ontology

This concludes our all too brief exploration of the prior and present unity of objective Science and perfectly subjective nondual Spirit that is its primordial ground—formless, timeless, selfless boundless whole of physical/mental/spiritual nondual *kosmos*—vast whole of Reality Being Itself. That all subsuming aboriginal 'supreme source' or Basis (*gzhi rigpa*) of all spacetime reality is mereologically (part-whole relations) the ground for our foundational integral panpsychic *Dzogchen* Kosmopsychic Noetic Quantum Ontology.

The multiplicity of spacetime parts are perforce included in the singular primordial whole that embraces and pervades them. Mereologically, where there are parts, there is a greater whole. Just so, where there is a whole, there are constituting parts subsumed within it. That constitutes a mereological 'proof' of the always present lucid Presence of that primordial 'groundless ground', by whatever grand name or concept.

Hence, the aboriginal ground state of all phenomenal reality, that great all inclusive whole, necessarily ultimately exists. As good a 'proof' for the existence of post-theistic, non-anthropomorphic, non-creator, trans-conceptual nondual primordial Godhead as we are likely to encounter.

Moreover, by leaving behind or surrendering even that conceptual fabrication and engaging contemplative direct experience (*yogi pratyaksa*) of that non-conceptual nondual luminous ground we may have direct certainty of it. As Longchenpa (2001) told,

> Rest in that ground of being where everything is the spacious expanse of awakened mind—Samantabhadra [*dharmakaya Buddha*] ... This brings about natural rest in the state [of Presence] that cannot be [conceptually] reified as anything ... All phenomena are timelessly free in awakened mind, equally existent and equally nonexistent ... Awareness is 'basic space' [*chöying*] because whatever manifests occurs within that single state of equalness. It is 'the ground of being' ... It is 'the vast expanse of being' ... It is 'awakened mind', like space, primordially pure ... Everything is subsumed and completely pure within awakened mind ... So awareness—awakened mind is always spontaneously present as the basic space, or ground, of all phenomena ... The entire universe of appearances and possibilities does not stray from the expanse of awakened mind ... timelessly free ... There is primordial freedom in that unborn expanse, the single state of evenness—vast expanse of timeless awakening.

Just so, objective *Science* (grounded in physics) arises in the relative-conventional domain of inherently but not ultimately subjective Quantum Field Theory (QFT/QED) which itself arises—along with everything else—in the perfectly subjective noetic nondual whole itself, primordial *Spirit* ground—*dharmakaya, kadag,* Tao, nondual *Nirguna Brahman,* infinite *Ein Sof, Abba* God the non-theistic nondual Primordial Father of Jesus the Christ.

Thus does the physical and mental phenomena of our spacetime cosmos, including our science and philosophy about it, arise and participate in its ontologically prior all embracing *kosmos*, its formless timeless ultimate source condition.

For Tibetan Buddhists that vast infinite boundless emptiness whole is Basic Space (*chöying, dharmadhatu*)—all embracing *dharmakaya* ground. That indwelling always already present luminous Spirit Presence (*rigpa, vidya, christos*) then manifests through the centrist Mahayana/Vajrayana Middle Way Prasangika Madhyamaka teaching vehicle—conceptual causal foundation of highest acausal nondual *Ati Dzogchen, The Great Perfection* view and practice. The blissful clarity of that innermost Presence is the urgent 'grounding relation'—grounding by subsumption—that is primary cause or modality of human happiness: *relative* human flourishing (*eudiamonia, felicitas*), and harmless *ultimate* Happiness Itself (*paramananda, mahasukha, beatitudo*).

At the end of our journey we were introduced to the essential Four *Ati Dzogchen* Yogas—1) mindful quiescent *shamatha, Buddha's 'mindfulness of breathing*; 2) analytic penetrating *samadhi* insight of *vipashyana*; 3) formless timeless primordial ground that is *kadag*; and 4) natural spontaneous primordial Presence of that ground or *lhundrub*. The prior and present unity of these four *Dzogchen* Yogas constitutes the essence of the Tibetan Vajrayana Secret Mantra spiritual Path.

We have learned in these pages that our understanding of the ontic prior and phenomenally present unity of objective Science and its noetic nondual perfectly subjective Spirit ground requires that we conceptually recognize, then contemplatively realize that unity of the Science of dualistic *relative* Quantum Field Theory with *ultimate* Spirit ground in which it arises. That Basic Space (*chöying*) 'ground of being' as expressed in subtlest *Ati Dzogchen*, "the heart essence of all spiritual teaching". [Longchenpa] *Dzogchen* naturally embraces the highest nondual teachings of each noetic path of our great Primordial Wisdom Tradition—Hindu, Buddhist, Taoist, and Abrahamic Hebrew, Christian, and Islam.

We now understand that the reality dimensions of objective conceptual spacetime Relative Truth (*samvriti satya*) and trans-conceptual, contemplative, perfectly subjective Ultimate Truth (*paramartha satya*) in which it all arises and is instantiated are, in an integral noetic view, an indivisible prior yet present one truth unity-equality (*dzog, samatajnana*).

That compassionate knowing-feeling love-wisdom *buddic* mind—naturally occurring timeless, selfless 'unborn awareness', utterly lucid all embracing 'Basic Space' (*chöying*) buddha nature of naturally *awakened mind*—is 'spontaneously present' for human beings as acausal nondual primordial wisdom (*jnana, yeshe,* gnosis) which naturally embraces its practical expression as cause and effect 'discriminating wisdom' (*prajna, sherab*). These two compassionate faces of wisdom are the root causes of our altruistic human happiness, both relative human flourishing and Happiness Itself, harmless ultimate happiness that cannot be lost.

We have in these chapters explored the amazing Quantum Field Theory of modern physics and cosmology, but as well the 'paraconsistent' intuitionist logical mathematical foundation of a 'post-empirical', post-quantum centrist Middle Way integral Noetic Quantum Ontology—a conscious finite awareness portal, cognitive bridge into the infinite timeless, selfless, formless all subsuming nondual noetic primordial emptiness ground of not only quantum formalist mathematics, but of all this arising physical and mental form—including all of us.

We have here discovered that such a panpsychic *Dzogchen Kosmopsychic* Noetic Quantum Ontology is required should we desire to fathom the meaning of the inherent subjectivity of the quantum theory beyond its arcane mathematical formalism and the prevailing metaphysical ontic biases that lie hidden in our deep cultural background 'global web of belief' (Quine 1969)—the classical ideology of Scientific Local Realism/ Physicalism.

Relative nonlocal entangled quantum 'ZPE zero point vacuum energy' of the Unified Quantum Vacuum or quantum emptiness, and *ultimate* Middle Way boundless Buddhist emptiness, emerge as nominally real spacetime phenomena within that aboriginal primordial ground that is the vast whole of Reality Bring Itself. As Buddha told, "Form is empty; emptiness is form." What is to become spacetime form is primordially enfolded in its prior nondual formless, timeless, selfless, emptiness Ultimate Truth dimension groundless Spirit ground, and naturally spontaneously unfolds and 'descends' into the Relative Truth dimension of space and time. The luminous present indwelling Presence of that *kosmic* process—that whole—is who we actually are now, our individual 'supreme identity'. That is the great Spirit gift of being here in form.

That nondual ultimate emptiness ground of everything is 'groundless' because it transcends all dualistic concepts and beliefs about such a 'nondual ultimate ground'. Middle Way founder Nagarjuna told that Buddhist emptiness is itself utterly "empty of any iota of intrinsic ultimate existence." That is known as the "emptiness of emptiness". "Buddhist emptiness is established by human conceptual minds." [H.H. Dalai Lama]

Our subtle dualistic concepts about nonduality are indeed a philosophical trap. With little or no direct realization (*yogi pratyaksa*) of the natural equality of that conceptually uncorrupted 'emptiness of emptiness' (*shunyata shunyata*)—the vivid clarity of the natural state absent any self-other dichotomy, indeed absent any cognitive reference frame whatsoever—we remain encaged in the primal ignorance (*avidya, marigpa, ajnana, hamartia/*sin) of our apocryphal concepts and beliefs about non-conceptual, nondual Spirit. The perfectly subjective Spirit ground of objective Science is decidedly not conceptual.

That said, Einstein's colleague quantum physicist David Bohm has told it well, "The vast implicate order of the one enfolded vast unbroken whole is the ground for the existence of everything." That is the completion of physics' prodigious Relativistic Quantum Field Theory (relativistic quantum electrodynamics or QED) as it opens into and adorns the nondual infinite ultimate source or Spirit ground in which, or in whom, all dualistic relative Science arises, participates, and is instantiated.

That vast primordial awareness-consciousness boundless whole is the metaphysical foundation, the 'grounding relation' for Jon Schaffer's (2010) post-micropsychic 'Priority Monism Cosmopsychicism' that I have described as a panpsychic all subsuming *Dzogchen Kosmopsychism* in which a centrist integral Noetic Quantum Ontology naturally arises.

We have seen that our perennial Two Truths trope—*relative,* dualistic, causal objective Science (form) and its perfectly subjective nondual acausal *ultimate* Spirit (emptiness)—is an ultimate unified invariant *one truth unity* (*dzog*), an ontic prior and phenomenally present nondual unity—the all embracing *Perfect Sphere of Dzogchen.* And yes, we have come to know and realize this all subsuming noetic unity via the contemplative practice of 'mindfulness of breathing' (*shamatha*) and the direct penetrating insight of meditative contemplative analysis (*vipashyana*).

Our *noetic cognitive doublet*: 1) relative, exoteric, objective, conceptual, mental, scientific; and 2) noetic, greater esoteric, higher mental, contemplative, spiritual, even ultimate perfectly subjective nondual. These two cognitive modalities that are conceptual objective Science and perfectly subjective nondual Spirit unified at last! In practice a real balancing act, to be sure. But good to know as we consciously engage this difficult joyous precious love-wisdom mind life path we've been given as esteemed guests of the beautiful phenomenal world on this precious little blue planet of ours.

So please practice and remain present to that always present enlightened awareness Presence of your 'already accomplished' indwelling love-wisdom Buddha nature/Christ mind. Upon the mindful love-wisdom *jnanaprana* breath in the belly, place and maintain your awareness-attention upon That.

Human happiness arises from conscious placement of your moment to moment attention-awareness. I have come to call that urgent cognitive process *awareness management.* We do have this choice. So stay mindful (*shamatha*) and aware (*vipashyana*) of That, your 'supreme identity' that is compassionate Presence of the primordial 'supreme source' in whom everything is embraced and enfolded. Relax often into that trans-conceptual quiescent aboriginal awareness-consciousness ground in whom this all unfolds and abides. Rest here always in that peaceful luminous numinous 'Basic Space' (*dharmdhatu*)—*buddic dharmakaya* ground. Feel and know That now, beyond your concepts and beliefs about it. *Tat Tvam Asi.* That I Am! That is original Buddha nature of your mind! Perfect just as it is.

As the Buddha of this present era told so long ago. "Rest your weary mind and let it be as it is; all things are perfect exactly as they are." That is our 'innermost secret' ultimate truth. Now that you know, arise and do some good. That will make you happy now.

> May all beings be free of suffering and the causes of suffering. May all beings have happiness and the causes of happiness.

Emaho! Mahasukaho!

Appendix A: Let It Be: Brief Course in Basic Mindfulness Meditation

Enjoy the clear bright space between your thoughts.

Happiness Arises From Your Present Mind State!

Awareness Management. Therefore, train your mind in happiness: peace, free of the habitual thinking of self-ego-I with its unhappy fear, anger, hatred, and pride. Mindfulness meditation is after all a conscious finite portal into infinite Basic Space—peace, boundless primordial awareness whole of everything arising herein—bright love-wisdom mind Presence of That, always already present within you now. Train your mind in *placement of awareness/attention* upon that aspect, or imprint, or *Presence* of you in this present moment now. 'Mindfulness of breathing' is the meditation that accomplishes this open secret of human happiness. Below are Ten Steps that will make you happy now. If you choose to practice it. It's easier than you think.

Begin by sitting in a chair, your back straight, hands in your lap, legs uncrossed, feet flat on the floor. Or sit on a cushion, legs crossed. Or consciously walk in a peaceful meadow. That is known as 'walking meditation'. Alternate between sitting and walking meditation.

1. **Thank You!**

Experience deep thanks for the great gift of your life, just as it is now. Accept yourself—all your positive and negative experience—exactly as you are, here and now. Feel *your selfless good will intention to benefit living beings. That is the primary cause of your own human happiness.*

Lower your gaze so that your neck is straight. Relax jaw, neck, gut. Feel the *prana* spirit wind upon the breath in your belly. Now *place your attention* behind your forehead. Close your eyes, raise your eyebrows. This will produce alpha/theta brain rhythm, the peace response, replacing stressful 'fight or flight' beta rhythm. Feel a subtle focused fullness in the forebrain. Let the crown of your head open as light streams in from above and meets the *prana* life energy rising upon each breath. *Feel* it pervade your entire body-mind—and deep into the earth. Now rest for a few moments in that quiet peaceful Presence of you.

2. Attention!

Now, gather the 'wild horse of the mind' by **placement of attention/ awareness** *on your breath. Be present to your breath as it rises and falls in your belly.* Let your mantra prayer begin. Softly recite *OM AH HUM* (see below). This is your 'alpha mantra breath': 5 seconds in; 7 seconds out through pursed lips (12 seconds). Do it 3 to 9 times (36 to 108 seconds). Let your mantra prayer continue, either consciously, or in the background, day and night.

Each breath feel your busy mind settle into its quiet natural state of wakefulness; your clear light love-wisdom mind Presence—*that aspect of you that is utterly connected with the great source of everything—your safe place, beyond all thoughts, concepts, beliefs; free of judgment, fear, anger, hatred, guilt, pride; free of self-ego-I. No need to think about it. Open and feel it!* Be *that stillness. Now say to the busy mind, "Peace, be still". Say to the frightened grasping self, "Peace, I Am".*

Thoughts, questions, feelings naturally arise. Briefly greet them. Negative or positive thinking, planning, mind wandering, worry/ anxiety, anger: label whatever arises "distraction". Then surrender it all on the out-breath. Or let it flow by on vast empty space of the sky, like a cloud, leaving no trace. *Again and again return attention to the breath.* After three minutes or so open your eyes slightly and breathe normally, mouth closed.

As you settle into, and rest in your selfless *wisdom mind Presence,* your breath will naturally be slow and gentle. Enjoy this feeling of delight within you. Feel your connectedness to everything. No need to create it; or grasp at it. Mindful Presence upon the breath is always already present—your "Supreme Identity". Who Am I? *That I Am!*

3. **In-Breath**

Open to receive luminous purifying 'life-force energy', sustainer of all life. It has many names. In the East this energy is *prana* or *ch'i* (spirit/breath). For the West it is *pneuma*/Holy Spirit, the very 'breath of life', 'bio energy', the subtle voice of gross physical light/energy/form ($E=mc^2$) arising from formless, non-conceptual, spacious unbounded whole; vast primordial awareness-consciousness ground itself in whom this all arises. *Breathe,* you are alive! Open and receive. Feel it pervade every space of your body-mind.

4. **Out-Breath**

Release thoughts, feelings, past, future, all self-ego-I grasping. Feel your stability deep in Mother Earth. Whatever arises—thoughts, feelings, doubts, happy or not—release it all on the out-breath. Surrender it all. Witness it all dissolve as you return to your breath, again and again. *Let it be just as it is* in this peaceful luminous sky-like space of your mind.

Please consider this well: Thoughts are only thoughts. They come and they go in dependence upon your present mind state. Thoughts are not a solid reality! You are now learning to choose *your realities by choosing your present mind state. All of the love-wisdom Masters of our great Primordial Wisdom Tradition have taught this great liberating freedom to be happy right here now.*

So, as thoughts and feelings retreat, *feel* your selfless, natural clear light *love-wisdom mind Presence*—peace, clarity, subtle bliss. From this natural spacious mind state the kind, compassionate *activity* of love spontaneously arises in your mind stream—the very secret and primary cause of human happiness. Place your *attention* on that. Let it be so now.

So it is, that deep blissful peace which you desire rides the breath. Remain close to the breath. When distracted by fear/anxiety, anger, or self-doubt—simply return to already present Presence of the nondual primordial ground upon the breath, again and again. When your mind is filled with this light of love-wisdom mind Presence, there is little room for the negative stuff. Practice that and be happy. Now, rest naturally in That for awhile.

5. **Presence**

Breathe Now. Open your heart and mind and feel your always present indwelling love-wisdom mind Presence of vast awareness whole in whom this all arises. It's right here! That you are now! Subtle Presence of That may be directly *experienced, prior to thinking, as luminous clear-light mind essence—the very Christ-Buddha Nature of Mind, beyond any name, concept, or belief.*

Now experience this *prana* spirit life-energy at the crown of your head. Feel it stream in from above upon each breath. Open your heart to receive. Feel it pervade your entire body-mind. Let it flow downward throughout your head, throat, chest, back, *hara* center in the belly, pelvis; then deep into Mother Earth. Feel your fearless stability in Earth. Release any negative thoughts and emotions that are ready to go. That is your 'full body scan'.

Let this energy of Presence penetrate any discomfort—that self-contraction from your natural life-energy flow: physical tension and pain, sense desire, grief, doubt, guilt, fear/anxiety, anger/hostility, harsh judgments of self and others. Patient love and wisdom heal fear and anger. Your alpha mantra breath is your touchstone to being happy now.

Now experience the emotional lift as any and all presently activated 'attachment and aversion' are inundated by Presence of clear light life energy. *Be* for a moment with whatever arises—attractive or aversive. Then surrender it all on the out-breath. Know now you are free of it. Let this light penetrate and pervade space of your entire emotional and physical body-mind: brain, nervous systems, heart, organs, cells, the very atomic structure of your physical/emotional/spiritual being. Now, rest in this feeling of delight within you. "Let it be as it is, and rest your weary mind, all things are perfect exactly as they are." (Buddha) "That which you seek…the Kingdom of God…is already present within you…and it is spread upon the face of the Earth, but you do not see it." [Jesus the Christ, Luke 17]

With each breath *feel* healing life energy Presence fill and overflow into your subtle energy field, this light of you that embraces and pervades your whole body-mind. Awaken to this 'basic goodness' that you are, prior to cultural skeptical 'global web of belief'. But don't *believe* it. It's beyond belief. *Feel* it. Now self-ego-I is at peace. Rest fearlessly in That.

6. **Wisdom Mind is a Choice**

"What you are is what you have been; what you will be is what you do now." [Gautama the Buddha] This bright basic space upon the breath is your natural wakefulness—your primordial love-wisdom mind Presence. *Choose* to be that space/peace, here and now, beyond ego: no past nor future; no attachment nor aversion; no true nor false; no judgment at all—just for this moment. No need to think, try or do anything. *Know that your clear-light mind is already awake, kind and wise. Rest in That, each breath. Let it be as it is; calm and clear.*

Love-wisdom mind *practice* is your Path to liberation from egocentric ignorance and delusion, root cause of human suffering. Stay with it. Your self-ego-I may resist. Notice the bogus excuses. This *choice* is kind, karma free *relative* human flourishing that does no harm; and *ultimate* happiness-liberation from suffering; the happiness that cannot be lost.

Thus is human happiness very much an awareness management skill set! Happiness arises, not so much from desirable stuff, but from the choice of your *placement of awareness/attention* upon your breath, in this present moment now! No belief, no leap of faith, no authority but your own is required. Simply settle your mind, open your heart, and be fully present to your alpha mantra breath now. That is your *connection* to peace and happiness already present within you. That is the foundation of your love-wisdom mind practice of the Path. *What is your mind? That is your mind. Feel that now, beyond your thoughts about it. Rest in That.*

7. **Refuge**

Now you know this precious Basic Space/peace of your *love-wisdom mind Presence.* Take refuge in it often. Breath by breath purify, pacify, stabilize, beautify your mind; a most courageous act; your most urgent activity. Make mindful breathing a priority, *'brief moments; many times',* all day, all night. Soon it becomes a quiet conscious continuity of awareness. Who am I? Feeling *Presence* of that vast whole—'*Tat Tvam Asi*; That I AM', without a single exception. You have never been separate from That! Feel it upon each breath. *That is the View. That is the Teaching. That is the Practice. It's like coming home.*

8. **Compassion Meditation**

By this good generated by each mindful breath make this aspiration for the benefit of all living beings: "*May all beings be free of suffering, and the causes of suffering. May all beings have happiness, and the causes of happiness*". That powerful mantra prayer is as well, your *Compassion Meditation* when practiced for a few minutes. It moves your attention from self to others. 'Come and see' what it does for your present state of happiness.

Is not your happiness already linked to the happiness of others? We're all in this reality boat together. Accomplish your own happiness through compassionate thought, intention, and action to benefit living beings. It's called altruism. In the East it's *bodhicitta.* It's the magic metric for a good life. So arise, and do some good. It will make you happy now.

9. **Real Practice**

Practice requires patience and courage. Patience is the antidote to anger, which arises from fear. It takes courage to face fear. Practice 15 minutes or more upon rising and retiring; and many '36 seconds of bliss' alpha mantra breaths during the day. 'Brief moments many times'. Peace is always here, between your thoughts, each mindful breath.

Take refuge in your love-wisdom mind Presence often. Feel it at your heart before sleep; and all night long. Be present while eating, walking, working, loving. Lovingly accept yourself as a mother accepts her child. No blame. Anxious, angry? No time? *Take three OM AH HUM belly breaths now!* Go ahead and do it now. Your goal is *not* peace and happiness in some ideal future mind state. *Make the practice itself your goal*—each mindful breath. "Mindfulness of breathing is the foundation of all wisdom and happiness." [Buddha]

10. The Five Benefits of Mindfulness Meditation Are Always Already Present

1) Body-Mind Relaxation experienced as profound peace, forgiveness, healing.
2) Non-Conceptuality: beyond self-ego-I thinking, concept, belief, fear/anger.
3) Clarity: clear mental and perceptual acuity, luminosity, vividness, wakefulness.
4) Deep Appreciation and acceptance of your life, and yourself, just as you are.
5) Wisdom Mind Presence: happiness expressed as kind, compassionate action.

Good Sleep

Engage your alpha mantra breath for a few minutes near your bed just before sleep. Now continue to recite *OM AH HUM* silently, on your back, hands over your solar plexus, or at your side, palms down. Settle into your clear light love-wisdom mind Presence. Feel That.

Now begin your *full body scan. Feel* the gentle peace of *prana spirit wind* life energy throughout your entire body mind. Your crown center is open. Feel Presence of Buddha or Christ three feet above your crown. Breath by mindful breath receive this love and peace from above through your crown and into your head; neck, shoulders, chest, arms and hands; then belly and back, pelvic area, legs and feet, and deep into Mother Earth. Let this life-light prana energy pervade your body-mind all the way down to the subatomic level.

Relax into this light. And rest in it. Let any obstruction to energy flow—tension, pain, worry, anger—flow away on the out-breath, and out through your hands and feet. "Rest your weary mind and let it be as it is." Feel life energy *prana* peace pervade your entire body-mind. Now say quietly, "May all beings be free of suffering and the causes of suffering. May all beings be happy, and have the causes of happiness." Thus do you go beyond 'self' to the quiet peace of Christ/Buddha selfless 'noself' Presence of the *ultimate* primordial ground of everything that arises here in this all too real *relative* space and time.

As your breath naturally becomes slow and regular, let your *OM AH HUM* mantra prayer settle into your awareness background as you assume your normal sleeping position. Rest in That. Let this spirit breath of yours be your love-wisdom lullaby and goodnight.

OM AH HUM: Our Three Reality Dimensions

Use this powerful mantra prayer as a touchstone in your practice—all that you think and do is practice—to instantly connect to and protect your primordial love-wisdom mind Presence. Let it be always in your awareness foreground or background. Free your mind by reciting it three or four times daily 108 times while walking, or sitting. (Get a 108 bead mala.) These three 'Buddha Bodies' are one prior and present indivisible *one truth unity.*

OM is *dharmakaya* dimension, formless, timeless, selfless empty 'Basic Space', all-pervading, always present primordial awareness ground of all arising spacetime phenomena, vast unbounded whole of Reality Being Itself. *AH* is *sambhogakaya* dimension, like the sun in empty space; Logos; selfless, lucid clear light awareness—light bridge into form. *HUM* is dream-like display of *nirmanakaya* form dimension—Buddha mind acting in time as love-wisdom *Presence* of *OM*—always already present now within you; light-form gift naturally expressing itself as skillful loving *bodhicitta*—thought, intention and action to benefit all living beings.

What is your mind? That (*tathata*) is your mind. Who am I? *I AM OM AH HUM*: body, voice, and mind of all the Buddhas and wisdom masters of the Three Times—past, present, future—instant connection to That! Three Gates to peace. Feel it purify your cause/effect karma. Don't think. Feel. *The benefit of 'mindfulness of breathing' is immeasurable.*

Now you know the 'innermost secret' of human happiness. Please consider it well. If you desire to be free and happy, then choose to practice it. You do have that choice. Now that you know, arise and do some good. It will make you happy, not in the future, but here and now. [Excerpted from *Mindfulness Meditation: The Complete Guide*, 2022, David Paul Boaz]

David Paul Boaz Dechen Wangdu: coppermount.org; davidpaulboaz.org 8.1.23

Appendix B: The Neuroscience, Logic, and Metaphysics of Mindfulness Meditation

> All the happiness in this world comes from compassionate service to others; all the suffering comes from serving oneself.
>
> —Shantideva

We have seen that human beings being here in time desire, require and deserve some semblance of happiness and well being. We have as well seen that these happiness mind states are already innately present within the human heart and mind. We access them through the conscious 'placement of awareness/attention' upon our always present love-wisdom mind *Presence* of the formless primordial ground in whom this all arises.

Therefore, both neuroscientifically, and metaphysically, human happiness is a function of one's here now *present* mind state; a blatantly obvious relative conventional truth. As our attentional awareness is consciously placed upon our always already present inherent innermost peace—that luminous 'I Am That I Am' Presence that we actually are—we are happy. If our present awareness is mired in fear and anger, we are unhappy. So the proper question as to human happiness is this: how do we accomplish such happy, peaceful mind states? That is to say, how do we learn "placement of attention" upon such already present happiness mind states? [*Appendix A*] We have examined the contemplative considerations. Let us now very briefly explore the pertinent data of neuroscience.

The Neuroscience of Mindfulness Meditation. Over 50 years of Western world neuroscientific research has demonstrated that the Buddha's "mindfulness of breathing" practice supports the following noble objectives: preventive medicine, chronic pain management, stress reduction (recovery from stressed induced cortisol production which is known to cause many psychophysical and physical symptoms), and psycho-emotional healing, learning, interpersonal relationships, and reports of happiness and well being.

Meditation supports parasympathetic function (alpha and theta brain rhythm "rest and digest" approach behavior), and thus enhances immune function, while tending to tonify sympathetic (beta rhythm "fight or flight" aversion/avoidance behavior) activation which suppresses immune function. [Porges 2014; Siegel 2013; Wallace 2009]

Neuroscientist Richard Davidson at the University of Wisconsin has shown that beginning meditators in a corporate business setting developed stronger immune systems than controls, as evidenced by statistically significant resistance to respiratory infections.

In another Davidson study, novice meditators reported being happier than non-meditating controls, experience more positive emotions including kindness, fewer stressful emotions, and an enhanced feeling of well being. [Davidson 2017]

The left and medial prefrontal cortex of the brain activated in mindfulness meditation has been shown to dampen response to negative emotional mind states, while enhancing positive states. Just so, the right prefrontal cortex is activated in negative states. The amygdala is responsible for such negative internal states as fear and anger, which are then expressed through egocentric, often destructive and self destructive behavior.

In other words, the left and medial prefrontal cortex generate alpha, theta and 25-42 hertz gamma brain oscillations which mitigate and pacify the negative neural signals from the right prefrontal cortex and amydgala which inwardly manifest as fear and its flipside—anger, hostility and aggression—which then manifest outwardly as adventitious human evil: alienation, hostility, aggression, despotism, ethnic hatred and endless war. [Siegel 2013; Begley 2007; Wallace 2007, 2009]

This process of aggression is observed clinically in stroke patients. Patients with damage to the "happy" left prefrontal cortex are generally more irritable than those with damage to the "angry" right ventromedial prefrontal cortex, who are often relatively calm, even serene [Sheng, Boaz in Hanson 2014].

Meanwhile, the frontoparietal control module network manages brain's multiple modules allowing it to function as a consciousness unity (executive function and cognitive control), while the default mode module, which spans the same lobes as the frontoparietal network, is linked to such cognitive functions as often habitual 'self-referencing' or 'selfing', introspective thought and subjective feeling, passive emotional listening and learning, emotional processing, memory retrieval and 'theory of mind' (empathetic emotional connection to others). [*Scientific American* November 2014; July 2019]

Mindfulness meditation meta-research, over thousands of studies, has demonstrated the following positive psycho-emotional outcomes: enhanced immune system function, respiratory function, post-surgical healing, blood pressure reduction, reduced age related brain atrophy, reduced symptoms of dementia and Alzheimer's disease, reduction in symptoms of anxiety, clinical depression, bipolar disorder I and II, obsessive compulsive disorder, attention deficit disorder, post-traumatic stress disorder, chronic pain, post-stroke symptoms, reduction of prison violence and recidivism, improvement of school grades, and much more. [Begley 2007; Sheng/Hanson 2014; Wallace 2009; Siegel 2013]

Mindfulness Based Cognitive Therapy (MBCT) has been shown to be as effective as antidepressants in treatment of clinical depression. [Lancet Vol. 386; Kuyken 2015]

The Neuroscience of Meditation and Our Experience of Self

We've surveyed the neurobiological influences of mindfulness meditation on human behavior. How do these influences effect our sense of self-ego-I; brain structure and function; relative human flourishing and relative happiness (*eudiamonia, felicitas*); and the ultimate happiness of liberation/enlightenment (*paramananda, mahasukha, beatitudo*)?

Both Zen Masters and neuroscientists agree, "mindfulness of breathing" ('focused attention meditation'), and 'compassion meditation' both facilitate 1) a beneficial shift of attention from obsessive, usually fraught *self-referential thinking* and concern for 'I, Me, Mine'; which 2) bestows a sense of inner peace and self-acceptance; which 3) reduces anxiety and anger toward self and others; which 4) enhances altruistic compassionate thought, intention and action for the benefit of living beings (*bodhicitta*); 5) enhancing individual well being and happiness. How then shall we understand this contemplative process in the scientific gloss of neurobiology?

The unfocused ruminating wandering mind, under sway of the brain's 'default mode network'—the medial prefrontal cortex (MPFC) and posterior cingulate cortex (PCC)—significantly increase *self-referential attention*—'selfing'—with its nearly always present fear/anxiety, anger/hostility, greed/pride, and negative judgments about self, which are then projected onto others. The micro-cognitive result in the individual is stress and unhappiness. The macro-cognitive result in the human sociocultural cognosphere is alienation, ethnic hatred, despotism, and endless war.

Scientific meta-research, synthesizing data from thousands of research projects since 1970, reveal substantial benefit in each of the three primary classes of meditation practice. These include 1) *shamatha mindfulness focused attention meditation*, usually upon the breath, or on an object image/vision of the Buddha, or of the Christ, or of the spiritual master; 2) *open monitoring mindfulness meditation*, witnessing whatever arises in awareness without grasping, rejecting, or judging; and 3) *loving-kindness compassion meditation*, feeling our natural empathy and caring for living beings. All three classes resulted in beneficial outcomes by conclusively reducing or deactivating processing in some physical structures, while enhancing activity in others. How is this so?

Proven Benefits of Mindfulness Meditation. 1) Reduced processing in the default mode network (PCC and MPFC) of the obsessive "selfing" wandering mind; which 2) reduced self-ego-I self-referential processing—habitual attention and concern about I-Me-Mine with its secondary anxiety, anger and ill-will mind states; 3) reduced activity in, and reduced size of the amygdala which is responsible for fear and anger ('fight or flight'); 4) reduced stress related cortisol production by the adrenal cortex while blocking its circulation throughout the upper body upon the autonomic CN-X vagus nerve ('rest and digest'); 5) enhanced beneficial brain alpha, theta, and high amplitude gamma band oscillations (25 to 42 hertz), while reducing excessive, often obsessive beta activity; 6) reduced activity in the right prefrontal cortex which is active in fear, anger, and ill-will mind states; 7) greatly increased left prefrontal cortex processing which enhances feelings of altruism, compassion and forgiveness toward self and others; 8) induced increased long term frontal cortex gyrification (neuroplasticity), which proved to be permanent, even when contemplative practice ceases. [Siegel 2013; Porges 2014; Begley 2007; Davidson 2017; Wallace 2007, 2009; *Scientific American* November, 2014]

The no longer surprising result of this neuroscientific meta-research includes 1) greatly reduced preoccupation with self-ego-I and its obsessive narcissistic self-narrative; 2) reduced psycho-emotional stress; 3) induced and enhanced subjective feelings of connection, well being, good will; and 4) subjective reports of enhanced happiness.

Thus does mindfulness meditation train the "wild horse of the mind" in the *placement of attention,* and continued focus of attentional awareness upon immediate, non-conceptual, present moment to moment sensory/feeling experience, upon the mindful breath—our eternal here now connection—while shifting self-referencing attention away from chronic unfocused wandering mind with its obsessive attachment to self-ego-I, and toward altruistic compassionate thought, intention, and action for the benefit of all living beings (*bodhichitta, karuna, patheos, hesed/lovingkindness, rahmah*).

We begin to realize that *mindfulness is an act of love*, a continuity of the mythopoetic ultimate primordial Great Love compassionately expressing itself in and through this relative gift of human form, for the benefit of all human and nonhuman beings, including our precious Mother Earth.

Awareness Management: Knowing-Feeling Presence of the Primordial Ground

Clearly, meditation reduces or suspends habitual, often obsessive self-referential thinking ('selfing') that causes the terrible suffering secondary to our pervasive sense of a fearful lonely separate self adrift in a hostile, dangerous, meaningless cosmos. And all of that accomplished through a program of mind training in present moment, trans-conceptual 'felt sense' *feeling awareness* upon the breath—mindful placement and maintenance of attentional awareness upon the mindful breath in the belly (wisdom *jnanaprana*) which settles the frantic, fearful, obsessively thinking 'monkey mind'. Thus does dualistic human awareness enter in, merge and awaken to its own indwelling, always already present Presence of nondual primordial awareness-consciousness whole itself, numinous formless *ultimate* ground of all *relative* space-time form—the very 'Nature of Mind' (*cittata, sems nyid, buddha-jnana*), nondual love-wisdom mind Presence of That, by whatever grand name. I have come to call that cognitive process *awareness management.*

Yes, neuroscientific meta-research demonstrates the profound value of mindfulness meditation—beginning with *shamatha* calm abiding, and loving-kindness compassion meditation—in support of human flourishing and happiness.

Indeed, there is a "mindfulness revolution" now abroad in the Western mind and its culture. It's alive and well in most of our institutions—education, medicine, psychology, the social sciences, business, government, military, corrections, even organized religion which has grown apart from its foundation in the contemplative mythos of the great noetic (body, mind, spirit unity) Primordial Wisdom Tradition of our beloved *Homo sapiens.*

Choosing Reality. On this neurobiological view then, human happiness is very much dependent upon an *awareness management skill set*—where, when, and how we *choose* to place our conscious awareness. In short, both happiness and unhappiness are the result of present placement of our cognitive attention/awareness in this present moment here and now. And this can be learned from the wisdom injunctions of those who know—the qualified mindfulness meditation teachers and their mentor/masters.

Self and Noself. Cognitive neuroscience has identified two ways of experiencing the self—two modes of self-reference: 1) *narrative focus upon self,* our urgent all consuming life story-drama about ourselves; and 2) *experiential focus upon self,* bodily proprioceptive sense experience, including trans-conceptual direct contemplative-spiritual feeling experience.

These two modes are hypothesized by neuroscientists to be neurologically distinct. Recent contemplative research with H.H. Dalai Lama's highly skilled Buddhist meditating monks (Davidson 2017) has shown these two modalities of self experience are in fact a neurological and phenomenological unity. Self-ego-I (*atman*) is here contemplatively (trans-conceptual compassion meditation) surrendered to the *buddic* always present luminous numinous wisdom mind 'noself' (*anatman*)—bright indwelling primordial Buddha mind Presence of That (*tathata*). As Gautama the Buddha of this present age told, "Noself is the true refuge of self." Self-destructive obsessive self referencing or 'selfing' is here surrendered to the 'clear light innermost secret' always present Presence of *buddic anatman.* [H.H. Dalai Lama 2009, 2000; Davidson 2017; Siegel 2013; Boaz 2020 Ch. V]

Yes. Volumes of research have demonstrated that in both meditators and non-meditators the *experiential focus* mode involving non-conceptual "mindfulness of breathing" as the Buddha called it, profoundly reduced egocentric narrative self-referential activity in the MPFC and PCC of the brain's default mode network.

For highly skilled meditators habitual fantasy-reverie self-referential thinking of the contemplatively untrained mind is absent during sitting meditation, and for varying periods of time following formal sitting meditation. Here, processing activity of the default mode network is nearly quiescent. [Siegel 2013] These skilled practitioners abide in a compassionate, calm post-meditation meditative mind state most of the time. This quiescent state persists through some sleep states. The subjective experience of such a stable neurological state of mind is known to such practitioners as the 'secret' happiness of the 'yogi's bliss'.

In short, 'advanced' meditators have demonstrated in hundreds of studies (Davidson 2017; Begley 2007; Siegel 2013) the capacity to maintain such stable direct non-conceptual contemplative mind states (*yogi pratyaksa*) with their corresponding brain rhythms (alpha, theta, gamma) in post-meditation activities—while "hewing wood and carrying water", and driving, talking, loving, and selfless creative thinking and planning.

Therefore, meditation practice for established meditators facilitates the *choice* of a fluent cognitive ambulation from conceptual self narrative modes to a selfless, peaceful, non-conceptual experiential mode, almost at will. The result is calm abiding quiescent peace of mind, and a felt sense of happy, blissful connection and interdependence with all living things, including our Mother Earth; and indeed, with the *ultimate* unbroken boundless whole of *kosmos* itself—even as inexorable human adversity continues to arise in the *relative* spacetime world of conditioned lived experience.

Clearly, the neuroscientific implications of meditation for the reduction of human suffering and for human happiness are profound. Mindfulness meditation and loving-kindness meditation offer skillful regulation of negative emotional response to life's endless adversity by transforming the painful narcissistic self-narrative into quiescent, peaceful, and altruistic states of mind, and their corresponding life stages. [Boaz 2023]

As we learn the practice of "mindfulness of breathing", we learn to place our present moment to moment awareness—our *attention*—upon our direct trans-conceptual wisdom mind *feeling experience.* Thus do we connect with an aspect of ourselves—indeed a Presence—that is selfless, profound, and directly experiential, beyond and embracing our habitual often narcissistic discursive concepts and beliefs.

We now begin to see that mindfulness meditation permits observing our thoughts without judging or identifying with them. We come to understand that we need not believe and defend our adventitious dreary and destructive negative fearful angry ego-centric thoughts and feelings; stress is reduced; and human happiness is enhanced. Thoughts and feelings are seen to be inherently evanescent, ever changing, and impermanent. They have only the power that we choose to give them. Clearly, we do have that choice. Perhaps we should take them less seriously, and with a bit of self-effacing humor. Perhaps after all we are not the egocentric center around which the universe revolves. Yes, that prodigious healing process is enlightened *awareness management.*

The Psycho-Social Benefits of Mindfulness Practice

As to my own, not especially astute experience working many years in a psychiatric corrections setting, I have personally introduced and guided hundreds of inmate clients with various psychiatric diagnoses in "mindfulness of breathing" (*shamatha, shiné, sati, smrti*) meditation practice. These include schizophrenic, delusional, dissociative, anxiety, mood, personality, and autism spectrum disorders, often with secondary substance abuse. I have taught Buddhist, Christian, Hindu, Taoist, Jewish, Islamic, Native American, and secular based mindfulness meditation to individuals of all capacities, including many staff psychiatrists who have seen its value and wish to practice it themselves, and teach it to their incarcerated clients. In all cases of those subjects with two or more weeks of actual, *vis a vis* reported practice, symptomatic relief was reported by the student-client—from reduced anxiety, anger, medication, and sleep disorder, to profound compassionate, emotional-spiritual healing transformation. I have trained many to teach it to fellow inmates, and to family members.

Even a few students with 'antisocial personality disorder' diagnosis (amoral sociopathy, psychopathy) learned to quiet 'the wild horse' of habitually thinking mind and gain some inner peace. Of course, those with ego-dystonic diagnoses (ego desire to heal) fared better those with ego-syntonic diagnoses (ego denial of any need to heal).

Student subjects in most cases were greatly relieved that their actual "true" identity was not their pathological psychiatric diagnosis, but far more profound. That is to say, my students came to understand that his or her 'supreme identity', by whatever name, is always already their innermost primordial love-wisdom mind Presence, to wit, the "I Am That I Am" Presence of Moses and the prophets, and the indwelling *christos* Presence of Jesus the Christ; the wisdom depth of Buddha nature/Buddha mind Presence (*vidya, rigpa, jnana*); Atman Presence that is one with Brahman; *Tao-chia* Presence of primordial Tao.

My student inmates are encouraged to conceptually unpack their non-conceptual meditation and contemplative prayer insights in the semiotic gloss of the wisdom tradition of which they are most familiar and comfortable—including secular "Scientism", the proto-religious belief system founded in the "global web of belief" of Metaphysical Scientific Materialism that has now colonized the Western mind and culture. While these secular students often, but not always remained skeptical of the esoteric notion of an innermost Presence of God, or of Christ, or of Buddha, they nonetheless benefitted from the exoteric secular quiescent 'mindfulness of breathing' practices of Contemplative Science.

'Clinical progress' was viewed here as client reported reduction in objective symptoms—anxiety and panic, sleep disorder, anger, voices, reduced medication—and perhaps more importantly, first person subjective reports of enhanced connections with others, and enhanced happiness and well being. These benefits proceeded fluently from two mindful awareness cognitive set points. These points included: 1) exoteric conscious aware mindful breathing ("focused attention meditation", and "compassion meditation"); and 2) our perennial esoteric noetic Primordial Wisdom Tradition contemplative teaching as to an *ultimate dimension* or primordial ground, "innermost esoteric", indwelling love-wisdom mind emotional-spiritual Presence—by whatever grand name—that co-exists, on the accord of these traditions, with a troubled and suffering self-ego-I living in this *relative dimension* of time and space. [Boaz 2022 *Appendix D*, "Light From the County Jail"]

Excerpted from Boaz 2022 *Mindfulness Meditation: The Complete Guide*,
coppermount.org; davidpaulboaz.org 8.1.23

Bibliography

Ajahn Brahm. 2006. *Mindfulness, Bliss, and Beyond.* New York: Wisdom.

Almas, A.H. 2008. *The Unfolding Now.* Boston: Shambala.

Anam, Thubten. 2009. *No Self, No Problem: Awakening to Our True Nature.* Boston: Shambhala.

Begley, Sharon. 2007. *Train Your Mind, Change Your Brain.* New York: Ballantine.

Boaz, David Paul. 2020. *The Teaching of the Buddha: Being Happy Now.* Cardiff: Waterside.

________________. 2021. *Buddhist Dzogchen: Being Happiness Itself.* Cardiff: Waterside.

________________. 2022. *Mindfulness Meditation: The Complete Guide.* Cardiff: Waterside.

________________. 2023. *The Noetic Revolution: Toward an Integral Science of Matter, Mind and Spirit.* Cardiff: Waterside.

Carroll, Sean. 2019. *Something Deeply Hidden.* New York: Dutton.

____________. 2003. *Spacetime and Geometry: An Introduction to General Relativity.* NY: Addison.

Chalmers, David J. 1996. *The Conscious Mind.* New York: Oxford Press.

Chögyam Trungpa. 2015. *Mindfulness in Action.* Boston: Shambhala.

Dzogchen Ponlop Rinpoche. 2002. *Penetrating Wisdom.* New York: Snow Lion.

________________. 2003. *Wild Awakening: Heart of Mahamudra and Dzogchen.* Boston: Shambhala.

Davidson, Richard. 2017. *Altered Traits: Science Reveals How Meditation Changes Your Mind,* Brain, *and Body.* New York: Avery Publishing.

Garfield, Jay. 2015. *Engaging Buddhism: Why It Matters to Philosophy.* New York: Oxford Press.

Gen Lamrimpa; Wallace, Alan. 1992. *Calming the Mind: Tibetan Teachings on Cultivating Meditative Quiescence.* New York: Snow Lion.

Gunaratara, Henepola. 2011. *Mindfulness in Plain English.* Boston: Wisdom.

His Holiness the Dalai Lama. 2007. *Mind in Comfort and Ease.* (Longchen Rabjam's *Finding Comfort and Ease in Meditation on the Great Perfection*). Boston: Wisdom.

_______________. 1998; 2009. *The Art of Happiness.* New York: Riverhead Books.

Klein; Anne. 2006. *Unbounded Wholeness: Dzogchen, Bön, and the Logic of the Nonconceptual.*

Longchen Rabjam. 2001. *The Basic Space of Phenomena.* California: Padma Publishing.

Mipham, Jamgon. 2007. *White Lotus.* Padmakara Translation Group. Boston: Shambhala.

________________. 2006. *Fundamental Mind.* Khetsun Tsongpo; Jeffrey Hopkins: Snow Lion.

Namgyal, Dakpo Tashi. *Clarifying the Natural State.* Hong Kong: Rangjung Yeshe.

Porges, Stephen. 2014. *Polyvagal Theory.* New York: Norton.

Quine, Willard Van Orman. 1969. *Ontological Relativity and Other Essays.* New York: Columbia.

Scientific American. November 2014.

Shantideva. 1997. *A Guide to the Bodhisattva's Way of Life.* Translated by B. Alan and Vesna Wallace. New York: Snow Lion.

Shechen Gyaltsap Gyurmé Pema Namgyal. 2020 by the Padmakara Translation Group. *Practicing the Great Perfection.* Boulder: Shambhala Publications.

Sheng, Chuan, Ed. *Exploring Buddhism and Science.* Singapore: Buddhist College of Singapore.

Siegel, Ronald D. 2013. *Mindfulness and Psychotherapy, Second Edition.* New York: Guilford Press.

Steinhardt, Paul and Turok, Neil. 2006. *Endless Universe.* New York: Doubleday.

Surya Das, Lama. 1992. *Awakening the Buddha Within.* New York: Broadway.

Suzuki Roshi. 1970, 2020. *Zen Mind, Beginner's Mind.* New York: Weatherhill.

Thanissaro Bhikkhu. 2015. *The Karma of Mindfulness.* Valley Center, CA: Metta Forest Monastery

Thich Nhat Hanh. *Miracle of Mindfulness.* New York: Beacon Press.

Tsoknyi Rinpoche. 1998. *Carefree Dignity.* Hong Kong: Rangjung Yeshe.

Tulku Urgyen Rinpoche. 1995. *Rainbow Painting.* Hong Kong: Rangjung Yeshe.

________________. 2000. *As It Is.* Hong: Rangjung Yeshe.

Wallace, B. Alan. 2007. *Contemplative Science.* New York: Columbia University Press.

__________. 2009. *Mind in the Balance: Meditation in Science, Buddhism, and Christianity.* New York: Columbia Univ. Press.

__________. 2003. *Buddhism and Science.* New York: Columbia University Press

__________. 2012. *Meditations of a Buddhist Skeptic.* New York: Columbia University Press.

Wilber, Ken. 2017. *The Religion of the Future.* Boston: Shambhala.

__________. 2006. *Integral Spirituality.* Boston: Shambhala.

Zajonc, Arthur. 2004. *The New Physics and Cosmology: Dialogues with the Dalai Lama.* New York: Oxford University Press.

Zeilinger, Anton. 1999. "A Foundational Principle for Quantum Mechanics". *Foundations of Physics,* 29 (4): 631-643.

www.ingramcontent.com/pod-product-compliance
Lightning Source LLC
Chambersburg PA
CBHW022051210326
41519CB00054B/307

* 9 7 8 1 9 6 0 5 8 3 7 6 5 *